# Cogeneration and Small Power Production Manual

Fifth Edition

# COGENERATION
## & Small Power Production Manual

**Fifth Edition**

**Scott A. Spiewak**
**Larry Weiss**

Published by
**THE FAIRMONT PRESS, INC.**
700 Indian Trail
Lilburn, GA 30247

**Library of Congress Cataloging-in-Publication Data**

Spiewak, Scott A., 1954-
Cogeneration & small power production manual / Scott A. Spiewak, Larry Weiss. -- 5th ed.
p. cm.
Added title page title: Cogeneration and small power production manual.
ISBN 0-88173-270-2
1. Cogeneration of electric power and heat. I. Weiss, Larry, 1949-. II. Title. III. Title: Cogeneration and small power production manual.
TK1041.S64 1997 333.793--dc21 97-7759
CIP

*Cogeneration & Small Power Production Manual By Scott A. Spiewak, Larry Weiss. -- 5th ed.*


Published by The Fairmont Press, Inc.
700 Indian Trail
Lilburn, GA 30247

Printed in the United States of America

10 9 8 7 6 5 4 3 2 1

0-88173-270-2 FP

0-13-760141-7 PH



Distributed by Prentice Hall PTR
Prentice-Hall, Inc.
A Simon & Schuster Company
Upper Saddle River, NJ 07458

Prentice-Hall International (UK) Limited, London
Prentice-Hall of Australia Pty. Limited, Sydney
Prentice-Hall Canada Inc., Toronto
Prentice-Hall Hispanoamericana, S.A., Mexico
Prentice-Hall of India Private Limited, New Delhi
Prentice-Hall of Japan, Inc., Tokyo
Simon & Schuster Asia Pte. Ltd., Singapore
Editora Prentice-Hall do Brasil, Ltda., Rio de Janeiro

# TABLE OF CONTENTS

## SECTION C
## ENGINEERING AND COGENERATION

## APPENDIX



## SECTION D
## FINANCE AND COGENERATION

## SECTION E
## POWER CONTRACTING AND PROCUREMENT

## SECTION F
## NEW ENTRANTS INTO THE COMPETITIVE POWER MARKETPLACE



## SECTION G
## TRANSMISSION

SECTION H
UTILITY STANDBY RATES



SECTION I
CONCLUSION

# EDITORIAL BOARD

**Larry Weiss** is President of Com Power Group. After graduating from the University of Virginia and University of Georgia Law School, he worked at the Federal Energy Administration and Nuclear Regulatory Commission before entering private practice in 1977. Mr. Weiss has more than a decade of experience in the negotiation of power sales agreements, wheeling agreements, and related documents for independents, utilities and industrial users. He also has assisted fuel suppliers in bidding fuel contracts, negotiated rate reduction agreements for large industrial users, and developed and helped implement a resource acquisition program for several utilities. Tel. (703) 486-2751; Fax (703) 486-8036.

**Harvey N. Morris** is President of CPM Energy, Inc., an electric and natural gas retailer. He has been a consultant to industry, commerce and governments for over 30 years. His clients have included both the U.S. and British governments, the National Academy of Sciences, OPEC and many of the Fortune 500 companies. Mr. Morris has been directly involved in over 50 major cogeneration projects, and has also been retained by a number of utilities across the U.S. to provide consulting services relating to cogeneration. In 1978 he was voted Energy Engineer of the Year by the AEE at the First World Energy Congress. He has served on the Presidents Energy Advisory Council, the Board of Governors of the American Association of Engineering Societies and as 1981 President of the Association of Energy Engineers. Tel. (800) 828-9427; Fax (908) 542-8655.

**Scott Spiewak** is General Counsel to CPM Energy, Inc. Mr. Spiewak holds a J.D. from Boston University School of Law and an LL.M. from the Georgetown University Law Center. He spent five years with the Federal Energy Administration and the U.S. Department of Energy before founding the predecessor to CPM, Cogen Power Marketing, in 1982. Tel. (201) 784-5349; Fax (201) 767-1928; E-mail spiewak1@soho.ios.com.

**Michael Schnell** has been a senior associate with CPM Energy, Inc. since 1995. His specialty is retail sales of natural gas and electricity. Mr. Schnell also did the research for the Power Marketing Association reference, "Utility Standby Rates", from which the chapter of the same name of this text is excerpted. Tel. (201) 928-1163; Fax (201) 928-1230.

# PREFACE

The electric market is rapidly changing. The generation sector is becoming truly competitive. The transmission system is well on its way to becoming a common carrier system open to all utilities. Many now believe that retail wheeling is just around the corner. If so, the electric industry, like the gas industry, will become a truly competitive market.

These changes have substantially increased opportunities for cogeneration.

Cogeneration has been a profitable business opportunity since the passage of the Public Utility Regulatory Policies Act in 1978. During the last fifteen years, developers have used this legislation as a wedge to open up the generation sector to market forces. The Energy Policy Act of 1992 reaffirmed the nation's commitment to a competitive electric generation market.

Just as significantly, ratepayers have suddenly become consumers which their local utilities can no longer take for granted. Those who feel they are being gouged by their local utility now have the option of putting up their own on-site generating unit or buying their electric needs from a third party developer who is willing to make the capital investment. Even a credible threat to cogenerate may make a local utility more receptive to discounting the end user's electric bill.

The *Cogeneration & Small Power Production Manual* begins with a discussion of the manner in which electric utilities set their rates and the impact this has on the cogeneration decision. It then turns to the legal rights of cogenerators: the right to sell power to their local utility and the right to receive power when their facility is down. The next sections discuss engineering and financial aspects relating to project development.

The *Cogeneration & Small Power Production Manual* then discusses procurement issues and focuses on what on-site generators and utilities should look for in purchasing power from a development team. It then examines new types of cogenerators – independent power producers and affiliated power producers. Finally, the Cogeneration & Small Power Manual looks at the ability of a cogenerator to move power from one utility's service territory to another (the so-called "wheeling issue").

The *Cogeneration & Small Power Production Manual* is designed for the plant manager or others with the ability to initiate corporate investment decisions. It is not intended exclusively for the expert engineer or attorney, although both may benefit from it. Rather, the Manual will provide the manager with detailed, timely and understandable materials which may serve as the basis of a cogeneration feasibility study.

# Section A

## How Electric Rates Are Set: Their Impact on the Cogeneration Decision

# Section A

## How Electric Rates Are Set: Their Impact on the Cogeneration Decision

When one considers the cogeneration option, it is usually because you believe you might be able to generate electricity more cheaply than your local utility or because you perceive an opportunity to profit by selling the power you generate. These opportunities generally exist because of the behavior of the local utility.

### INTRODUCTION TO UTILITY RATE REGULATION

Utilities are best understood in the context of their environment. Since the early part of the century, electric utilities have been regulated as "natural monopolies". This treatment followed from the belief that consumers would benefit from avoiding unnecessary duplication of plant and equipment. This view, in turn, depended on the notion that the industry manifested economies of scale. In return for being granted exclusive franchises to serve specific areas, individual electric utilities were regulated to ensure that that they fulfilled their obligation to serve all customers and did not charge an unreasonable rate for their service. This trade-off is often referred to as "the regulatory bargain".

Regulation is unquestionably the single most important influence on utility behavior. Regulators determine how much utility shareholders may earn, what expenses they may recoup from ratepayers, the level of rates for various classes of customers, and the amount of flexibility a utility has to modify those rates in certain circumstances. Increasingly, regulators also are influencing utility strategy through the so-called "integrated resource planning" process.

While state laws differ significantly, all state regulatory regimes have two common characteristics. They first seek to determine what is called the "utility's revenue requirements" After determining the appropriate level of revenues, regulators then seek to divide those requirements among broad classes of customers based on cost-of-service principles.

In retrospect, this regulatory model rested on two assumptions: (a) new technologies would significantly reduce the cost of power in both nominal and real terms and (b) there was relatively little difference in the price of power among neighboring utilities. This was the experience of the industry from its founding through the early 1970s. Unfortunately, however, the experience since the early 1970s have not been in accordance with these assumptions, and the traditional regulatory model has had difficulty coping with these changed circumstances. The focus on revenues often means electric utilities charges rates which are unrelated to the value of their product– at times higher, at times lower. And rate structure regulation has proven a mechanism for discrimination against unfavored (i.e., nonvoting) classes of ratepayers, further broadening the chasm between value and price.

Much of this book is devoted to the ramifications of these developments. Before turning to those ramifications, some understanding of the current regulatory model is essential.

## DETERMINING REVENUE REQUIREMENTS

In theory, revenue regulation is a simple process. It consists of a utility's permissible expenses plus a return on its capital investment. This calculation yields what is called a utility's total revenue requirements.

Put another way, regulators determine a utility's revenue requirements by making three separate determinations:

(1) what are the utility's *operating expenses*, including maintenance, depreciation, and all taxes, appropriately incurred to provide service;

(2) what utility property provides the service for which rates are charged and thus represents *the rate base* on which a return should be earned; and

(3) what percentage figure should be applied to the rate base to establish the *rate of return* to which investors are reasonably entitled.

We shall now examine each of these determinations separately.

### Operating Expenses

Operating costs represent the largest percentage of a utility's total revenue requirement. Many of these costs are determined by normal competitive factors (wages, salaries, fuel and maintenance) or by various levels of government (taxes). Others are determined by the individual firms (expenditures on advertising, research and development, charitable contributions, and purchases from affiliated companies) or by the regulatory commissions (annual depreciation rates). For a typical electric company, operating expenses average about 59% of revenues, depreciation about 9%, and taxes about 14%.

Public utility commissions (PUCs) have broad authority to examine the reasonableness of operating expenses. It is a long-standing principle of regulation that such expenses will be disallowed if imprudently incurred. When an expenditure is disallowed, it, in effect, is charged to a utility's stockholders rather than its customers. In regulatory terminology, expenditures allowed for ratemaking purposes are charged "above-the-line;" expenditures disallowed in determining new operating income are charged "below-the-line".

Regulators may control operating expenses in two broad ways: (1) by disallowing improper charges already incurred and (2) by prohibiting the charges before they are incurred.

Operating expenses which traditionally have proven especially controversial include fuel costs, dealings with affiliated companies and advertising expenses. Increasingly, PUCs are inquiring into the reasonableness of expenses associated with plant outages and replacement power.

Several areas in this regard are worthy of greater attention.

Depreciation is an annual charge to operating costs, designed to reimburse those supplying the capital used to purchase the related assets. The general rule for valuating depreciable property is the original cost of the asset rather than its replacement value. The service life is usually determined by a combination of a review of past retirement experience and engineering estimates of future service lives. Three methods are used for distributing the depreciation charge over the expected service life of the property: straight line, sinking fund, and accelerated.

Utilities are permitted to pass through the taxes they pay to consumers. Two accounting procedures are used: "flow through" and "normalization". The former method permits a utility to collect these taxes as they are due. Normalization, which is the method followed in most jurisdictions, spreads tax savings brought about by the use of accelerated depreciation or the investment tax credit over the life of the property. A utility using this method collects taxes as if it were following a straight line depreciation method. Although normalization does not increase net income, it does add to cash flow.

Taxes are part of total operating expenses, and therefore are recovered from the ratepayer. In many areas of the country, utilities have for years served as surrogate tax collectors. The utility is taxed, the taxes are allowable expenses, and are passed through to the ratepayers who really pay the tax.

Most states have adopted fuel adjustment clauses to protect the utility from sudden changes in fuel costs. Because fuel prices can change quickly and drastically, the utility could suffer if it had to wait for a general rate increase to recover the higher costs. Without this provision a utility might be requesting rate changes almost monthly. Recently, some have argued that the automatic nature of the fuel clause pass-on provides utility management with little incentive to control fuel costs. The concept of the fuel adjustment clause is under scrutiny, and a few regulatory agencies have adopted incentive arrangements to encourage efficient operating procedures.

Finally, purchased power costs which a utility incurs from purchasing power from either another utility or a cogenerator is treated as an operating expense.

### Rate Base

Determination of the rate base is one of the most important and most difficult problems in ratemaking.

From a utility perspective, the rate base determines a company's earnings. Since operating costs are flowed through, its earnings are dependent on the size of the rate base and the return permitted on it.

At the risk of over-simplification one may calculate a utility's rate base by using the following formula:

| | |
|---|---|
| + | net plant |
| + | other assets of value |
| – | depreciation |
| – | customer contributed capital |
| = | rate base |

Net plant includes a utility's investment in generation, transmission and distribution facilities. Other assets of value include working capital and property held for future use. Customer contributed capital is deferred taxes and customer advances.

The three most controversial items in determining the rate base are:

(1) the method used to valuate the items in the rate base,

(2) the treatment which should be accorded excess capacity and overbudget plants, and

(3) the treatment which should be accorded construction work in progress (CWIP).

The large majority of states (37) use original cost in determining the value of assets in a utility's net plant. While there are number of methods used in determining original cost, it essentially consists of those costs prudently incurred in constructing a facility.

Ten states follow an approach called "fair value". This figure is somewhere between the cost of reproducing the facility at today's prices and its original cost.

Three states consider all evidence submitted.

While the difference between the two methods have occupied a great deal of academic attention, the difference is not as substantial as one might think. Five studies on the subject have concluded that when commissions adopt fair value rate bases, they partially offset the effect of the larger bases by granting lower percentage rates of return. Nevertheless, the combined effect of the larger rate base and the lower rate of return is to permit higher levels of earnings in fair value jurisdictions. A 1962 study found that the average rate of return for telephone companies was approximately 12% higher for fair value companies than they were for original cost companies. This gap may have narrowed in the interim, although no recent studies have made a conclusive comparison.

A major regulatory issue is the rate base treatment which should be accorded new plant might not be currently needed and which cost far more than originally projected.

Until recently, most utilities could reasonably expect that all plant costs would be put in the rate base soon after the plant was completed. The controlling legal standard is that net plant is placed in the rate base once it becomes "used and useful". The commissions have adopted two alternatives to determine what is used and useful: (1) a "prudent management" test, which permits the plant in the rate base if the decision to construct it was prudent, and (2) an "economics benefits" test, which permits the plant in the rate base to the extent that it provides an economic benefit.

States differ on whether construction-work-in-progress (CWIP) should be included in the rate base. This issue was really not all that important until after the oil embargo. Until that time deferral periods were short, interest expenses were minimal and there was relative certainty that the capacity would be needed. As the deferral period has lengthened because of longer construction periods, CWIP began accounting for a greater percentage of utility earnings. In the mid-Eighties it constituted about half of the industry's earnings, though the percentage had significantly declined by the early 1990s.

Finally, it is important to note that the rate base is reduced by the amount of depreciation taken each year. As a result, the fixed costs associated with a powerplant is very high soon after it is commissioned, but decreases each year. Within ten or fifteen years, the cost is below any reasonable market value of the plant.

**Rate of Return**

For regulatory purposes, the rate of return is the amount of money earned by a public utility, over and above operating costs, expressed as a percentage of the rate base. In other words, the rate of return includes interest on long-term debt, dividends on preferred stock, and earnings on common stock (including surplus on retained earnings). The important point to note is that rate of return includes profit (in the traditional sense), as well as interest on debt capital and dividends on preferred stock.

As with most aspects of rate regulation, the rate of return determination is conceptually simple: the Commission determines the cost of each segment of the capital structure and adds the results to determine the overall cost of capital, or rate of return. (See Table 1.) Determining the cost of debt and preferred stock is fairly easy because the required interest charges and dividends are fixed. Determining the return on common equity, however, is a long and complicated process.

The allowance for rate of return has had four phases. Between 1915 and 1929, regulators granted returns in the 7% to 8% range. As interest rates declined during the Depression,

**TABLE 1**

A utility's capitalization is half debt, half equity. The $500 of debt is borrowed at 8%; the regulatory agency finds the cost of equity is 13%. The utility cost of capital is:

| | Amount | Cost | Weighted Cost |
|---|---|---|---|
| Debt | $ 500 | 0.08 | $ 40 |
| Common stock | $ 500 | 0.13 | $ 65 |
| | $1000 | | $105 |

The rate of return is 105/1000, or 10.5%

returns were generally in the 5.5% to 6% range and remained there until the mid-1960s. Because of strong business conditions, inflation, and sharply rising interest rates that began in the mid-1960s, rate of allowed return began rising. In 1984, for example, the average authorized return on equity was 15.42%. As interest rates fell in the mid- to late 1980s, returns began declining again. By 1993, returns were generally in the 10% to 11% range.

The Supreme Court decisions have set three relevant economic criteria in setting the return on common equity: financial integrity, capital attraction, and comparable earnings. Stated another way, the two leading Supreme Court cases – *Bluefield Water Works* (1923) and *Hope Natural Gas Co.* (1944) – found:

(1) the utility can charge rates sufficiently high to maintain its financial integrity;

(2) the utility's rates may cover all legitimate expenses including the cost of capital; and

(3) the utility should be able to earn a return at a rate that is comparable to companies of comparable risk.

PUCs have used a number of standards to determine the permissible rate of return on common equity. Each has a superficial attraction, but all have a significant problems. What follows is an extremely brief overview of these standards.

The market-determined standard relies upon stock market transactions and estimates of investor expectations. Three major approaches have been employed: earnings-price ratio, the discounted cash flow model (DCF), and the capital asset pricing model.

The earnings-price ratio approach holds that the cost of equity capital to a utility is equal to the ratio of current earnings per share. Thus, if a utility's annual earnings are $5 per share and the average market price of its common stock for that same period is $38, the earnings-price ratio is 13.16%. (The actual figure would be 14.62% to account for 10% underpricing to allow for the costs involved in the issuance of new stock.)

The discounted cash flow model represents an attempt to estimate the equity investors' capitalization rate. Mathematically, the discount rate equals the current dividend divided by the current market price per share plus the expected growth rate in dividends per share. Thus, if the stock of a particular utility pays a $3 dividend, which is expected to grow at a rate of 6% per year, and if investors are willing to pay $38 for the stock, the required return on equity (assuming a 10% allowance for underpricing) is 14.77%.

The capital asset pricing model holds that cost of equity capital is equivalent to that on a riskless security plus a risk premium related to the risk inherent in a particular utility's stock. The formula is that the total return equals the risk free return plus the utility's relevant market risk (its beta) times the difference between the stock market return and the risk free return. Thus, assuming a stock market return of 13.9%, a risk free return (i.e. Treasury bills) of 7.8%, and a beta of 1.057, the cost of equity capital would be 14.25%.

Another standard which has been used is comparable earnings. In essence, this standard is the reported returns on the book equity of comparable enterprises. The problem is determining what constitutes a comparable business. Some commissions have taken the term quite literally and found a utility should earn a return on book equity that is close to the return being earned by similar utilities. The circular nature of this reasoning is obvious.

In recent years a number of PUCs have adopted what has been called financial integrity standards in response to specific problems, especially for plummeting interest coverage ratios and the fall of stock prices to levels below book value. One commentator has identified three such approaches.

The first approach is to set a return to produce a necessary interest ratio. Where a company needs to obtain financing and regulators believe that a 2.5 interest coverage ratio will permit it to do so, the regulators may start with that ratio and work backward.

A second approach has been to provide equity with a return set at a fixed amount above current debt costs. Some PUCs will set the proper return on equity at a specific number of percentage points above long term bond yields.

A third approach is to provide a return on equity designed to keep the common stock price at or slightly above book value. Because stock sold below book value dilutes the value of existing shareholders, many regulators believe that earnings should be high enough to keep the stock at or slightly above book value.

Before leaving the subject of rate of return, one should note Commissions have broad discretion in setting the rate of return. Perhaps the best example is the authority of regulators to substitute a hypothetical capitalization ratio for the real one. A PUC may decide that a utility which had no debt was being unduly conservative, particularly where the cost of debt was figured to be 8% and the cost of equity was found to be 12%. As a result, it may determine the rate of return based on a ratio of 50% debt and 50% equity on the theory that consumers in the state should not pay for the utility's conservatism.

## THE RATE STRUCTURE

The preceding three subsections concerned determination of the rate level. The end result is the total revenue requirement. The next step is the rate structure – the distribution of these revenue requirements among the utility's different classes of customers.

Here, as we have seen elsewhere, the legal standards governing the regulatory commissions are broad. Each specific rate must be "just and reasonable". Further, "undue" or "unjust" discrimination among customers is prohibited. The setting of rate structures was, prior to the 1970s, left primarily to the utilities, on the theory that they were fairly neutral when it came to balancing equities between customers. However, in the past two decades, utility commissions have become more involved in the rate setting process.

Utilities do not charge a uniform price for each kWh of electricity which they sell. In general, the prices customers pay for electricity should be an accurate reflection of the actual cost of providing the service. But the actual cost of generating electricity may differ from hour to hour within a given day, and during different seasons of the year. Different customer classes have different load profiles, some with more constant demand, others with a disproportionate share of their demand during peak periods. Price differentiation on the basis of customer classification is therefore necessary for both equity and, indirectly, for the efficiency of the system.

From our perspective, the rate structure is important for several reasons.

First, the rate structure may serve a variety of purposes. It may be designed to bring about redistribution of wealth (by charging business users substantially higher rates than its other consumers), to promote economic efficiency, to serve social policies (e.g. lifeline rates), or just to assume that each customer group provides a certain rate of return.

Second, the rate structure may not produce the gross revenues which will permit the utility to generate its authorized rate of return. Rates are based on past costs and sales. This can lead to revenue deficiencies when those assumptions are incorrect. This problem has been especially pronounced during periods of high inflation.

Third, a rate structure which places a disproportionate share of a utility's revenue requirements on its industrial customers creates incentives for on-site generation.

In the following segments we shall examine (1) the functional and causal aspects of cost and price variations, (2) basic rate principles, (3) rate structure design and criteria, (4) economic pricing principles, and (5) several computational methodologies employed in allocating cost.

**Functional and Causal Aspects of Cost**

Costs are categorized so that one can more accurately compute specific costs for each customer class, and then set prices based on the actual cost of service. This cost categorization process involves looking at a utility's cost from both a functional and a causal point of view. Functionally, costs reflect the dollars that must be spent to perform a certain function. Looked at causally, cost reflects the respective cost responsibilities of the different customers who require these functions so they may be provided with electricity.

The four functional cost categories include generation cost, transmission cost, distribution cost, and customer cost:

- *Generation Cost:* Those capital and operating expenses incurred in building a generating station that will transform other forms of energy into electric energy.
- *Transmission Cost:* Those capital and operating expenses incurred in the process of transporting electric energy in bulk from a source or sources of supply to other principal parts of the system, or to other utilities.
- *Distribution Cost:* Those capital and operating expenses incurred in the process of distributing electric energy from convenient points on the transmission or bulk power system to the customer.
- *Customer Cost:* Those capital and operating expenses incurred by the utility in taking on another customer. These usually includes such costs as services and metering, as well as accounting and sales promotion expenses.

The three causal cost categories include demand related cost, energy related cost, and customer related cost.

- *Demand (Capacity) Cost:* The capital and operating expenses incurred by the utility on behalf of an individual customer in providing sufficient capacity (and a large enough generation, transmission and distribution system) to meet the maximum demand of that customer when needed. Demand costs vary with the kilowatts (kW) of power demanded. These costs are typically charged to the customer in the form of dollars per kW.

- *Energy Cost:* The operating expenses incurred by a utility on behalf of an individual customer in providing that customer with a kilowatt hour (kWh) of electricity. Energy costs vary directly with kWh used and are primarily fuel costs. These costs are usually charged to customers as cents per kWh. Sometimes rates are structured so that the energy component includes hidden demand charges. This can make the energy charge appear to be inordinately high, particularly when fuel costs are falling.
- *Customer Cost:* The capital and operating expenses incurred by the utility on behalf of an individual customer relating primarily to the number and size (usage) of the customer. Customer cost does not vary significantly with the amount of power used. The "causal" customer cost category is pretty well analogous to the "functional" customer cost category. It is generally expressed in terms of a fixed number of dollars per month.

The next step is to show how functional cost and causal cost may be interrelated. Cost due primarily to the capital cost of generating facilities is classified as "capacity costs," while operating expenses may be divided between capacity related costs and energy related costs. Fuel costs are almost always allocated entirely to the energy cost category. The capital and operating expenses associated with transmission costs are usually considered capacity related costs. The capital and operating expenses associated with distribution costs are divided between capacity related costs and customer related costs. Customer related capital and operating expenses vary proportionately with the number of customers served and have little if any relationship to the kWh of energy used by the customer, or even to kW energy demands.

Utility rate analysts use three load factor formulas to derive estimations of actual cost variations per kWh in producing electricity based on capacity: one for the system, another for the customer, and yet a third to determine the system's diversity. For illustrative purposes, let us take a hypothetical example of an electric utility's daily load curve. In this example, plant capacity is 9,000 kW, maximum kW peak demand is 7,000 kW, and total production during the day is equal to 117,000 kWh (or an average of 4875 kw per hour).

The system load factor is the ratio of average load to peak demand. Peak demand is the maximum demand placed on the system at any given time. The System Load Factor, in the example, then, would be expressed as:

$$4875 \text{ kW}/7000 \text{ kW} = 70\%$$

The customer load factor is the ratio of the customer's average consumption to its maximum consumption. Say our first customer, Customer A, used a maximum of 2000 kW, but averaged only 1250 kW use. In this example, Customer A's load factor would be:

$$1250\text{kW}/2000\text{kW} = 62.5\%$$

Customer A's 2000 kW peak usage is referred to as its "noncoincident peak demand". This is to distinguish it from its "coincident peak demand". Coincident peak demand is Customer A's demand during system peak periods. If A's coincident peak demand were also 2000 kW, then A would be a relatively high-cost customer. This is because new plant would be needed to serve A (due to the fact that A's needs peak just when the system is already subject to the most stress), but A doesn't make very good use of plant, with a load factor of only 62.5%, well below the average load factor of 70% for the customers on the system. A must be charged more per kWh, in order to support that idled plant.

The System's Diversity Factor is the ratio of the maximum non-coincident demands of all customers to the system's peak demand. In our example, assume Customer A is joined by two other customers, B and C, and that Customer A, B and C's maximum non-coincident demands are 2,000, 2,500 and 3,000 kW, respectively. Here, assuming system peak demand is 7,000 kW, the diversity factor for this example would be expressed as follows:

$$7500/7000 = 1.07$$

The higher the system diversity factor, the better. This is because customers complement each other's usage. If A is not using plant, B is using it. When A requires the plant back, B no longer does so.

If the utility managers can raise the system load factor, so that plant is used as much as possible, they can lower the average production cost per kWh. This is because the fixed costs associated with generation plant can then be spread over the maximum number of kWh's generated.

Assume the utility gets a request for an additional 1,000 kW of electricity from 12 midnight to 8 a.m. What would it cost the utility to deliver this much electricity during this off-peak time period? The only cost the utility would incur would be energy and customer costs. The capacity costs associated with the generating plant were already incurred in serving the normal, daytime peak loads.

The utility may argue that adhering strictly to the cost of service principle justifies charging lower rates to this off-peak customer. Further, if the utility allocates any capacity cost to this customer at all, theoretically, existing customers could benefit by receiving lower rates themselves. This is so because by spreading total capacity cost out over more kW of electricity, the total average cost per kWh will decrease, since the energy and customer cost remain constant per kWh. If the new customer actually used its maximum 1,000 kW throughout the time period (12 midnight to 8 a.m.), the utility's system load factor would increase.

In essence, by gaining this off-peak customer, the utility increased its diversity factor , which in turn increased the system load factor. It is important to note that large industrial or commercial customers, not residential ones, are the only ones likely to use significant kWh's in off-peak times. This is one of the justifications for offering lower rates to industrial facilities which have elastic load requirements during off-peak periods.

The utility could also argue that giving quantity discounts to large customers who have historically demonstrated off-peak usage patterns could increase the system load factor. How could the utility entice this off-peak customer to use 2,000 kW? Suppose the utility granted the customer a rate structure as follows:

| | |
|---|---|
| 0-1000 kWh | $.06/kWh |
| 1000+ kWh | $.03/kWh |

The customer could lower its average kWh fuel cost from $.06 to $.045/kWh if it used 2,000 kW continuously throughout the time period. But the only way the customer can achieve this quantity discount is by using more electricity.

Traditionally, utility rate analysts have argued that offering quantity discounts through a declining block rate structure is an effective device for increasing a utility's system load factor. As long as the $.045/kWh rate brings in enough revenue to cover the customer's energy and customer costs, and contributes something to the utility's capacity costs, the utility's average production cost per kWh will be reduced.

What all this comes down to is that a kWh of electricity produced at peak has a greater capacity cost associated with it than a kWh of electricity produced off-peak, unless the utility has plenty of surplus generating capacity.

### Cost Allocation Evaluation Techniques

There are two basic cost allocation techniques used by utilities in the U.S.: the Fully Distributed Cost Method and the Marginal Cost (or Incremental Cost) Method.

#### *Fully Distributed Costs*

In the fully distributed cost method, the utility rate analyst attempts to apportion the total cost of service accurately among the various customer classifications and also to apportion the capacity cost, energy cost, and customer cost within each customer classification. Energy cost can be apportioned among customers on the basis of actual consumption as registered by meters. Customer cost can likewise be accurately ascertained. Many customer costs, such as billing and metering, can be divided equally among all customers, because these kinds of customer costs do not change substantially with changes in consumption per kWh or per kW. But the rate analyst using the fully distributed cost approach does run into severe difficulty when he attempts to apportion capacity cost. In rate proceedings, it is usually the allocation of capacity costs around which major controversies take place.

The difficulty in capacity cost allocation is known as a "joint cost problem". In essence, joint costs are common costs (e.g., capital costs, salaries, taxes, etc.), which means that they cannot be causally related to a particular customer class. Joint costs do not change proportionately to changes in output per kWh or per kW, but decrease as output increases. Whatever method is used to allocate these common costs is at least to some extent arbitrary. It's like trying to determine what portion of the cost of raising a cow should be allocated to the hide and what should be allocated to the beef. In real life the market decides the price of the hide and the beef. In utility ratemaking, the utility commission makes the decision.

In the last half century, over 30 methods for allocating capacity costs have been developed. We shall briefly examine three of the more prominent methods: (1) the peak responsibility method, (2) the noncoincident peak method, and (3) the average excess demand method.

In the peak responsibility method, a utility's capacity costs are apportioned among customer classifications in proportion to each classification's percentage of the system peak. As illustrated in the example given, capacity cost would be allocated among Customers A, B and C in the following manner:

*Peak Responsibility Demand Method*

| Demand | Peak Maximum Demand by Class | Percent of kW Plant Capacity | Allocation of Customers kW |
|---|---|---|---|
| A | 1750 | 25.0% | 2250 |
| B | 2250 | 32.1% | 2889 |
| C | 3000 | 42.9% | 3861 |
| TOTAL | 7000 | 100% | 9000 |

The analyst first computes the average maximum peak kW demand for each customer during the peak demand period. He then determines each customer's percentage of the system peak and then multiplies that percentage by the system's total capacity.

Clearly, the major flaw in this approach is that any customer who does not consume electricity during a system's peaking period will not be allocated any capacity cost, no matter how much electricity the customer uses during off-peak hours.

In the noncoincident peak demand method, a utility's capacity costs are apportioned among customer classifications in proportion to each classification's maximum kW demand. For example, using the same data as above, capacity costs would be allocated between customers A, B and C as follows:

*Noncoincident Peak Demand Method*

| Customers | Peak Maximum kW Demand | Percent of kW Demand by Class | Allocation of Plant Capacity |
|---|---|---|---|
| A | 2000 | 26.7% | 2403 |
| B | 2500 | 33.3% | 2997 |
| C | 3000 | 40.0% | 3600 |
| TOTAL | 7500 | 100% | 9000 |

Using this methodology, the analyst first ascertains the maximum kW demand for each customer whenever that maximum demand occurs peak or off peak. Second, she determines each customer's percentage of the combined maximum kW demand, and then multiplies that percentage by the system's total capacity.

The major flaw in this approach is that it does not in any way account for the actual time that individual customers use the plant's facilities. One customer may have a maximum demand of 500 kW for only one hour, while another may have a maximum demand of 500 kW twenty-four hours a day. In the noncoincident peak demand method, each would be apportioned the same capacity cost. By the way, the peak responsibility method is subject to the same criticism.

The Average Excess Demand Method is an attempt to avoid the flaws inherent in both the peak responsibility method and the noncoincident peak demand method.

With the average excess demand method, a portion of the utility's capacity cost is allocated to each customer on the basis of each customer's average kW load, the remainder (the difference between maximum kW load and average kW loads) is allocated by the noncoincident peak demand method.

The Average Excess Demand Method applied to our example would yield the following results:

| -1 | -2 | -3 | -4 | -5 | -6 | -7 |
|---|---|---|---|---|---|---|
| End user | Load Factor Alloc | Max KW Demand | Ave Load Alloc | Load Alloc | Excess Load Alloc | Total |
| A | 62.50% | 2000 | 1250 | 750 | 1178 | 2428.4 |
| B | 58.30% | 2500 | 1458 | 1042 | 1637 | 3095.2 |
| C | 72.20% | 3000 | 2166 | 834 | 1310 | 3476.4 |
| TOTAL | | 7500 | 4874 | 2626 | 4126 | 9000 |

First, the analyst determines the average kW load per customer classification. If the customer load factor is known, this can be computed by multiplying each customer's maximum kW demand by its load factor (Column 2 x 3). This will yield the actual Average kW load capacity cost allocation to each customer (Column 4). Second, the analyst subtracts each customer's average kW load from its maximum kW load (Column 3 minus Column 4). This yields each customer's excess peak load (Column 5). Then the analyst multiplies the ratio of the system's total excess load (4126, which is the system capacity of 9000, minus the total in Column 4) to its total excess peak load (Column 5). This will yield the actual excess load capacity cost allocation (Column 6). For example, customer A's excess load allocation is 4126/2626 x 750 = 1178.4. Then the analyst adds the average kW load and the excess load capacity cost allocation, Columns 4 and 6, to obtain each customer's total capacity cost allocation (Column 7).

Of the three methods, most professional economists consider the Average Excess Demand Method to be the best. This approach comes closest to tracing the actual capacity cost customer responsibilities since it takes off-peak usage into consideration. However, all three of these fully distributed cost methods of allocating capacity costs are subject to a common criticism: they are all based on historical demand for electricity, which has been shaped by past pricing policies. Rates are based in part on the cost of constructing plants decades ago, with the cost of building new plants being averaged in as they become on line. Thus the actual cost of adding new capacity to the system is grossly understated in the rates charged to the consumer, encouraging what many economists consider to be uneconomic expansion of demand. These economists favor some form of marginal cost pricing, as the best way of achieving the most appropriate allocation of resources for the generation of electric power.

### Marginal Cost Pricing

The implicit assumption underlying the arguments in favor of some form of marginal cost pricing for utilities is that the competitive price system is the most efficient allocator of a society's resources, and that in the free market, competition would drive prices down to equal the marginal cost of production. In the theoretical model, firms are forced to operate at the minimum point on their average total cost curves, at the point where average total cost equals marginal cost. In addition, they would also set prices at this same point, due to competition, and thus price would also equal the firm's marginal cost. Since a firm's cost curve reflects the technical production conditions under which the industry operates, when operating at the minimum point on its average total cost curve, the firm is also operating in the most technically efficient manner, which means that resources are being used efficiently.

In this brief discussion we shall confine ourselves to examining the basic principles involved in marginal cost pricing including (1) definitional and cost/price relationships, (2) the case for and against short run marginal cost ratemaking and (3) the case for and against long run marginal cost ratemaking.

Marginal cost refers to the increase in a firm's total production cost due to increased output. When a firm increases its output in the short run by increasing its use of existing facilities, we call the change in the firm's total production cost its short run marginal cost (SRMC). When a firm increases its output by increasing plant capacity, either by addition to existing facilities or building entirely new facilities, changes in total production cost are called long run marginal cost (LRMC).

In the short run, the difference in marginal cost per kWh of electricity is due primarily to variable costs. Variable costs are those costs which change substantially in direct relationship to changes in the rate of output. In electric utility economics, variable costs consist primarily of energy costs. Fixed costs are those costs which do not change substantially with changes in output and in electric utility economics, fixed costs are analogous to capacity costs. In the short run, marginal costs fluctuate for the following reason:

Electric utilities use their most efficient generating units under normal operating conditions and hold their least efficient and obsolete generators in standby reserve. The average and marginal cost curves will turn up when the utility is forced to use these less efficient generators during on-peak periods. In essence, fuel costs per kWh greatly increase when using the older, obsolete generation equipment, causing the short run marginal cost to be substantially above the utility's average cost.

The concept of long run marginal costs differs from that of short run marginal costs, in that it takes into account the costs incurred in changing the amount of generating capacity.

Short run marginal cost pricing is more valuable if the primary concern is the efficient utilization of existing plant capacity. As such, it is useful in circumstances of severe overcapacity. Long run marginal cost pricing is of more use when a change in utility plant is foreseeable during a useful planning period. For example, if no change in plant is likely for the next forty years, long run incremental cost pricing is not terribly useful. However, if a need for new capacity is foreseeable in the next five years, then according to proponents of marginal cost pricing, it would be appropriate to price electric power at a level which reflects the cost of production from that new powerplant.

To do otherwise would lead to uneconomic use of resources. For example, if a company were planning to locate an aluminum reduction plant in an area with substantial hydroelectric resources, under fully distributed cost principles, that company would pay a rate reflecting the average costs of production in that service area. This would be so even if the serving utility had long since exhausted opportunities for new hydroelectric capacity, and would now have to construct an expensive coal fired plant to serve new load. It may be that another utility could provide the power required less expensively, perhaps through better utilization of existing plant. Nevertheless, new industry would continue to flock to the area with existing low cost power sources until it forced enough new construction to bring rates, on average, up to those charged in other service areas. This is obviously uneconomic, but politically, marginal cost pricing has been hard to accept.

The reason for this is as follows: if marginal costs are declining, then charging rates reflecting marginal costs will not provide the utility with sufficient revenue to support its older, less efficient plants. The shortfall would have to be made up in subsidies. If marginal costs are rising, then charging rates reflecting marginal costs will provide the utility with more revenue than is required for a reasonable rate of return. In a rising marginal cost situation, changing over from fully distributed cost principles to marginal cost principles would also require a large rate hike, which is politically unpalatable.

Even if marginal costs do not replace embedded costs in the calculation of rates to be paid by ratepayers, marginal cost principles may be embodied in the rate structure design. Rate structures are designed to recover from the ratepayer class that amount of the utility's total costs assignable to it under its methods of cost allocation.

A typical rate structure might call for the payment of a set fee each month, plus a flat kWh rate. This "flat" rate structure could look something like this:

| | |
|---|---|
| $5.00/Month | – Fixed charge |
| $0.05/kWh | – Energy charge |

Another variant is the declining block rate structure. The declining block rate structure, perfectly valid in circumstances in which increases in output result in decreased costs per unit, is a disaster in the opposite circumstances, encouraging increased use by discounting at a time when those very increases in use result in an increase in costs per unit. A declining block rate structure looks something like this:

| | |
|---|---|
| 1-500 kWh/Mo | $.05/kWh |
| 501-1000 kWh/Mo | $.04/kWh |
| 1000+ kWh/Mo | $.03/kWh |

An inverted block rate structure is the opposite of a declining block rate structure. It penalizes increased use of electricity, and is designed to reduce usage in circumstances of increasing electrical costs. However, while the inverted block rate structure is an improvement over flat or declining rate structures in circumstances of increasing costs, it is still a crude tool for controlling the need for new capacity expenditures. An inverted rate structure looks something like this:

| | |
|---|---|
| 1-500 kWh/Mo | $.03/kWh |
| 501-1000 kWh/Mo | $.04/kWh |
| 1000+ kWh/Mo | $.05/kWh |

These types of rates do nothing to discourage ratepayers from using electric power during peak periods, and recognize no difference in the value of electricity delivered at different times.

Most proponents of marginal cost pricing would not only favor the replacement of fully distributed actual costs with incremental costs of production in the allocation of total revenue to be collected among the various classes of ratepayers, they would also favor replacement of flat rates with rates which attribute to "peak" users, the costs attributable to the construction of new capacity. This is on the theory that it is only the peak users who are responsible for the need to construct new capacity–therefore only peak period users should have to pay for capacity.

Nevertheless, rate design is a separate procedure from establishment of the total revenue to be recovered by the utility. Even if capacity costs are established by traditional principles involving the original cost of equipment, the rate may be designed to recognize the different values of electricity delivered at different times. In this manner, rate design may be used as a load management tool, totally apart from incremental cost pricing principles although consonant with them.

This type of rate design is often referred to as time of use pricing, as the price per kWh varies depending upon the time of day that kWh is used.

A time of use rate can be fairly complex, and require metering equipment which is either unavailable, or uneconomic for smaller users of power. An example of a time of use rate structure is as follows:

$4 fixed charge per month

Summer:

| | |
|---|---|
| $.05/kWh | 10 a.m. to 2 p.m. |
| $.02/kWh | 2 p.m. to 10 a.m. |

Winter:

| | |
|---|---|
| $.04/kWh | 9 a.m. to 1 p.m. |
| | 5 p.m. to 7 p.m. |
| $.02/kWh | 1 p.m. to 5 p.m. |
| | 7 p.m. to 9 a.m. |

These rate structures can become far more complicated, with separate rates for peak and on peak periods, but also for shoulder periods. Multiple part rates may also be employed, with not only a fixed "customer" charge, but also separate "capacity" and "energy" components of the rate. In some regions there may also be differentiation based on whether service is at high or low tension service levels, and whether it is at primary or secondary distribution levels.

To the extent that rates may be designed to recover costs from the customers on whose behalf they are incurred, the rates are truer to the principles of marginal cost ratemaking.

While these principles are rarely applied to utility customers, marginal cost principles are required to be used under PURPA section 210 and the implementing regulations when establishing buyback rates for the purchase of power from cogenerators or small power producers, and the rates are often calculated first by the same rate analysts who have used the above principles in setting the rates charged by the utility. Thus, information concerning your utility's marginal costs is likely to be available. It can be very useful in determining whether the utility at least has the capability of negotiating lower rates than it currently charges, and in determining whether on-site generation is the optimal means of obtaining firm generating capacity in the future.

## HOW RATE REGULATION AFFECTS THE COGENERATION DECISION

Regulation has three very significant impacts from the perspective of the potential cogenerator. First, it determines whether the utility will have an interest in building new capacity itself or purchasing that capacity from someone else. Second, it is a significant factor in the decision as to whether on-site cogeneration is an economic alternative. Third, it is a major factor in determining whether a utility will have the flexibility to lower its rates for an industrial customer in an effort to defer the cogeneration option.

## BUY VERSUS BUILD

All else being equal, a utility which is relatively assured of earning more than its cost of capital has every incentive to build its own new capacity. This is the so-called "Averch, Johnson & Wellisz" (or AJW) effect.

The implications of the AJW effect are severalfold. Assuming that a utility can increase its earnings by adding new investment, the utility will have a tendency to overbuild. It will tend to resist to any proposals which would decrease demand or save capital. It also will have the tendency to gold-plate reliability standards so to increase investment.

On the other hand, a utility whose management questions its ability to earn a return on its investment sufficient to cover its cost of capital is likely to purchase power from others and seek to build its rate base through investment in transmission and distribution facilities (which are much less likely to be the cause of a regulatory disallowance). In this manner, shareholders eliminate the risk to earnings from new powerplant construction, while continuing to increase earnings (albeit at a slower pace than if the utility were to build new generating capacity).

As the experience of the 1980s illustrates, political factors will influence a commission to keep the allowable rate of return low or disallow some of the costs for new construction. This has discouraged utility investment in new generating capacity and given utilities an incentive to "find" that new capacity is not needed. Because of a utility's obligation to serve, it must either buy or build new capacity to serve its forecasted load. Low forecasts enable the company to avoid making that decision.

## DECISIONS REGARDING ON-SITE GENERATION

There are two motivations for on-site generation: (a) rates which make self-generation an attractive alternative and/or (b) a concern that the utility won't be able to fulfill its obligation to serve.

Previously, we have seen why electric utility rates bear only a passing relationship to the value of electricity. Simply put, there is no tie between the economic cost of power and the regulatory accounting method of pricing power. The total revenue requirements approach doesn't work very well in cases where the capital costs of new plant is above the marginal cost of new generating capacity – as is true in many areas of the country today.

In addition, regulators often have worsened the situation by either foisting off a major share of politically unpopular rate increases on industrial and large commercial users or using their rates to subsidize other popular programs (e.g., lifeline rates or DSM programs). In some cases this has significantly worsened the discrepancy between the rate that a utility must charge under its tariffs and the cost of alternative generation.

Finally, in a business which is subject to the free market, there is always an incentive to keep costs as low as possible. But since public utilities are permitted to collect enough revenue to cover total operating expenses plus a predetermined amount of earnings, there is little monetary or regulatory incentive to hold costs down – and the opposite can be true. Needless to say, an on-site generator buying fuel for its own powerplant, with money out of its own pocket, is likely to be more interested in negotiating a better deal for fuel than a utility with automatic fuel adjustment clauses in its tariffs.

Another reason to consider self-generation is concern that the utility is "low-balling" its demand forecasts. It is a well-known fact of life that industrials are the first to be interrupted when supplies are short.

When you see a utility rate which has large differences between peak and off peak prices, you have a tip-off that the utility is becoming capacity short. The higher peak period rates are designed to reflect the need for new construction, or the reliance on high cost purchased power or low efficiency plants during those periods. In these circumstances, it is worthwhile to consider building on-site generation, unless the local utility is credibly preparing for the future.

Of course, any decision on self-generation must consider not only the cost of power from the utility vis-à-vis that of an on-site facility but also the cost of partial requirements power (backup, supplementary and maintenance). These are discussed in Section B.

## COGENERATION DEFERRAL CONTRACTS

Many states permit their utilities to charge certain customers rates below their published tariffs in an effort to defer cogeneration.

This generally occurs in cases of "uneconomic bypass" – where a utility's marginal costs are favorable from a competitive standpoint, but its rates are unattractive because of the recovery of capital costs. In most cases the net amount of annual savings to the industrial customer which is considering installing a cogeneration facility is less, and sometimes, dramatically less, than the net annual loss of revenue to the electric utility.

The rationale for cogeneration deferral rates is that the utility and its other customers are better of with some contribution to fixed costs than none (as would be the case if the customer if left the system).

There are several forms of cogeneration deferral rates. The three most popular are discussed in the following paragraphs.

In some cases the utility discounts its rates designed to meet the customer's self-generation alternative. For instance, several utilities have signed contracts with industrials at a rate designed to match a cogeneration project's economics. In such cases the monthly bills mirrors the fixed and operating costs of an on-site generator.

In other cases utilities simply have restructured their rates to lower the cost of customers threatening to leave the system. A utility might offer a potential cogenerator interruptible service or incremental off-peak rates to make its rates more attractive.

Some states have allowed utilities to discount rates to any large customer when in the judgment of the utility it is economically feasible and practical for the customer to displace its service by using cogeneration. States have varying conditions for such contracts, although all require that the discounted rates shall be greater than its marginal customer, energy and capacity charges.

As a result, many industrial and commercial customers have used the threat of cogeneration to get rate breaks from their utilities. Obviously, this strategy is most likely to work if the threat to cogenerate is credible. This means the customer often must undertake the legal, economic and engineering studies necessary to convince the utility that it is seriously exploring the self-generation option.

# Section B

## Legal Rights of Cogenerators

# Section B
## Legal Rights of Cogenerators

The Public Utility Regulatory Policies Act of 1978 (PURPA) gives various legal rights to owners of cogeneration facilities and small power production facilities meeting certain standards and which are not owned by electric utilities. The net effect of these rights is to make on-site generation a viable alternative for large industrial users of steam and to open up the generation sector of the electric industry to competition.

Cogeneration facilities simultaneously produce two forms of useful energy, such as electric power and steam. Cogeneration facilities use significantly less fuel to produce electricity and steam (or other forms of energy) than would be needed to produce the two separately. Thus, by using fuels more efficiently, cogeneration facilities can make a significant contribution to the nation's effort to conserve its energy resources.

Small power production facilities use biomass, waste, or renewable resources (including wind, solar and water) to produce electric power. Reliance on these sources of energy can reduce the need to consume traditional fossil fuels to produce electric power.

### PUBLIC UTILITY REGULATORY POLICIES ACT

Prior to the enactment of PURPA, a cogenerator or small power producer seeking to establish interconnected operations with a utility faced three major obstacles. First, a utility was not generally required to purchase the electric output. Second, some utilities charged discriminatorily high rates for back-up service to cogenerators and small power producers. Third, a cogenerator or small power producer which provided electricity to a utility's grid ran the risk of being considered an electric utility, thus becoming subject to state and Federal regulation as an electric utility.

Sections 201 and 210 of PURPA are designed to remove these obstacles. Each electric utility is required under section 210 to offer to purchase available electric energy from cogeneration and small power production facilities that have obtained qualifying status under section 201 of PURPA at rates equal to, or less than, the utility's avoided cost. Section 210 also requires electric utilities to provide electric service to qualifying facilities at rates which are "just and reasonable, in the public interest, and which do not discriminate against cogenerators and small power producers". Section 210(e) of PURPA provides that the Commission can exempt qualifying facilities from various state and Federal laws.

### FEDERAL AND STATE ROLES

PURPA gave the Federal Energy Regulatory Commission (FERC) considerable discretion in the manner in which PURPA should be implemented, leaving it up to the Commission to determine the qualifications of facilities meeting the standards of the legislation and the rate at which utilities should be required to purchase power from a "qualifying facility" (i.e., one which meets the standards determined by the Commission). Qualifying facilities are also referred to as QFs.

The implementation of FERC's rules, which were promulgated on March 20, 1980, is left up to the individual states. States have implemented PURPA through the enactment of laws or regulations at the state level and by application on a case-by-case basis.

Review and enforcement of state rules can occur either in state courts or at the FERC. As a practical matter, FERC has been extremely reluctant to overrule state decisions implementing PURPA.

## REQUIREMENTS FOR QUALIFICATION UNDER PURPA

### Cogeneration

PURPA defines a "qualifying cogeneration facility" as a cogeneration facility which (i) the Commission determines, by rule, meets such requirements respecting minimum size, fuel use, and fuel efficiency) as the Commission may, by rule prescribe; and (ii) is owned by a person not primarily engaged in the generation or sale of electric power (other than electric power solely from cogeneration facilities or small power production facilities). A cogeneration facility is defined as a facility which produces (i) electric energy, and (ii) steam or forms of useful energy (such as heat) which are used for industrial, commercial, heating, or cooling purposes.

FERC's regulations implementing this language require that topping-cycle facilities produce at least five percent of its total energy output in the form of useful thermal energy output. Compliance with this standard is to be based on estimated annual energy output.

They also include efficiency standards for new oil- and gas-fired topping cycle cogeneration facilities. Any topping-cycle cogeneration facility using oil or gas as an energy input must each calendar year meet the following standard: the useful power output plus one-half the useful thermal energy output of the facility not be less than 42.5 percent of the energy input of natural gas and oil to the facility. Note this standard weighs thermal energy with only half the value of electricity.

Efficiency standards for bottoming cycle units are limited to cases of oil or natural gas supplemental firing. The standard specifies that the useful power output of the bottoming cycle must, during any calendar year, be no less than 45 percent of the energy input of natural gas and oil for supplementary firing. The fuels used in the thermal process "upstream" from the bottoming-cycle facility's power production system are not considered in this efficiency test.

FERC rules also prohibit an electric utility or electric utility holding company from owning more than 50 percent of a qualifying cogeneration facility (see discussion infra).

### Small Power Production Facilities

A small power production facility is defined as a facility which produces (i) electric energy solely by the use as a primary energy source, of biomass, waste, renewable resources, geothermal resources, or any combination thereof; and (ii) has a power production capacity which, together with any other facilities located at the same site (as determined by the Commission), is not greater than 80 megawatts.

Legislation passed in 1990 removed the size limit on eligible solar, wind, waste and geothermal small power production facilities for which notice or application for Commission certification is submitted to the Commission by December 31, 1994, and for which construction generally commences by December 31, 1999.

FERC rules implementing the 80 megawatt limitation provision provide that the capacity of all facilities which use the same energy resource, are owned by the same person, and are located within one mile of each other be added together. Hydroelectric facilities (within this distance) are considered to be located at the same site only if the facilities use water from the same impoundment for power generation. This provision for hydroelectric facilities is necessary because use of the one-mile rule alone might discourage the development of facilities on separate waterways which are within one mile of each other, or of closely-spaced impoundments on an individual stream.

Requirements for qualifying small power production facilities provide that the primary energy source of the facility must be biomass, waste, renewable resources, or any combination thereof, and that more than 75 percent of the total energy input must be from these sources. This requirement is not intended to force small power producers to continually monitor the energy input, but rather to require that reasonable estimates based on sampling methods be used.

Qualifying small power production facilities using biomass as a primary energy source are treated differently than are facilities using other resources for purposes of exemption from the Public Utility Holding Company Act and certain State law and regulation under section 210(e) of PURPA. FERC regulation specifies that any primary energy source which, on the basis of its energy content, is more than 50 percent biomass shall be considered biomass. In other words, a qualifying facility may be considered biomass-fired if, on an estimated annual basis, at least half the energy input, exclusive of fossil fuel use, is biomass.

Much more than half of the energy content in municipal solid waste is due to "organic material not derived from fossil fuels," or "biomass" under the Commission's definitions. Thus, a small power production facility fired with municipal solid waste may be considered a biomass facility. The same treatment applies to facilities fired with forest-industry residues, sewage sludge, or peat.

Another aspect of what constitutes "primary energy source" is a specification of what fuels may be used in addition to the primary energy source for purposes of ignition, startup, testing, flame stabilization and control, and during equipment outages and emergencies.

Section 3(17)(B) of the Federal Power Act, as amended by PURPA, provides

> "Primary energy source" means the fuel or fuels used for the generation of electric energy except that such term does not include, as determined under rules prescribed by the Commission, in consultation with the Secretary of Energy:
>
> (i) The minimum amounts of fuel required for ignition, startup, testing, flame stabilization, and control uses, and
>
> (ii) The minimum amounts of fuel required to alleviate or prevent
>
> (I) Unanticipated equipment outages, and
>
> (II) Emergencies, directly affecting the public health, safety, or welfare, which would result from electric power outages.

Under FERC rules, no more than 25 percent of the total energy input during any calendar year may consist of fossil fuels–namely oil, natural gas, and coal.

With this simple rule, a qualifying facility can use up to the allowed quantity of fossil fuel for purposes specified in the statute. The standard does require that a small power producer be able to estimate the energy content of the primary energy source. However, for some energy sources, municipal solid waste in particular, energy content is not constant. In such cases, reasonable estimates will suffice. Finally, it should be noted that the fossil fuel limitation applies only to small power production facilities and not cogeneration facilities.

Here, as in the case of a cogenerator, FERC rules prohibit an electric utility or electric utility holding company from owning more than 50 percent of a qualifying facility.

### Ownership Criteria

As noted previously, PURPA provides that a qualifying small power production facility or cogeneration facility must be owned by a person not primarily engaged in the generation or sale of electric power (other than electric power solely from cogeneration facilities or small power production facilities). Regarding this provision, the PURPA Conference Report states that:

> ...(e)lectric utilities may participate in an entity which owns such (qualifying small power production or cogeneration) facilities with other persons, and such entity could qualify under these definitions.

The test in this case is whether the entity which owns the facility is primarily engaged in the generation or sale of electric power other than in connection with its ownership of the cogeneration facilities or small power production facilities.

Thus, either directly or through a subsidiary company, an electric utility could participate in the ownership of a qualifying cogeneration or small power production facility.

However, FERC rules provide that if, based on the proportion of ownership by electric utilities, public utility holding companies, or subsidiaries of either, more than 50 percent of the entity which owns the cogeneration or small power production facility is comprised of these electric utility interests, then the facilities are not qualifying facilities.

Note that nothing limits the extent of utility ownership or operation of cogeneration or small power production facilities. The Conference Report reads:

> ...it is also the intention of the conferees that the definition of "qualifying cogeneration facility" and "qualifying small power production facility" will not be construed as prohibiting or discouraging electric utilities from cogenerating.

Utilities may not, however, qualify for the benefits under PURPA if their ownership interests exceed the limits set forth in the rules.

## PROCEDURES FOR OBTAINING QUALIFYING STATUS

A cogeneration or small power production facility which meets the criteria for qualification is a qualifying facility. There is no mandatory case-by-case qualification procedure.

However, any owner or operator of a qualifying facility must furnish notice to the Commission. The contents of the notice must contain the information required of an applicant for qualifying status.

An optional procedure is provided whereby the owner of a small power production facility may, should it prove desirable, file an application with FERC for certification that the facility is a qualifying facility. The application must contain enough information to enable the Commission to make an accurate finding that the facility should be certified.

Each application must contain the name and address of the applicant and the location of the facility, a brief description of the facility including a statement indicating whether such facility is a small power production facility or a cogeneration facility, the primary energy source to be used by the facility, the rated power production, and the nature and amount of ownership interest held by electric utilities, or public utility holding companies, or by any person owned by either.

Applications by owners or operators of small power production facilities must also contain the location of the facility in relation to any other small power production facilities within one mile of the facility owned by the applicant which use the same energy resources, and information identifying any planned usage of natural gas, oil or coal.

An application by a cogenerator must contain the date installation of the facility commenced, a description of the cogeneration of the facility, including whether the facility is a topping or bottoming cycle facility, and sufficient information to determine that any applicable efficiency or operating requirements have been met.

Within 90 days of the filing of a complete application for qualifying status, the Commission will issue an order granting or denying the application, extending the time for issuance of an order, or setting the matter for hearing. If no order is issued within 90 days of the filing of the application, it shall be deemed to have been granted.

An electric utility is not required to purchase electric energy from a facility with a design capacity of 500 kilowatts or more until 90 days after the facility notifies the utility that it is a qualifying facility, or 90 days after the facility has applied to the FERC.

The Commission may revoke the qualifying status of a facility if it ceases to comply with any of the statements contained in its application for Commission certification.

While the application procedures described above are optional, certification by FERC is sometimes required to finance a cogeneration project.

## RIGHT TO SELL POWER TO UTILITIES

Most QFs will want to sell power to utilities at some point. Even if the QF is used primarily for on-site generation, its owner will want to be able to generate power if the processing plant is down for any reason. Of course, the right to sell power, and the rates received, is extremely significant for the owner of a QF built primarily to sell power to a utility.

### Statutory language

Section 210(b) of PURPA provides that in requiring utilities to purchase electric energy from a qualifying facility, FERC must ensure that the rates for the purchase are just and reasonable to the electric consumers of the purchasing utility, in the public interest, and nondiscriminatory to qualifying facilities, but that they not exceed the incremental costs of alternative electric energy (the costs of energy to the utility, which, but for the purchase, the utility would generate itself or purchase from another source).

The use of the word incremental is significant. Under the principles of economic dispatch, utilities generally turn on last and turn off first their generating units with the highest running cost. At any given time, an economically dispatched utility can avoid operating its highest-cost units as a result of making a purchase from a qualifying facility. The utility's avoided incremental costs and not average system costs are used to calculate avoided costs. With regard to capacity, if a purchase from a qualifying facility permits the utility to avoid the addition of new capacity, then the avoided cost of the new capacity and not the average embedded system cost of capacity should be used.

FERC's rules implementing PURPA provide that electric utilities must purchase electric energy and capacity made available by qualifying cogenerators and small power producers at a rate reflecting the cost that the purchasing utility can avoid as a result of obtaining energy and capacity from these sources, rather than generating an equivalent amount of energy itself or purchasing the energy or capacity from other suppliers. To enable potential cogenerators and small power producers to estimate these avoided costs, the rules require electric utilities to furnish data concerning present and future costs of energy and capacity on their systems.

FERC defined "avoided costs" as the costs to an electric utility of energy or capacity or both which, but for the purchase from a qualifying facility, the electric utility would generate or construct itself or purchase from another source. It includes both the fixed and the running costs on an electric utility system which can be avoided by obtaining energy or capacity from a qualifying facility.

FERC listed the following factors which should be taken into account in setting avoided costs:

> The costs which an electric utility can avoid by making such purchases. These generally can be classified as "energy" costs or "capacity" costs. Energy costs are the variable costs associated with the production of electric energy (kilowatt-hours). They represent the cost of fuel and some operating and maintenance expenses. Capacity costs are the costs associated with providing the capability to deliver energy; they consist primarily of the capital costs of facilities.
>
> The availability of capacity or energy from a QF during the system's daily and seasonal peak periods, including:
>
> - the ability of the utility to dispatch the QF;
> - the expected or demonstrated reliability of the QF;
> - the terms of the contract, including the duration of the obligation, termination notice requirement and sanctions for non-compliance;
> - the extent to which scheduled outages of the QF can be usefully coordinated with scheduled outages of the utility's facility;
> - the usefulness of energy and capacity supplied from a QF during system emergencies, including its ability to separate its load from its generation;
> - the individual and aggregate value of energy and capacity from QFs on the electric utility's system; and
> - the smaller capacity increments and the shorter lead times available with additions of capacity from QFs.
>
> The costs or savings resulting from variations in line losses from those that would have existed in the absence of purchases from a QF, if the purchasing electric utility generated an equivalent amount of energy itself or purchased an equivalent amount of electric energy or capacity.

FERC left it up to the individual state regulatory authorities to determine the avoided costs for utilities under their jurisdiction.

## STATE DETERMINATION OF AVOIDED COSTS

The translation of the principle of avoided capacity costs from theory into practice has proven to be an extremely difficult exercise, being, by definition, a determination based on estimation and forecasting of future occurrences. Accordingly, FERC has left the states and nonregulated utilities "flexibility for experimentation and accommodation of special circumstances" with regard to implementation of rates for purchases. Therefore, to the extent that a method of calculating the value of capacity from qualifying facilities reasonably accounts for the utility's avoided costs, and does not fail to provide the required encouragement of cogeneration and small power production, FERC said it would be considered as satisfactorily implementing the Commission's rules.

States followed a variety of approaches in setting avoided cost.

California followed a negotiated approach which took the capital cost of a new utility-constructed gas turbine and the incremental heat rate of the state's utility marginal units. Energy costs for the first ten years would be paid on the basis of the state's forecast of future gas prices, irrespective of the fuel used and the actual price of gas.

New York followed the differential requirements approach. The differential revenue requirement approach uses load forecasts with an expansion planning model to develop expansion plans both with and without the estimated output of alternative energy facilities. The resulting two expansion plans are then used as inputs for two runs of the utility financial planning model which yields the utility's projected revenue requirement both with and without the existence of purchases from QFs. The difference in the present value revenue requirements of these two expansion plans is the avoided revenue requirement made possible by the expected output from alternative energy facilities. This avoided revenue requirement would encompass both projected avoided energy and capacity costs as well as other factors.

Florida and Ohio adopted the specific unit approach. This approach uses the expected capacity and running costs of a future baseload unit (generally a coal-fired unit) to estimate the long-run avoided costs of capacity and energy. The capacity costs are annualized over the expected life of the generation facility to yield an annual capacity cost per kW. The process of annualizing the total capacity costs may involve either the application of a carrying charge to the total capacity cost per kW or the use of present value and annuity calculations.

Other approaches also were used: the purchased power approach, the "reverse-the-meter" approach, and the simplified average incremental cost approach, among others.

None of these approaches proved very satisfactory. Various problems were soon identified with the manner in which state determinations were made.

First, administrative determinations regarding utilities' avoided costs have often been cumbersome, slow and inconsistent.

Second, some state commissions have implemented the avoided cost rule by considering only the cost that the utility would incur by producing itself the power to be purchased from a QF. This procedure may not accurately identify the utility's avoided cost because it may not consider the utility's opportunities to purchase lower cost electric energy from alternative sources.

Third, computer models involved forecasts bearing no relationship to either real world developments or a cogenerator's costs. Administrative determinations of avoided costs necessarily placed the focus on the utility's marginal cost of production, not on a cogenerator's actual costs of construction and operation.

As a result, most states that now need new capacity have turned to some form of competitive bidding as an alternative means of determining avoided costs.

Here, as with administrative determinations, FERC has left the states considerable discretion as to the manner in which bidding programs are to be implemented. In 1988 FERC did issue a notice of proposed rulemaking to provide some uniformity, but the Commission withdrew this document after it was widely criticized.

Consequently, state programs have continued to evolve and vary widely from state to state. A history of the evolution of competitive bidding programs is contained in Section E.

However, administrative determinations of avoided costs are still important for on-site generators who are looking for an occasional home for their power.

## ARBITRAGE

FERC rules differentiate between "old" and "new" production in connection with simultaneous purchases and sales. The rules require an electric utility to purchase at its avoided cost the total output of a facility, construction of which was commenced after the date of issuance of these rules, even if the utility simultaneously sells energy to the facility at its retail rate. The effect of this rule is to separate the production aspect of a qualifying facility from its consumption function. Under this approach, the electric output of a facility is viewed independently of its electric needs. Thus, if a cogeneration facility produces five megawatts, and consumes three megawatts, it is treated the same as another qualifying facility that produces five megawatts, and that is located next to a factory that uses three megawatts.

The rules thus permit a cogenerator, at its option, to indulge in the fiction that the electricity it generates for its own use is simultaneously purchased by a utility and sold back to it. The purpose for imposing this "paper transaction" on utilities is to permit cogenerators to take advantage of the difference between a utility's real rates, typically set by the state commissions on the basis of embedded or average costs, and cogeneration purchase rates set under FERC rules on the very dissimilar basis of incremental cost.

While a utility's incremental costs will vary, often widely, over any period of time, the rates it charges its customers typically do not vary over that time period, but are based on an average of the utility's costs. Furthermore, the utility's rates are based not on the incremental costs of producing the last, most expensive kilowatt hour, but rather on the average cost of all kilowatt hours. Consequently, by proceeding under the Commission's bookkeeping fiction, the cogenerator is able to require the utility to pay to it the difference between the utility's incremental cost at any given time and its average cost-based rates for electricity.

This effect, sometimes referred to as the FERC/PURPA "arbitrage" provision, can increase the potential incentive for cogeneration in particular, because it not only gives the qualifying facility the benefit of "avoided cost" prices for its surplus generation, but also makes it possible for the cogenerator to receive a payment roughly equal to the differential between the utilities' rates and its incremental costs for its entire electric output, even if it is almost all used on site.

One should note, however, that in the current economic climate, the "arbitrage" procedure often is unusable, as avoided costs today are often less than average costs. The reason for this however, is not because the electric utility industry has again become a declining cost industry, but rather because "capacity" payments are not available in many states. Capacity payments are only made to cogenerators whose operation permits a utility to avoid the construction of new capacity. Today, with much capacity lying idle in the recession, new capacity plans are being canceled. Thus, cogenerators are only being paid avoided "energy" cost payments in many areas. Avoided energy costs alone, even incremental energy costs, generally will not exceed energy and capacity costs together, even though the energy and capacity costs used in computing rates paid by users is based on average costs.

## NET ENERGY BILLING

A simple alternative method of billing, more commonly used for calculation of purchases and sales from and to smaller cogenerators, is the net energy billing procedure. Under this procedure, no attempt is made to calculate avoided costs. Instead, the meter is set to run backwards when electricity is being generated in excess of plant requirements. In circumstances in which avoided costs exceed retail rates, this would not be a desirable method of dealing with a utility. However, if avoided costs are below retail rates, this method could result in a premium to cogenerators above avoided cost payments. Particularly if there is no time of use metering, a cogenerator could purchase high cost, peak power, and replace it with low value, off peak power.

## PERIODS DURING WHICH PURCHASES ARE NOT REQUIRED

Finally, one should note an electric utility will not be required to purchase energy and capacity from qualifying facilities during periods in which such purchases will result in net increased operating costs to the electric utility. This condition can occur during light loading periods. If a utility operating only base load units during these periods were forced to cut back output from the units in order to accommodate purchases from qualifying facilities, these base load units might not be able to increase their output level rapidly when the system demand later increased. As a result, the utility would be required to utilize less efficient, higher cost units with faster start-up to meet the demand than would have been supplied by the less expensive base load unit had it been permitted to operate at a constant output.

The result of such a transaction would be that rather than avoiding costs as a result of the purchase from a qualifying facility, the purchasing electric utility would incur greater costs than it would have had it not purchased energy or capacity from the qualifying facility. A strict application of the avoided cost principle would assess these additional costs as negative avoided costs which must be reimbursed by the qualifying facility.

In order to avoid the anomalous result of forcing a qualifying facility to pay an electric utility for purchasing its output, FERC rules provide that any electric utility which seeks to cease purchasing from qualifying facilities must notify each affected qualifying facility prior to the occurrence of such a period, in time for the qualifying facility to cease delivery of energy or capacity to the electric utility. This notification can be accomplished in any reasonable manner determined by the state regulatory authority.

Any claim by an electric utility that such a light loading period will occur or has occurred is subject to such verification by its State regulatory authority as the state authority determines necessary or appropriate either before or after its occurrence. Moreover, an electric utility which fails to provide adequate notice or which incorrectly identifies such a period will be required to reimburse the qualifying facility for energy or capacity supplied as if such a light loading period had not occurred.

This does not override contractual or other legally enforceable obligations incurred by the electric utility to purchase from a qualifying facility. In such arrangements, the established rate is based on the recognition that the value of the purchase will vary with the changes in the utility's operating costs. These variations ordinarily are taken into account, and the resulting rate represents the average value of the purchase over the duration of the obligation. The occurrence of such periods may similarly be taken into account in determining rates for purchases.

## RIGHT TO PARTIAL REQUIREMENTS SERVICE

One way utilities used to deter on-site generation was to refuse to sell power to any customer which self-generated, or to make those rates so high as to make self-generation economically unattractive. Some utilities even refused to interconnect with the self-generator.

PURPA sought to remedy both these problems.

## OBLIGATION TO INTERCONNECT WITH QUALIFYING FACILITIES

Section 210(a) of PURPA requires that each electric utility offer to sell electric energy to qualifying facilities. State law ordinarily sets out the obligation of an electric utility to provide service to customers located within its service territory.

However, it is possible that a qualifying facility located outside the service area of an electric utility might require back-up, maintenance, or other types of power.

State and local law limit the authority of some electric utilities to construct lines outside of their service area. Accordingly, while FERC requires electric utilities to serve any qualifying facility, an electric utility is only required to construct lines or other facilities to the extent authorized by State or local law. As a result, a qualifying facility outside the service area of a utility may be required to build its line into the service area of the utility.

Under FERC regulations, an electric utility must make any interconnections with a qualifying facility which may be necessary to permit purchases from or sales to the qualifying facility. A state regulatory authority or nonregulated electric utility must enforce this requirement as part of its implementation of FERC rules.

Under general Federal Power Act principles, an electric utility cannot be required to interconnect until appropriate notice has been given, and interested parties have been given an opportunity to request a hearing. Before such an order is issued, FERC must first find that the interconnection:

- is in the public interest;
- would (a) encourage overall conservation of energy or capital; or (b) optimize the efficiency of the use of facilities and resources; or (c) would improve a utility system's reliability;
- is not likely to result in a reasonably ascertainable uncompensated economic loss for any utility or cogenerator;
- will not place an undue burden on any party;
- will not unreasonably impair the reliability of the utility; and
- will not impair the utility's ability to render adequate service to its customers.

### SUPPLEMENTARY, BACKUP, INTERRUPTIBLE AND MAINTENANCE POWER

PURPA requires utilities to sell power to QFs at a rate which is (1) just and reasonable and in the public interest, and (2) nondiscriminatory.

FERC rules defined this obligation as giving qualifying facilities the right to demand supplementary, backup, interruptible and maintenance power from utilities.

Supplementary power is electric energy or capacity used by a facility in addition to that which it ordinarily generates on its own. Thus, a cogeneration facility with a capacity of ten megawatts might require five more megawatts from a utility on a continuing basis to meet its electric load of fifteen megawatts. The five megawatts supplied by the electric utility would normally be provided as supplementary power.

Back-up or maintenance service provided by an electric utility replaces energy or capacity which a qualifying facility ordinarily supplies to itself. FERC rules authorize certain facilities to purchase and sell simultaneously. The amount of energy or capacity provided by an electric utility to meet the load of a facility which simultaneously purchases and sells will vary only in accordance with changes in the facility's load; interruptions in the facility's generation will be manifested as variations in purchases from the facility. In such a case, sales to the qualifying facility will not be backup or maintenance service, but will be similar to the full requirements service that would be provided if the facility were a non-generating customer .

Back-up power is electric energy or capacity available to replace energy generated by a facility's own generation equipment during an unscheduled outage. In the example provided above, a cogeneration facility might contract with an electric utility for the utility to have available ten megawatts, should the cogenerator's units experience an outage.

Maintenance power is electric energy or capacity supplied during scheduled outages of the qualifying facility. By prearrangement, a utility can agree to provide such energy during periods when the utility's other load is low, thereby avoiding the imposition of large demands on the utility during peak periods.

Interruptible power is electric energy or capacity supplied to a qualifying facility subject to interruption by the electric utility under specified conditions. Many utilities have utilized interruptible service to avoid expensive investment in new capacity that would otherwise be necessary to assure adequate reserves at times of peak demand. Under this approach utilities assure the adequacy of reserves by arranging to reduce peak demand, rather than by adding capacity. Interruptible service is therefore normally provided at a lower rate than noninterruptible service.

Supplementary, backup and maintenance rates (which collectively often are called "standby service") are superficially similar to any of the service tariffs for which a utility has the primary responsibility to design rates. However, these classes of service are unique in a critical way.

With other tariffs, utilities generally have little incentive to discriminate between classes of customers. So long as they receive enough income to fulfill their total revenue requirement, they may be treated as neutral arbiters.

However, in serving customers with on-site generation, utilities are potentially aiding their competitors. Thus, more than in other areas of rate design, the role of the utility commission in reviewing the equity of utility partial requirements service tariffs is crucial.

## FERC GUIDANCE ON PARTIAL REQUIREMENTS SERVICE TARIFFS

The basic problem of setting rates for partial requirements service is one of ensuring that a utility recovers all its costs for providing that service without overcharging. This is an especially difficult task for standby service: a utility must be "on call" to provide such service without knowing when it will be required to provide the power and how much power it will be obligated to supply. Most customers have fairly steady demand patterns from day to day or week to week. However, the demand of self-generators will be erratic since unit outages are not predictable. Therefore, the rate design for the demand charges must incorporate pricing features than can follow the customer's use.

FERC regulations and case law give some guidance in establishing partial requirements service.

The regulations provide that rates for sales of backup or maintenance power shall not be based upon an assumption (unless supported by factual data) that forced outages or other reductions in electric output by all QFs on an electric utility's system will occur simultaneously, or during the system peak, or both. They also state these rates should take into account the extent to which scheduled outages of the QFs can be usefully coordinated with scheduled outages of the utility's facilities.

In Oglethorpe Power, a 1986 case, FERC read PURPA as requiring that utilities offer standby service to cogenerators on an interruptible basis. It provided two possible outs to this requirement. First, a waiver could be sought provided the state jumped through certain procedural hoops. Second, the interruptible tariff can be set at a level which reflects no savings, i.e., a rate could be the same for interruptible as for non-interruptible service. If, for example, a utility had excess supply, the rates could be the same since there is no advantage to the utility and its ratepayers in offering interruptible service.

In 1988 the Commission rejected the Florida Commission's partial requirements tariff on the grounds that it failed to offer interruptible rates (without offering a justification for its action) and that Florida didn't establish separate rates for standby and maintenance service. It also required Florida to justify to a state court a provision that self-generating customers pay a special fee designed to cover utility costs associated with transmission and distribution facilities which the utility must maintain to provide service when needed (a so-called "minimum reservation charge"). Finally, FERC criticized, but did not overturn, a "ratchet" for usage above the amounts of power reserved. (Ratchets are an increase in fixed charges for a period of time.

Several things were notable about this decision.

First, FERC required states to establish separate rates for standby and maintenance service. Theoretically, it should be possible to reduce the costs of providing maintenance service by scheduling maintenance for times when the utility has low marginal costs of operation. The Florida Commission rolled standby and maintenance service into a single tariff, stating that as a practical matter it was difficult to distinguish between the costs of the two types of service – i.e, there was no consistent, verifiable benefit to scheduling downtime.

Second, the minimum reservation charge for self-generators was upheld in theory. Full requirements customers are not always charged this fee, but instead may pay for their share of these costs through per kWh charges. FERC said that, while the minimum reservation charge is discriminatory on its face, it may not be discriminatory in its impact.

## STATE COMMISSIONS AND PARTIAL REQUIREMENTS SERVICE

State decisions have revealed three key issues in establishing these tariffs for partial requirements service.

First, what will be the availability of the units for which partial requirements services will be requested? A corollary issue is how often, and for how long, will service be rendered?

PURPA and FERC regulations require that rates for QFs must recognize the special characteristics of self-generators' operations and that the same principles used for other customers are to be applied to pricing QF services. Establishment of a traditional rate schedule depends upon a cost of service study, using defined usage patterns, to establish the revenue target for the class. Data drawn from billing histories is then used to design a rate that will collect the appropriate revenue.

In the early days of PURPA there was a great deal of controversy about the availability of cogeneration facilities. As experience has indicated that these facilities have extremely high availability rates (often higher than 90%), the contentiousness of this issue has declined somewhat.

A second issue that has proven troublesome is: What utility facilities are required to provide access to auxiliary services, to what degree will they be utilized, and how is the cost to be recovered?

Here, there is a need to distinguish between generation, transmission and distribution.

We have discussed most of the issues related to allocating the costs of generation in Section A and partial requirements service is usually determined in accordance with the functional and causal principles in use in each state. Some would argue that demand charges for backup or maintenance generation should be minimal or nonexistent where a utility has excess capacity, or where these services are interruptible.

Pennsylvania has decided that nonpeak maintenance power should be priced the same as peak interruptible maintenance power. On the other hand, peak noninterruptible maintenance power should be priced the same as firm power. (One should note that maintenance power is generally provided by contract specifying that it not be used during system peak period.)

Standby rates generally are set based on a probabilistic assessment that a self-generator will need power during the peak period. They also may be tailored to the individual customer. Some standby customers may have a higher likelihood of requiring power on system peak than others. For example, a customer which depends on windmills may have a greater likelihood of requiring service during the peak period demand, if peak generally occurs on windless summer days.

The demand charge for the generation component of supplementary power is generally the same as that for firm customers taking a similar level of power.

The transmission and distribution functions generally are poorly understood by customers. Many feel that if they have no use for these systems in a particular month, they should not pay anything. Yet the systems must be able to handle the full load of the self-generator whenever the unit goes off-line. The costs of both the transmission system and the distribution are related to the maximum power demanded and not to the amount of energy delivered over time.

Many states impose a reservation charge similar to that used in Florida to cover all or a large fraction of the costs of transmission and distribution. Others cover it through the demand charge for partial requirements service imposed on self-generators.

The third issue is: How appropriate is the use of demand ratchets?

In setting rates for backup service, the utility usually will propose a demand charge for the quantity of power which the customer may need. We have discussed the setting of the appropriate demand charge. The issue here is establishing the quantity of power to which the demand charge will apply (the so-called "billing demand").

Many backup rates have a minimum billing demand of a small percentage of the customer's potential needs. The billing demand is the highest of (a) the minimum billing demand or (b) a percentage of the highest of the maximum demand established during an on-peak period in the preceding year or (c) a lower percentage of the maximum demand established during the off-peak period of the billing month. Demand is determined by the highest usage during any fifteen minute period.

Some have argued that these so-called "demand ratchets" overly penalize a cogenerator for a one-time shutdown. As a result, it often is necessary to look at the totality of the backup rate structure. If a utility has a relatively low demand charge for backup power, regulators are likely to approve a formula yielding a higher billing demand.

## COMING ISSUES IN PARTIAL REQUIREMENTS SERVICE

As more competition enters the electric industry, new issues are likely to come to the fore. This section addresses some of them.

- *Unbundling of services.* Most utility services are bundled – i.e., all components of service are covered by a single price structure. Though the notion of partial requirements service involves some unbundling, most utilities still bundle generation and transmission costs. As individual customers gain freedom to seek out alternative suppliers, rates will have to include a separate transmission component in addition to distribution. Also they are likely to be some charge for regulating services (e.g., reactive power).

- *Rights of the returning customer.* Many cogenerators use natural gas. If gas should again become relatively expensive, many may seek to return to the system. This raises a number of issues, such as their right to return, the notice they should have to give, and the price at which the utility is obligated to serve them. For example, should the returning cogenerator be assured service at rolled in average costs, or only at marginal cost? Would such treatment be illegal discrimination?

- *Appropriate demand charge.* Is any demand charge appropriate for the utility with excess capacity? If a utility has been recovering demand-related costs in its energy charges, might it actually charge cogenerators a higher demand charge than the ordinary customer? Under what circumstances?

- *Rate stability for cogenerators.* Attractive backup rates are the equivalent of a big sign saying "cogenerators welcome". What happens when a utility no longer needs additional capacity? Should there be vintaging of rates for cogenerators, much as there is for avoided cost rates? Contracts rather than tariffs?

- *Need for multiple cogenerator classes.* Different cogenerators are likely to have different cost of service characteristics – due to reliability, fuel choice, operating economics, etc. Should these cogenerators receive separate treatment? Might they be covered by formula in a single tariff which accounts for operating characteristics?

## RESPONSIBILITIES OF COGENERATORS

### Interconnection Costs

Each qualifying facility must reimburse any electric utility which purchases capacity or energy from the qualifying facility for any interconnection costs. The interconnection costs of a facility which is already interconnected with the utility for purposes of sales are limited to any additional expenses incurred by the utility to permit purchases.

State commissions have the responsibility and authority to ensure that the interconnection requirements are reasonable, and that associated costs are legitimately incurred.

For qualifying facilities with a design capacity of 100 kw or less, interconnection costs may be assessed on a class basis, and the standard rates for purchases established for classes of facilities of this size might incorporate these costs. State commissions may also determine interconnection costs for qualifying facilities with a design capacity of more than 100 kw on either a class average or individual basis.

There has been some controversy regarding the manner of reimbursement of the utilities for interconnection costs. Potential owners and developers of qualifying facilities generally prefer the costs be amortized on a reasonable basis, because paying a large lump sum payment would be a considerable obstacle to the program. Electric utilities generally prefer payment up front, although several permit amortization for credit-worthy facilities. FERC has left the manner of reimbursements (which may include amortization over a reasonable period of time) to the states.

### System Emergencies

Except as provided under section 202(c) of the Federal Power Act (FERC's emergency authority), no qualifying facility may be compelled to provide energy or capacity to an electric utility during an emergency beyond the extent provided by agreement between the qualifying facility and the utility.

A qualifying facility is not required to make available all of its generation to the utility during a system emergency. Such a requirement might interrupt industrial processes, with resulting damage to equipment and manufactured goods. Many industries install their own generating equipment in order to ensure that, even during a system emergency, their supply is not interrupted. To put this supply in jeopardy could result in the discouragement of interconnected operation and a resultant discouragement of cogeneration and small power production. FERC therefore provides that the qualifying facility's obligation to provide energy and capacity in emergencies be established through contract.

In order to receive full credit for capacity, a qualifying facility must offer energy and capacity during system emergencies to the same extent that it has agreed to provide energy and capacity during non-emergency situations. For example, a 30 megawatt cogenerator may require 20 megawatts for its industrial purposes, and thus may contract to provide 10 megawatts of capacity to the purchasing utility. During an emergency, the cogenerator must provide the 10 MW megawatts contracted for to the utility; it need not disrupt its industrial processes by supplying its full capability of 30 megawatts. Of course, if it should so desire, a cogenerator could contractually agree to supply the full 30 megawatts during system emergencies. The availability of such additional backup capacity should increase utility system reliability, and should be accounted for in the utility's rates for purchases from the cogenerator.

An electric utility may discontinue purchases from a qualifying facility during a system emergency if such purchases would contribute to the emergency. In addition, during system emergencies, a qualifying facility must be treated on a nondiscriminatory basis in any load shedding program – i.e., on the same basis that other customers of a similar class with similar load characteristics are treated with regard to interruption of service.

Credit for capacity will also take into account the ability of the qualifying facility to separate its load and generation during system emergencies. However, the qualifying facility may well be eligible for some capacity credit even if it cannot separate its load and generation.

## EXEMPTION FROM FEDERAL AND STATE LAWS AND REGULATIONS

Section 210(e) of PURPA states that FERC shall prescribe rules under which qualifying facilities are exempt, in part, from the Federal Power Act, from the Public Utility Holding Company Act of 1935, from the State laws and regulations respecting the rates, or respecting the financial or organizational structure of electric utilities, or from any combination of the foregoing, if FERC determines such exemption is necessary to encourage cogeneration and small power production. The Commission has made liberal use of its exemption authority in order to remove the disincentive of utility-type regulation.

Originally, Section 210(e)(2) of PURPA provided that FERC was not authorized to exempt small power production facilities of 30 to 80 megawatt capacity from these laws. An exception was made for small power production facilities using biomass as a primary energy source. Such facilities between 30 and 80 megawatts may be exempted from the Public Utility Holding Company Act of 1935 and from State laws and regulations but may not be exempted from the Federal Power Act.

Legislation passed in 1990 removed the size limit on eligible solar, wind, waste and geothermal small power production facilities for which notice or application for Commission certification is submitted to the Commission by December 31, 1994, and for which construction generally commences by December 31, 1999.

Section 210(e)(3)(C) of PURPA provides that no qualifying facility may be exempted from any license or permit requirement under Part I of the Federal Power Act dealing with hydropower. FERC issued simplified procedures for obtaining water power licenses for hydroelectric projects of 5 megawatts or less.

Cogeneration and small power production facilities could be the subject of an order under section 202(c) of the Federal Power Act requiring them to provide energy if the Economic Regulatory Administration determines that an emergency situation exists. Because application of this section is limited to emergency situations and is not affected by the fact that a facility attains qualifying status or engages in interchanges with an electric utility, FERC qualifying facilities are not exempted from section 202(c) of the Act..

FERC regulations exempt QFs from:

- all rate regulation under the Federal Power Act;
- all Federal Power Act regulation relating to the securities of public utilities;
- PUHCA (to the extent permitted by law); and
- State laws and regulations which would conflict with the state's implementation of the Commission's rules under section 210.

## RETAIL SALES

A retail sale occurs when the owner of a cogeneration facility seeks to sell power to an unaffiliated end-user. No retail sale occurs when the owner sells power to a utility or the owner and end-user are the same entity.

Many cogeneration facilities are owned by third party investors who plan to sell steam and electric power to the owner of an industrial site which allows for a feasible cogeneration project. PURPA does not prohibit these sales, (although neither does it protect them) as will be seen in the cases below.

**Federal Regulation**

The FERC has for some time held that qualifying status under PURPA may not be denied to a cogeneration facility which is owned by a third party developer which sells all of its electric output to an end user. The leading case here was decided in 1984. Here PRI Energy Systems, Inc sold individual cogeneration units to various individuals and business concerns that are able to use the heat and power produced. PRI retained ownership of the facility and sold energy to the users. The local utility Hawaiian Electric Company, Inc. (HECO) challenged PRI's certification as a qualifying facility under PURPA on the grounds PRI intended to engage only in direct retail sales of electricity to end users

In deciding the case in favor of PRI, the Commission reasoned that under the statute, a cogeneration facility is qualifying if it meets the requirements established by the Commission respecting (1) fuel use, (2) fuel efficiency, (3) reliability and (4) ownership. None of the criteria involve a consideration of the type of purchaser to whom the sale may be made by the cogeneration facility.

In another significant case, Alcon challenged the Puerto Rico Electric Power Authority (PREPA) for its refusal to sell backup power to third party financed cogeneration projects. PREPA agreed that O'Brien Engineering, the project developer, was entitled to partial requirements service. However, it said it had to obligation to Alcon, the purchaser of O'Brien's power. Alcon countered by arguing that regardless of the divided ownership, the industrial plant owned by Alcon, and the cogeneration facility, owned by O'Brien Engineering, were really a single facility, so that either of the co-owners should qualify to receive backup power from PREPA

FERC initially sided with PREPA, then reversed itself, citing the impact its original decision would have on PURPA's intention of encouraging cogeneration development. Saying it did not believe third party ownership was relevant in this case, it held that the right to backup power covers both the production and consumption functions, irrespective of whether they have the same ownership.

In the Fin-Lin case FERC held that a cogeneration plant which served both of its owners qualified for PURPA treatment, though the two owners were geographically separated and one of the facilities did not receive steam from the project.

In this case Union Carbide and Fina Oil and Chemical agreed to form a partnership to be called "Fin-Lin," that would establish and operate a cogeneration facility at a site adjacent to the Fina Plant. In 1987 the partners filed a joint application with FERC requesting QF status for the facility. The application stated the facility would provide Fina with electrical power and steam and would transmit electrical power, but not steam, to the Union Carbide plant located 1.7 miles away. Fin-Lin would oversee the facility's management, but Fina would run its day-to-day operations on behalf of Fin-Lin.

Over the arguments of the local utility, Gulf States Utilities, FERC ruled that the fact that a proposed cogeneration facility would be owned and operated by multiple owners and would produce energy to be sold among the owners was irrelevant to the Commission's consideration of the unit's QF status. Matters relating to the making and pricing of retail electric sales fall under state commission jurisdiction, the FERC said, and are of no consequence to federal certification inquiries.

The D.C. Circuit sent the case back to the Commission, saying it did not understand why the Commission had rejected Gulf States' argument that the distance between Union Carbide's plant and the cogeneration operation's physical location in combination with the fact that Union Carbide was taking only electricity, but not steam from the operation, made Union Carbide "essentially ... a retail purchaser of surplus electricity". It further questioned why Union Carbide was deemed an integral portion of the plant, thereby entitling it to obtain backup power from the utility.

On remand, the Commission gave four reasons why it had determined Union Carbide was an "integral component" of the cogeneration facility:

1. Union Carbide was a part owner of the power producing component as well as a consumer.
2. Union Carbide's plant 1.7 miles away is in "close proximity".
3. The power line used to transmit electrical power to Union Carbide is a private line indicating that the Union Carbide plant is a "part of an integral industrial operation".
4. There was a long-standing supplier-customer relationship between Union Carbide and Fina before they entered into this joint venture. The former sold the latter nitrogen.

This time the D.C. Circuit upheld FERC, though it acknowledged that it was not entirely comfortable with the Commission's reasoning. "[W]e certainly do not mean to suggest that the Commission is totally free to bestow QF status on all sorts of loose arrangement of cooperating plants – thereby undermining the customer base of public utilities." It questioned, for instance, whether multiple plants at an industrial park, could form a cogeneration joint venture, bound together only by relative propinquity and a shared objective of producing energy more cheaply than the local utility company. Nevertheless, it said these matters probably need to be decided on a case-by-case basis.

The decision clearly leaves the retail sales issue up to the states. It is generally agreed PURPA allows a state to block retail sales. If the state allows them, however, PURPA and FERC's regulations probably preclude them from regulating the rates of any retail sales in their jurisdiction which result from QFs selling power to end-users.

**State Regulation**

Most states which have decided this issue have done so on the basis of a state statute.

State statutes differ widely. Retail sales in Georgia and Florida are prohibited; California limits a QF to two retail customers. In New York they are permitted so long as the cogeneration unit has QF status, is under 80 MW, and the retail customer is at or near the site of the cogeneration plant. In Texas a cogenerator may make retail sales to a steam customer of the plant.

# Section C
## Engineering and Cogeneration

# Section C
# Engineering and Cogeneration

## SELECTING AN ENERGY EFFICIENT COGENERATION SYSTEM TO MEET INDUSTRIAL PROCESS NEEDS

In the face of increasing energy costs, does cogeneration really offer an effective way to conserve energy while reducing the energy costs per unit of product?

There is no simple answer. A cogeneration plant is capital intensive. The matching process is critical. Development of an optimum design concept is both plant-specific and site-specific.

Energy managers in some major plants may have access to a sizeable in-company engineering department experienced in conceptual design of power systems and able to analyze their performance in terms of energy and economy. Most energy managers, however, must seek outside assistance, first from consultants and later from architect and engineering firms.

The optimal end result in both cases is the same: A conceptual design of a cogeneration system which meets the needs of the industrial process and is economically viable.

A clear understanding of both the potential and limitations of such a system underlies the energy manager's ability to direct such a conceptual design while staying within acceptable cost limits. He must be acutely aware of the design's interaction with power and heat demand and with economic performance as a function of energy costs and escalation rates.

This discussion will address these questions from the viewpoint of the end user, addressing important technical aspects of cogeneration.

### Basic Energy Use in Manufacturing

At the most elemental level, an industry will use fossil fuel to produce heat and will buy electricity from the local utility. The fuel will be burned to produce hot gases for:

- direct applications, such as drying
- indirect applications, such as salt baths,
- use in steam generators (boilers), and
- use in combustion engines doing mechanical work, such as gas turbines.

In all of these conversion processes, only a part of the original heat content of the fuel is utilized. The remainder is rejected. Unless this reject heat is put to good use, it can contribute only to global entropy increase.

Combustion gases are carriers of energy and supply the conversion equipment with "available heat". This is the heat content which can be converted to work by reducing the temperature of the carrier medium, in this case, of the combustion gases. During the conversion process, the combustion gases get cooler and the amount of available heat is gradually reduced.

Ideally, useful energy can be extracted from the combustion gases until they reach ambient temperature. At that point, the heat content becomes "unavailable". In an ideal conversion process, an equivalent between work and the change in available heat is expressed:

$$1\text{kWH} = 3{,}413 \text{ Btu}$$

In the real, unideal, world, all conversion processes are afflicted with losses. Thus, an internal combustion engine will convert only one third of the available energy into useful work at the coupling. One third will be dissipated through the cooling system. One third will be expelled through exhaust gases.

A typical utility, for example, will generate electricity at an efficiency of about 34 percent or at a heat rate of 10,000 Btu per kW-hr.

The available heat of a medium is measured in Btu per pound. Since we are dealing primarily with industrial equipment such as heat exchangers and adiabatic power generators under steady flow conditions, the heat content of the medium involved is equal to the enthalpy h (in Btu/lb).

In an industrial process, the available heat can be utilized as process heat. The heat output is then measured in MBTU/hr or MW thermal ($MW_t$ in short). If the available energy is converted to power, it is normally measured in electrical terms as MW electrical or $MW_e$.

After power and/or process heat has been extracted from the combustion gases, a certain amount of reject heat is still available. The quality of the reject heat (or enthalpy) in BTU/lb varies according to the conversion process. It can be quite high in the case of gas turbine exhausts or it can be quite low, as with water returning from the cooling tower of a central power plant.

**Cogeneration: Use of Reject Heat**

Cogeneration is a process by which the reject heat of one process becomes a source of energy for a subsequent conversion process. If power is generated first and the reject heat is used as a process heat, the process is called "topping". If the reject heat of an industrial process is used secondarily to generate power, then we have a "bottoming" configuration.

In the bottoming configuration, the heat provided by the combustion gases goes entirely into raising the level of the heating medium to the required enthalpy level h1. The exhaust at a rather high level h2 is then further reduced by the generation of power to h3. No additional fuel is required by adding power generation through bottoming.

This is not the case in topping. The power generator must reject process heat at the required level, h1. This means that the working medium has to be raised now to a value of h0, which is higher than h1 and requires the burning of additional fuel.

Thus, power generation by bottoming is gratis in terms of fuel consumption. Power generation by topping requires the burning of additional fuel.

Steam is a very convenient heat transfer medium in many industrial processes. Thus, steam generators play a very important role in both topping and bottoming. The following remarks, therefore, will deal primarily with cogeneration options providing power and steam to industrial plants.

**Prime Mover Options**

The "prevailing state of the art" supplies us today with three major cycles for conversion of heat to energy, each with adequate warranties as to performance and integrity. They are the Rankine (steam turbine) cycle; the Brayton (combustion turbine) cycle; and the Diesel (diesel engine) cycle. A fourth, the Otto cycle, has been used successfully in Europe for residential and commercial buildings with very small heat demands.

The Rankine cycle is always a closed-cycle operation, in contrast to the others. (A closed-cycle, coal-fired gas turbine has been used in Europe, but due to the high price of European coal today, none are being built at this time.)

A closed-cycle operation has the important advantage of being able to use practically any type of fuel, including solid waste. Open-cycle machines are limited to liquid and gaseous fuels. Gas turbines generally burn natural gas or light distillates although residual oil can be used, after appropriate preparation, in the large capacity units.

The achievable power/heat ratio is a function of the highest temperature incurred by the working medium in the cycle. The peak temperature is lowest in the case of the steam turbine, highest in the diesel.

Selection of the appropriate prime mover is determined by: the power/heat ratio delivered and the incremental heat rate in Btu/kWh required to generate power.

The power/heat ratio is a function of many factors and must be determined on a case-by-case basis. The incremental heat rate (the additional amount of fuel needed to generate topping power) is cycle-specific and is critical to a system's energy efficiency.

Typical values illustrating the performance capabilities of various prime movers are cited in the accompanying table. This table also contains a combined-cycle configuration, consisting of a gas turbine trailed by a steam turbine which tops the process steam.

This table highlights the facts that: highest fuel utilization is achieved by the steam turbine, the lowest by the diesel; the combined-cycle outperforms other contenders as an energy-saver; and a steam turbine converts most of the source energy to steam, the diesel converts most to power. Thus, the diesel is a fine power generator, but a poor steam generator.

The table shows representative values only and should be used with care. The exact determination of the achievable power/heat ratio at design and part load for selected prime movers as a function of the demand characteristics is a critical step in any feasibility study and conceptual design work.

### The Matching Process

Ranges of the power/heat ratios achievable by the various prime movers show that a broad band of industries fall within the operating range of steam turbine, gas turbine, and combined cycle cogeneration. The choice of the appropriate cogeneration system in terms of energy is straightforward, provided that the power-heat ratio is constant or varies only over a limited range.

The choice is not at all straightforward where the coincidence between power and steam demand is poor. In some industries, for example, seasonal effects (such as high electricity demand, low heat demand in summer and the reverse in winter) play a major role.

Such cases leave a cogenerator with three options: 1) operate as an island, and meet both heat and power demands, 2) meet the power demand at all times and generate associated steam, and 3) meet the steam demand at all times and generate the associated electricity.

Tracking the system demand is, in most cases, the most favorable operating mode for an industrial cogeneration plant. In steam tracking, the sizing of the cogeneration equipment and the selection of its operational mode are the most critical elements.

### Integration with Production Plant

Cogeneration will be most energy- and cost-efficient where it is designed as an integrated subsystem of a completely new processing plant. The heat and mass balance can then be carefully adjusted to encompass all major mass and energy streams in the total plant, maximizing the utilization of source energy.

Design of an optimum cogenerator is far more difficult if it is retrofitted to an existing process plant. This is particularly true if power and steam demands are not in phase or do not change proportionately to each other.

It is highly inefficient and uneconomical to design a cogeneration plant to meet peak demands. Efficiency and economy are best for a cogeneration plant that operates at design point or capacity continuously for many hours during the year. Economy of scale favors plants which meet a large heat demand.

In retrofitting, it is essential to obtain a complete data base on the mass and energy flow through all major pieces of equipment. This establishes whether systems can be introduced to assist the cogenerating plant to operate at capacity for a high percentage of the time. Other energy systems often can be incorporated into the cogeneration concept.

### Data Base for Optimization

A data base which contains all of the parameters that have a major impact upon plant selection and performance is critical to the design of a cogeneration plant. These essential data fall into two major groups: the first deals with mass and heat flows, the second with economics.

From a technical viewpoint only, a well-designed data base for choosing the proper cogeneration option in the proper size will encompass these points:

#### Main Plant

Needed to develop load duration curves for both heat and power:

- Required heat content and mass flow for process heat stream(s)
- Power demand, with daily and seasonal variations
- Heat demand, with daily and seasonal variations
- Peak power and heat requirements
- Continuous demand component separate from temperatures dependent demand.

#### Other Plant Components

Need to investigate "smoothing" potential for heat demand curve.

- Typically: chilling equipment; hot water generating equipment; heat pumps; etc.

#### Fuel Data

For siting and operation considerations:

- Fuel availability and limitations, including combustible wastes
- Storage capability
- Dual fuel capability
- Environmental impact (SOx, NOx, particulates)

Table C-1

## CHARACTERISTICS OF COGENERATION TECHNOLOGY

| COGENERATION CHARACTERISTICS | UNITS | STEAM TURBINE (BACKPRESSURE) | GAS TURBINE | COMBINED CYCLE (BACKPRESSURE) | DIESEL |
|---|---|---|---|---|---|
| Power/steam ratio | kWh/million Btu steam | 50 | 200 | 250 | 400 |
| | Btu power/Btu steam | .171 | .683 | .853 | 1.365 |
| Incremental heat rate | Btu/kWh | 4700 | 5800 | 5500 | 6500 |
| | Btu fuel/Btu steam | .235 | 1.160 | 1.375 | 2.600 |
| Cogeneration | | | | | |
| fuel consumption | Btu fuel/Btu steam | 1.488 | 2.365 | 2.375 | 3.805 |
| power/fuel ratio | Btu power/Btu fuel | .115 | .289 | .359 | .359 |
| steam/fuel ratio | Btu steam/Btu fuel | .672 | .423 | .421 | .263 |
| fuel utilization | | 78.7 | 71.2 | 78.0 | 62.2 |
| Separate generation fuel consumption | Btu fuel/Btu steam | 1.705 | 3.205 | 3.705 | 5.205 |
| Savings by cogeneration fuel consumption | Btu fuel/Btu steam | .22 | .84 | 1.33 | 1.4 |
| | % | 12.9 | 26.2 | 35.9 | 26.9 |

DIAGRAM ILLUSTRATIONS OF
TOPPING CYCLE COGENERATION SYSTEMS

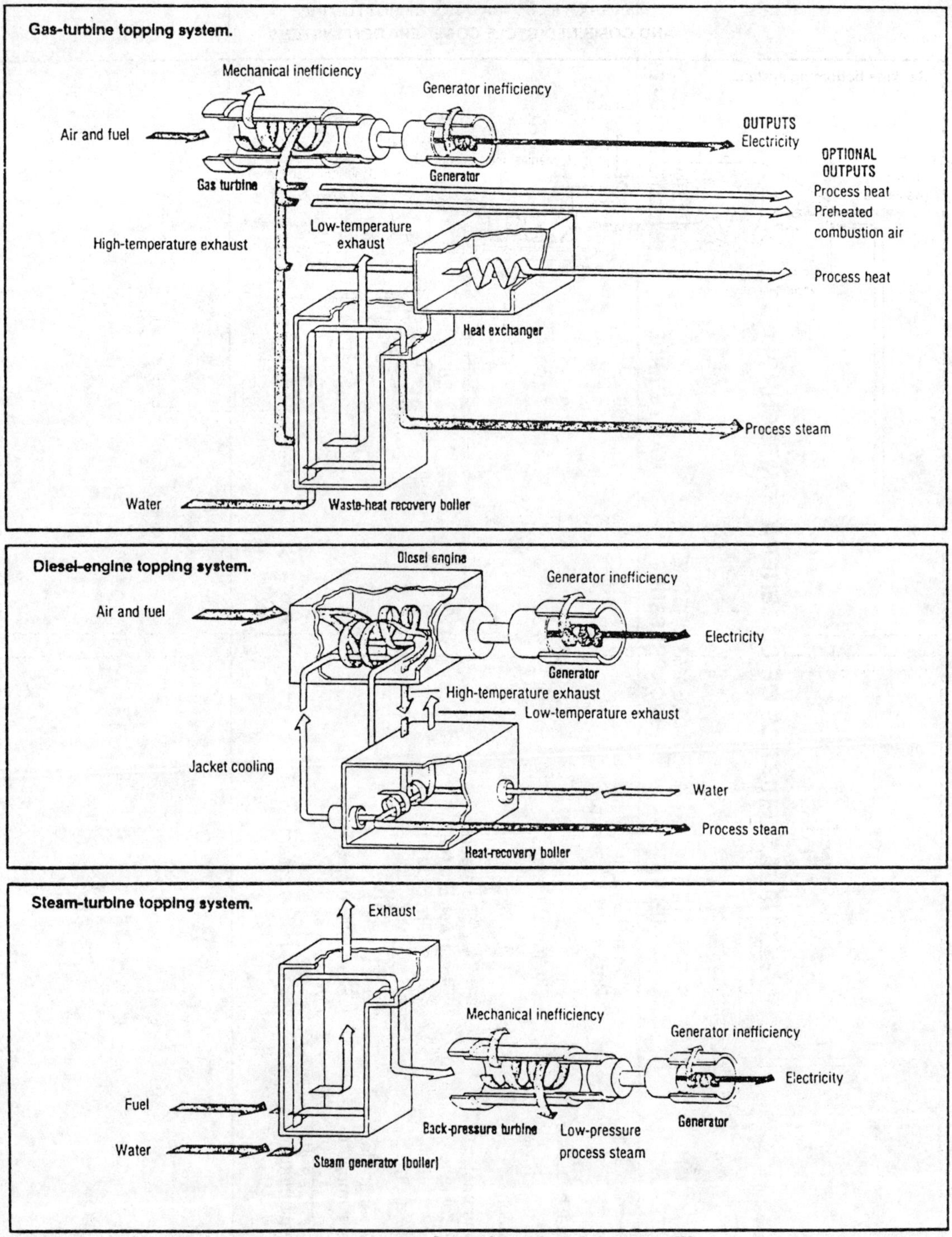


Source: Department of Energy-Resource Planning Associates, Inc. Cogeneration: Technical Concepts Trends Prospects, Sept. 1978

**Figure C-1**

DIAGRAM ILLUSTRATIONS OF BOTTOMING
AND COMBINED CYCLE COGENERATION SYSTEMS

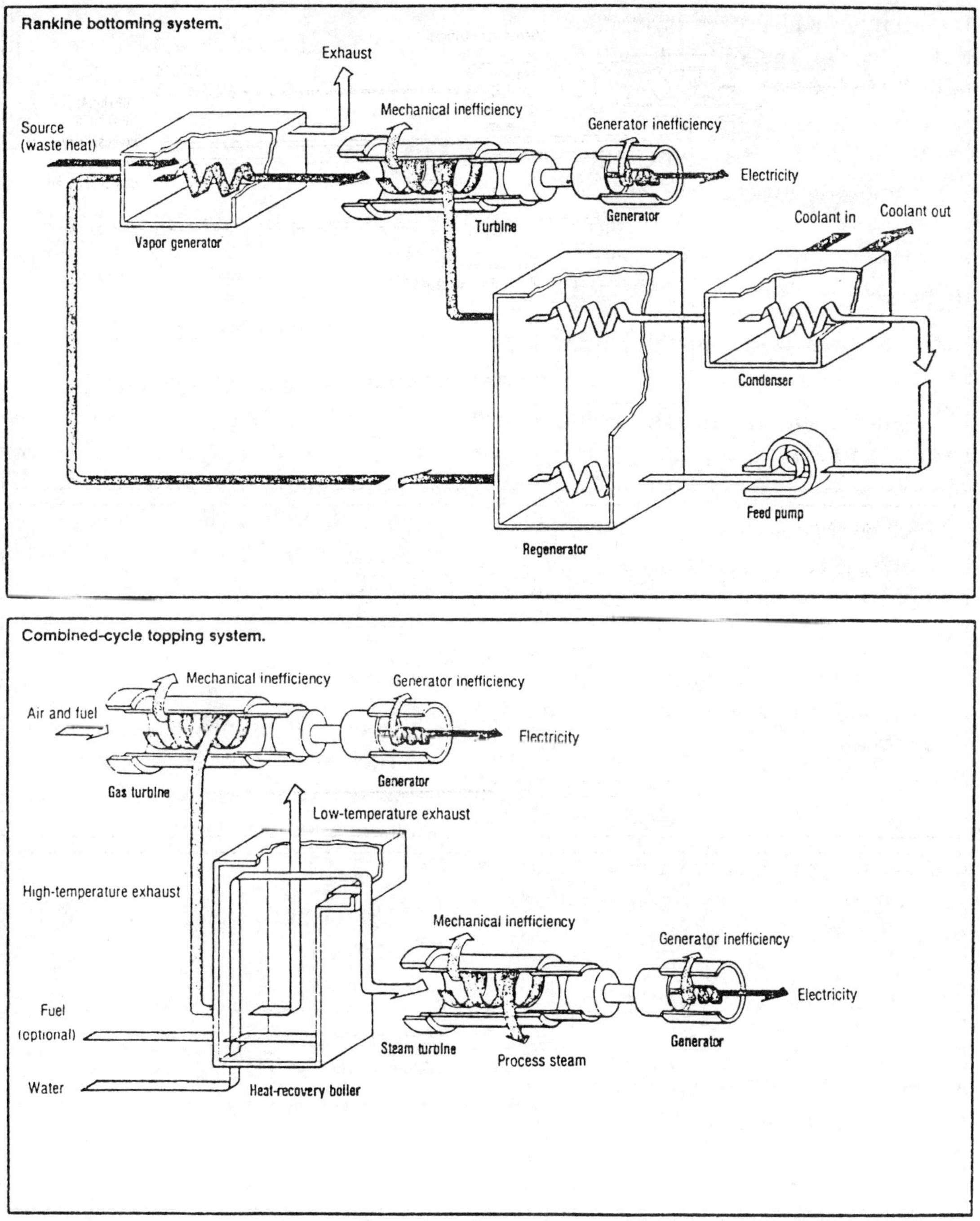


Source: Department of Energy-Resource Planning Associates, Inc. Cogeneration: Technical Concepts Trends Prospects, Sept. 1978

**Figure C-3**

## OPERATIONAL CONSIDERATIONS

This section considers some of the technical issues which are of concern in the operation of an electric system in parallel with (connected to) a qualifying small power production or cogeneration facility. The emphasis will be on the smaller generating facilities, which are usually connected to distribution lines and receive little individual engineering attention. Larger generating facilities for parallel operation require more detailed engineering attention and are more familiar, and therefore will not be given detailed attention in this section.

This section is organized into the following three parts:

- Part A – Is an introduction to technical questions and concerns which deserve attention.
- Part B – Gives an overview of the characteristics of the types of generation and interfacing apparatus which a qualifying might employ.
- Part C – Summarizes the main concerns of an electric utility in assessing the possible impact of qualifying facilities on the electric system.

See also, Appendix to Section D: Contracts – Parallel Operation of Customer Generation, for a description and diagrams of installation options.

## A. INTRODUCTION TO QUESTIONS AND CONCERNS

The concerns in this section are based upon the rules that require the utility to interconnect with qualifying facilities, to purchase energy and to provide the cogenerator with electric service (supplementary, back-up, interruptible and/or maintenance power) when requested.

To discuss the questions and operational concerns that arise as utilities begin to implement the FERC rules, it will be helpful to identify some of the key issues. In approximate order of importance, they appear to be as follows:

1: SAFETY

To avoid accidental energizing of facilities by independently owned generators, how much reliance can be placed upon isolation devices which are provided by the qualifying facility? What additional apparatus should be required for compliance with work rules and code requirements? What additional work procedures will be required for line repairs?

2: EQUIPMENT AND LINE PROTECTION

How do we assure that line sectionalizing devices will operate correctly with the independent generators operating in various combinations? How do we assure that power system apparatus, the generators and other consumer-owned property will be adequately protected against damage under normal and abnormal conditions?

3: VOLTAGE

What is the probability of problems with voltage flicker, voltage surges and/or voltage unbalance from operation of the generators? Who pays for correction of such problems if they occur?

4: HARMONICS, DISTORTION AND INTERFERENCE

If interference problems result from operation of the cogenerator, who is available to provide assistance? At whose expense? Should generators and converters be required to meet specific wave shape and noise objectives?

5: METERING

What metering practices can be employed to implement the rates for energy purchase and energy sales?

## B. CHARACTERISTICS OF GENERATORS AND INTERFACING APPARATUS

A small power producer may utilize wind, solar, biomass, geothermal, hydroelectric or other renewable energy resources. For cogeneration, the facilities may consist of turbine driven or engine driven generators utilizing steam waste heat or combustible gas. These facilities may be relatively predictable and constant (hydro, geothermal or waste heat from an industrial process), or as variable and unpredictable as the wind.

For any of these energy sources, the electrical apparatus feeding energy into the utility system will be one of three general types: a synchronous generator, commonly referred to as an alternator; an induction generator; or one of several descriptions of static converter to match a dc or variable frequency source to the 60 Hz utility system frequency. The differences in behavior of these "generators" are important to the electric utility.

### 1: GENERATORS FOR PARALLEL OPERATION

The generators used for ac power production are induction generators, which are widely used for intermediate or smaller sized generators, and synchronous generators, which are usually referred to as alternators. Alternators are used in most of the largest capacity installations, including giant steam turbine generators. However, small alternators including permanent magnet types are also appearing in wind energy systems with rated outputs as low as 1 to 10 kW.

#### (a) INDUCTION GENERATORS

An induction generator is like an induction motor, operated as a generator by being driven above synchronous speed. Induction generators are self-regulating in adjusting to varying torques and load conditions, and can act as motors for starting. They present a lagging reactive load to the electric system when generating as well as when acting as motors.

Induction generators, as used for parallel operation, depend on the utility system for control of speed and frequency and also for exciting current, unless capacitators are connected to provide excitation.

Induction generators have advantages over alternators in simplicity of control and start-up, and in some situations, in their ability to re-establish normal operation after electrical and mechanical transients. Also, induction generators automatically cease generating when the utility source is cut off (unless capacitors are connected), thus reducing the possibility of hazardous voltages being produced by the isolated generator.

#### (b) SYNCHRONOUS GENERATORS

Synchronous generators (alternators) must run, like electric clock motors, precisely in step with the electric system frequency. An alternator therefore requires special circuitry and controls for synchronizing before it connects to the electric utility source. Alternators can generate electricity at unity power factor or at leading power factor. Because they are self excited, they can operate independently of the central station source.

Alternators may also operate at variable speed and frequency, and this is done in some applications. The resultant variable-frequency output is then used directly in resistance loads such as heating, or rectified to charge batteries, or fed into a static converter for conversion to 60 Hz for parallel operation with a utility system.

#### (c) DIRECT CURRENT GENERATORS

Direct current generators can be used in complete dc battery systems or, without batteries, to serve loads such as water heating which provide their own energy storage. For ac applications, dc generators (with or without batteries) are used with static inverters to serve ac loads and/or feed energy into the electric utility system. With developing technology, dc "generators" may include photovoltaic or electrochemical sources (fuel cells) as well as rotating machines.

2: STATIC CONVERTERS

The modern converter is a solid state apparatus; without moving parts, which converts ac to dc, dc to ac, or ac to ac at another voltage and frequency. This type of apparatus is usually referred to as a rectifier when used only for ac to dc conversion, or as an inverter when used for dc to ac conversion.

The converter or inverter performs an important function in matching a variable-frequency or dc source to the 60 Hz utility frequency, and by regulating power flow. However, these benefits are bought at a price in terms of additional complexity, losses, reactive power requirements and/or quality (wave form) of the electrical output. These characteristics, and differences between types of converters, should be recognized in engineering decisions involving power production systems with or without converters and, if a converter is to be used, in specifying performance characteristics.

(a) Line Commutated Converters.

Line commutated converters are dependent on an ac source, such as the utility system, for their operation. They are not capable of operating by themselves. They require lagging reactive kvars which may be equal to or greater than the kW rating of the converter. The chopped-wave output of these converters may cause heating in motors, metering errors and electrical interference problems.

(b) Self Commutated Converters

Self commutated converters generally cost more than line commutated converters of equal capacity but are capable of producing a better output wave form, do not require a central station ac source, and can operate by themselves to supply standby power.

The term "synchronous inverter" may be applied to either a line commutated or self commutated device which is triggered by the utility source and therefore synchronizes its output with the utility.

In the part which follows, additional information will be given regarding harmonics, wave shape distortion and interference.

C. CONCERNS WITH PARALLEL OPERATION

The many questions about consumer-owned generation, revolve mostly around the issues relating to costs and benefits. What will be incurred costs by the owner of the qualifying facility to pay for controls, protective apparatus and wiring? What amounts are necessary for the utility to spend for metering, switching and protective apparatus in order to accommodate the power producer? These amounts to the extent that they exceed the utility's costs had they not obtained power from the cogenerator, are chargeable to the cogenerator.

Another issue, less often thought of, is complexity. How does the would-be owner of a cogenerator gain enough knowledge, about such things as siting and apparatus to decide correctly about type of equipment and where and how to install it or, indeed, whether any generator system will deliver the results he expects? Can he manage the operation of a power production facility?

The utility, meanwhile, has its own set of problems in deciding how to operate in parallel with uncontrolled generation while maintaining standards of service in all respects, and in providing terms for purchases and sales of energy that will meet the FERC's rules.

Considering these constraints, what are reasonable costs for production and other apparatus required because of generation by the cogenerator.

1: SAFETY

The first concern that is usually expressed by electric utility people about dispersed generation is whether an uncontrolled generator might possibly be able to energize an isolated section of line that has been cut off from the central station source. The scenario pictured is one in which a fuse or oil circuit recloser operates to isolate a section of line on which one or several generators are operating. The combination of generator(s), line capacitance and load may be such that the generators keep operating with voltages and frequency uncontrolled. The line section could remain energized at a voltage that presents unacceptable hazards to line crews and the public, and could cause damage to utility or consumer property including generators.

The scenario could be aggravated for a three phase generator if one phase of the three phase supply line were disconnected. The generator would concentrate its output on the isolated phase, trying to maintain voltage on the faulted phase. This phenomenon is sometimes experienced with three phase motors, but a generator has greater capability for sustained operation under these conditions.

Accidental energizing is possible with any generator that is capable of independent operation, and this includes alternators and self commutated converters which generate their own triggering signals. Induction generators can operate independently if sufficient capacitors are connected in the vicinity to provide excitation. Capacitors at an induction generator for power factor correction, or line capacitors on the isolated phase(s), may result in sustained output at a voltage and frequency determined by the circuit constants and generator speed. Line commutated converters are not regarded as capable of isolated operation, since they depend on the utility source for commutation (switching to provide ac) as well as for lagging kvars.

Commercially available wind energy systems generally have built-in protection circuitry to cause automatic shut down in response to conditions such as high or low voltage, overspeed, underspeed and frequency variations. A momentary loss of voltage, as from an automatic reclosing fuse or breaker, should result in automatic shut down. Similar features can be included in other small power production apparatus including any which employ a static converter. Unfortunately, however, independently owned apparatus can be modified or exchanged for other components, or protective apparatus can be bypassed or defeated. For example, the owner of a cogenerator experiencing an emergency for lack of electric power might bypass the protection circuitry in order to continue operation.

With safety involved, the utility must be conservative in requiring a disconnecting means under its control to assure isolation of the independent generator. Work rules may require visual verification or other positive assurance of disconnection before crews can work on a de-energized line, thus adding to operations and maintenance expenses.

2: EQUIPMENT AND LINE PROTECTION

Parallel generation introduces variables in distribution line protection in at least two ways. First, the coordination of fuses and other sectionalizing devices may be affected by generator contributions to fault currents. Secondly, consumer-owned generators need to be protected against abnormal power circuit conditions, including operation of utility line protective devices.

(a) Line Sectionalizing

Line sectionalizing studies should be checked to verify correct operation of sectionalizing devices over the range of conditions that could arise with various combinations of generators operating. Small generators, with rated outputs of a few kW or less, are not likely to contribute significantly to fault currents. However, cogeneration or small power production in higher ratings, such as tens of kW or more, may be significant in this respect.

(b) Coordination with Generator Protection

Generators in parallel with the utility system need protection against overload and against operation of sectionalizing devices which momentarily deenergize and then reenergize the line. For example, an oil circuit recloser (OCR) responds to a fault, such as a tree limb against a conductor, by deenergizing the line for approximately one-quarter to one second, and then reclosing to restore service in the event the fault was temporary. This operation may be repeated up to three times, depending on the recloser setting, before the OCR goes to "lock-out", leaving the line deenergized. Such alternative connecting and disconnecting can cause damage to the generator or the driving machine.

Additional problems may arise for a three phase generator if one or two phases, but not all phases from the central station source become disconnected. In such a situation, the generator might partially energize the disconnected phases, with possible damage to utilization equipment of the owner or his neighbor, or to the generator. It is hardly feasible to provide adequate protection in all such situations by means of apparatus on the utility facilities. To be fully effective, protection should be designed and installed as part of the power production facility.

(c) Quality of Voltage

Problems with flicker and with alternately bright and dimming lights are usually associated with varying loads and motors starting currents on the same power circuit. They may also be caused by generators with varying output.

Voltage problems are related quite directly to the size of the generator as compared with the capacity of the circuit to which it is connected. The greatest effects are on the premises of the owner of the generator or others serviced by the same transformer. The likelihood of problems from a single cogenerator can be minimized if the rated output is limited, for example, to 50 percent of the distribution transformer rating; or two to ten kW for single phase unit, depending on distance from the substation; or 25 to 100 kW for three phase units with the largest allowed only in locations near the substation. However, other problems with voltage control may arise as additional cogenerators are connected, if the aggregate capacity of generators becomes significant compared with total load.

If the total capacity of uncontrolled generators becomes large compared with minimum load on a feeder or substation, voltage regulators may be subject to more severe duty, requiring increased maintenance. A wider range of regulation, calling for additional line regulators may become necessary. Electric system operating conditions can vary from minimum load with all generators operating to full load with all generators off.

For reasons of voltage control and stability of operation, limitations are necessary on the maximum unit size and on the maximum total generating capacity that can be allowed on a line or substation without special analyses and possibly costly system improvements. The total rated capacity without special study might need to be restricted, for example, to 50 percent of the line or substation minimum load. Otherwise, the connection costs might need to include the cost of increased circuit capacity or a separate circuit for the power producers. Similarly, special analysis should be required prior to commitments, for service to individual cogenerators larger than those mentioned above, in order to anticipate connection costs that may prove to be prohibitive.

3. HARMONICS, DISTORTION AND INTERFERENCE

Alternating current electric systems and apparatus are designed to operate with a relatively smooth, undistorted wave form. Distortions, when they occur, are in the form of harmonics or multiples of the 60 Hz fundamental frequency; for example, the third, fifth and seventh (180, 300 and 420 Hz). Excessive levels of harmonic voltages and currents may cause errors in watthour meter registration, excessive heating of motors and overloading of capacitors. Lower levels of harmonics may interfere with telephone and other communication circuits (including voice, data and signaling or control circuits) or with ripple-type (audio frequency) load management signals on the power line conductors.

Interference from induction generators or alternators may be heard in a telephone receiver as high pitched, singing sounds, which are familiar to many who use telephones in or into rural areas. Strength of the interfering signals depends on design as well as size of the generators, and also on the locations at which they happen to be connected to the electric line and on whether the line is lightly or heavily loaded. Experience with motors of similar size should be helpful in anticipating whether an induction generator or alternator may cause troublesome interference.

Static converters are more potent sources of interference than rotating generators of equivalent size. The converters (or rectifiers or inverters) cause relatively severe waveform distortions because of the abrupt commutating or switching action which occurs in each half cycle of the voltage wave. The voltage waveform is smoothed when the converter output is connected to the utility line, because the capacity of the utility system is usually very large compared with that of the converter. However, the harmonic currents produced by the converter flow along the electric line and contribute the power circuit "influence" which is a measure of its tendency to cause interference in nearby communication facilities.

The characteristics of the converters that are commercially available in smaller sizes, such as those rated at 10 to 25 kW or less, are such that the harmonics produced are over the entire spectrum, from the lower order harmonics (3rd, 5th, 7th and possibly 2nd, 4th, 6th) through the audio range and into the RF spectrum. Filtering, if required, is likely to be difficult and costly because of the presence of the lower order harmonics.

Converter output wave form can be improved at additional cost by incorporating filtering or other changes in the converter. Pulse width modulation may be employed, utilizing multiple pulses rather than a single pulse each half cycle, to eliminate lower order harmonics and reduce the cost of filtering. Additional information, including suggested distortion limits, is given in IEEE Standard 519-1980, IEEE Guide for Harmonic Control and Reactive Compensation Static Power Converters.

A lightly loaded electric line is particularly vulnerable to interference problems if it is quite long (roughly, 10 miles or more) or if a capacitor is connected on the lightly loaded feeder. Figure C-3 is an illustration of conditions in which a small source of harmonics such as a 2 kW converter could interfere with the telephone circuit shown, or with a ripple control system on the electric line. In this illustration, the capacitor and the series reactance of the electric line cause a resonant condition that amplifies one or more harmonics, causing high harmonic currents between the capacitor and the utility power source and high harmonic voltage at the end of the line. Such a combination may result in a power circuit influence 100 times as great as that of the same circuit without a capacitor or with higher load density.

The electric utility is usually considered to be the party responsible when interference problems arise, since the interfering harmonics are flowing in its lines. Testing, analysis and corrective measures are costly and time consuming, particularly if the offending harmonics are from several sources. Corrective actions, if required, are the responsibility of the owner of the cogenerator, but the burden of proof, of testing, analysis and negotiation, falls upon the utility. Also, it is the utility that is usually called upon for advice and assistance and, if necessary, for the hard decision that a cogenerator causing difficulty must be disconnected.

However, the probability of difficulties due to interference is small, if the sources of harmonics are dispersed and of limited capacity compared with load on the feeder. To minimize difficulties, the following are suggested:

1. Present and prospective owners of cogenerators should be aware of the possibilities of interference problems, their responsibility and the choices of apparatus that might make a difference.

2. Be alert to the particular locations and conditions most likely to cause difficulty.

3. If a problem might result, or if one has occurred, request assistance from your local telephone company for testing and interpretation.

# INTERFERENCE FROM A SMALL SOURCE OF HARMONICS

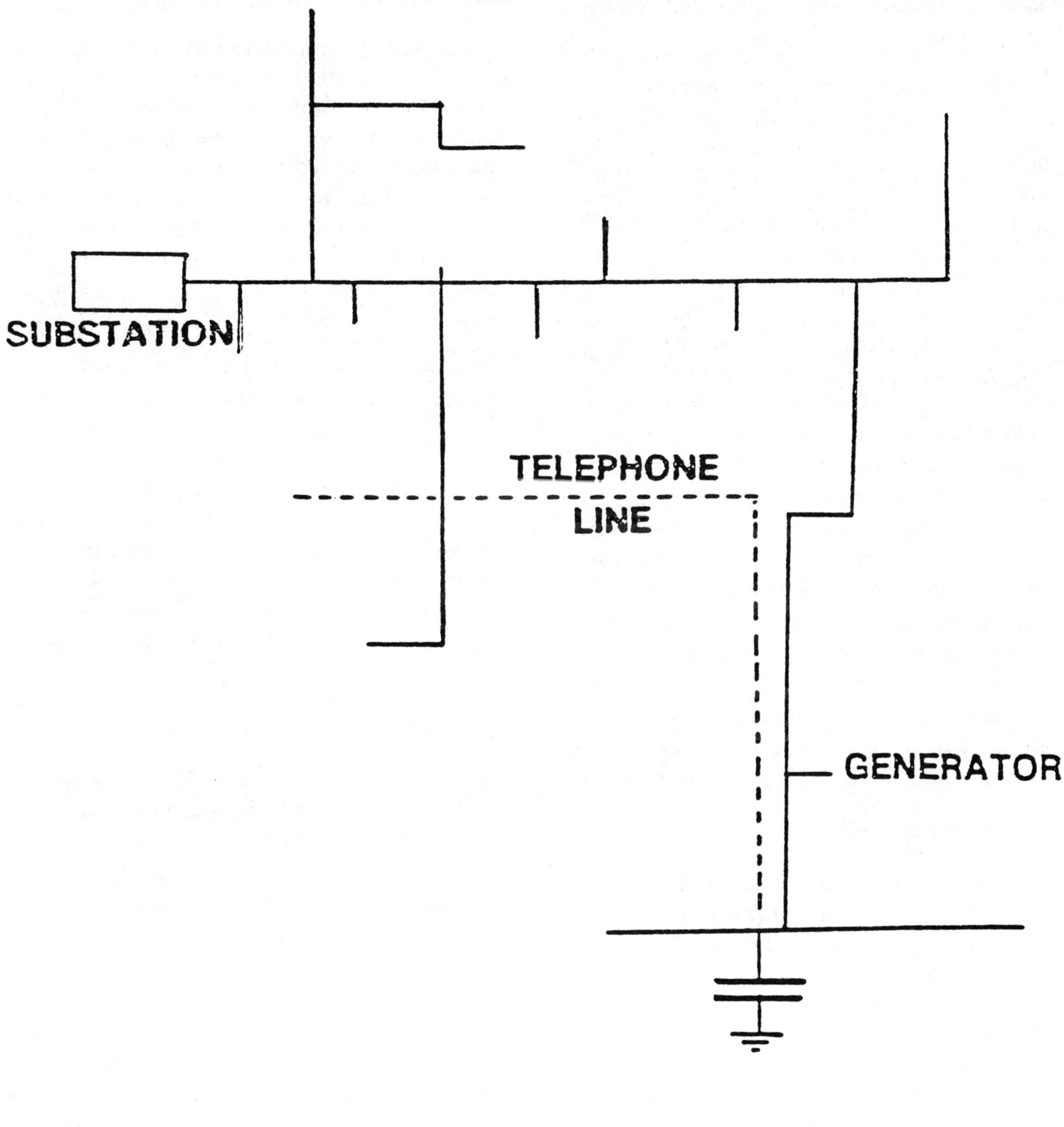


Figure C-3

4: METERING PRACTICES

Rates for the purchases of electric energy from a qualifying facility and rates for sales to the facility should be in accordance with the PURPA rules and FERC regulations described in Section A of this Manual.

Metering requirements are determined by the needs of both the cogeneration facility and the utility. The cogenerator may wish only to sell energy; or to simultaneously sell generated energy and purchase energy (gross sales and purchases); or to sell only surplus energy and purchase supplementary energy when needed (net energy purchase or sale). Metering should be recognized as the means for obtaining data which will permit purchase and sale of energy on terms that are fair and reasonable (not discriminatory), as well as being the "cash register" for sales and purchase.

Elaborate metering, while desirable for obtaining accurate information, can result in unacceptably high connection costs for a small cogeneration facility. On the other hand, when a simplified arrangement with only one or two watthour meters is used, it is important to recognize the significance of the quantities measured, and of other quantities such as load factor and power factor that are not being measured. Low load factor and low power factor tend to increase the costs of service and therefore may be significant in ratemaking.

(a) Net Energy Metering

The simplest metering connection for a consumer who is also a small power producer, is with one watthour meter connected as shown in Figure C-4. In this arrangement, the generator may supply part or all of the electrical load at the owner's service. Only the loads in excess of the generator output, including those when the generator is shut down, are supplied by the utility and registered by the watthour meter. When the generator's output is in excess of the owner's electrical load, then the surplus energy is fed back into the utility system and causes the watthour meter to rotate backward. At all times, the utility source controls the frequency, stabilizes the voltage, supplies back-up power and lagging kvars if required and accepts excess output from the generator when that is necessary.

(b) Surplus Energy Metering

To measure separately the surplus energy generated and the net kWh consumed, two meters may be connected where one measured net consumption whenever the amount used exceeds the amount generated. The other meter operates only to register the surplus energy delivered to the utility when the generator is producing electrical output that is not being used.

(c) Gross Energy Metering

For gross energy metering, two watthour meters may be connected to separately measure the kWh produced and kWh consumed, as in Figure C-5. This arrangement provides more complete data on the electrical energy generated and consumed. Additional metering for reactive power or power factor or for time-of-day may be added if desired.

(d) Engineered Metering

The simple watthour metering methods discussed above should be adequate for small qualifying facilities (under 100 kW), if enough data are available for establishing that the standard rates being applied are fair and nondiscriminatory. However, additional metered data may be needed even on the smaller qualifying facilities until some experience has been gained, to indicate cost-related characteristics for purpose of developing rates. On larger facilities, additional metered data are necessary for more accurate calculations of

# RECOMMENDED ELECTRICAL ARRANGEMENT I

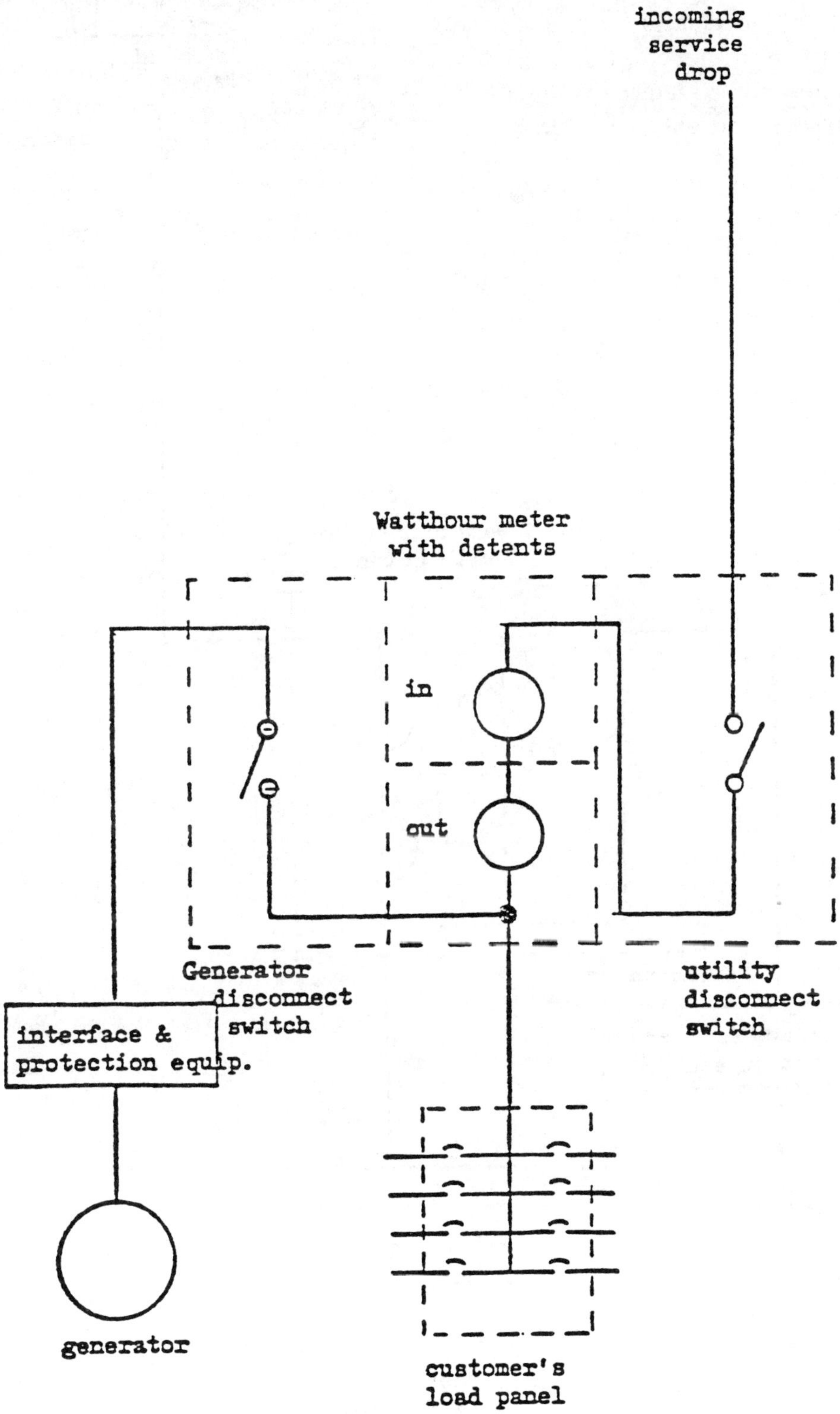


Figure C-4

# RECOMMENDED ELECTRICAL ARRANGEMENT II

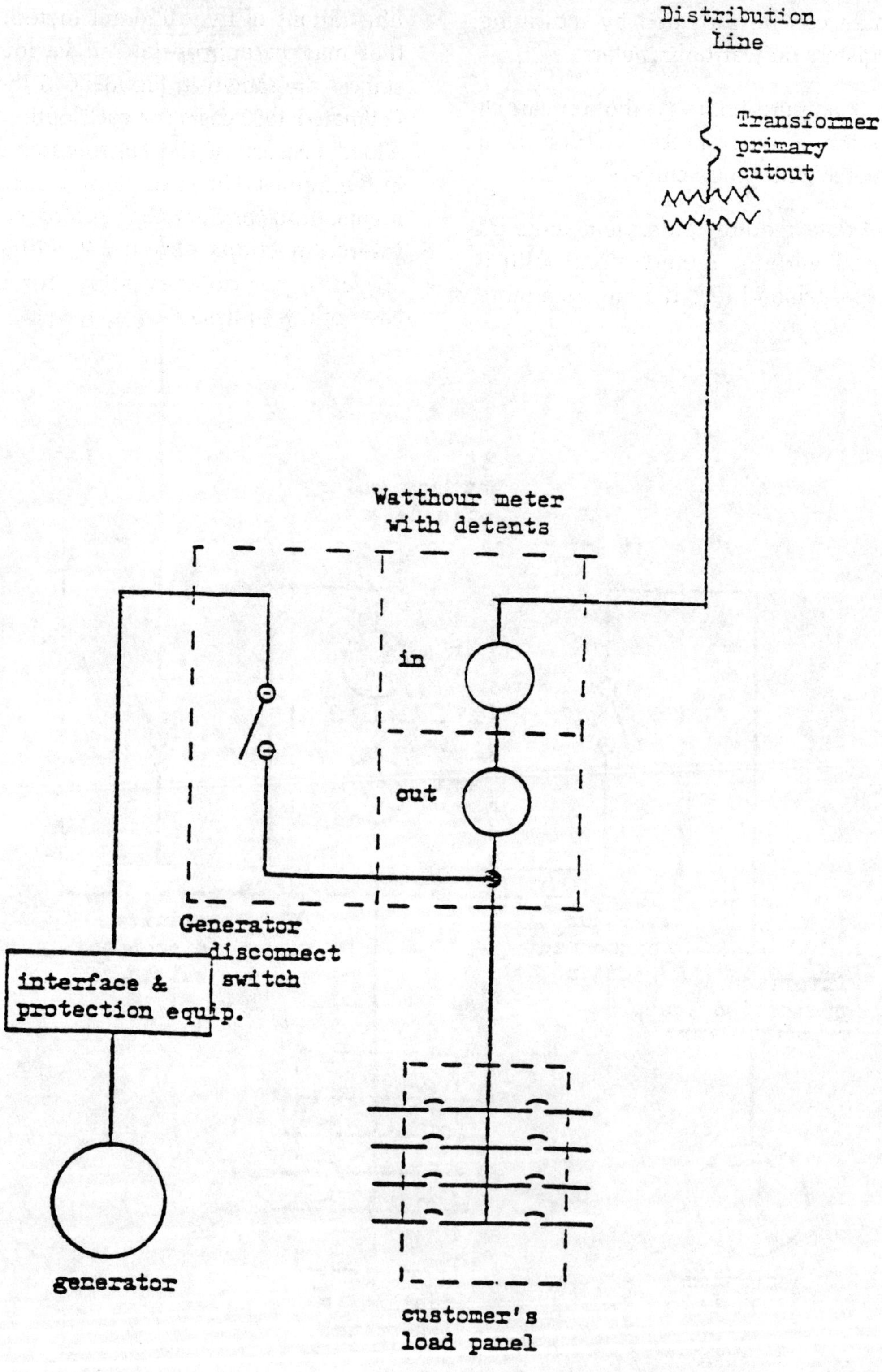


Figure C-5

amounts payed and amounts billed to the cogenerator.

Capacity data can be provided by indicating demand registers on watthour meters.

Reactive volt ampere hours (varhours) can be metered with a varh meter, which is a watthour meter with phase shifters.

Time-of-day data requires pulse generators on watthour and varhour meters along with a magnetic tape demand recorder, unless a more complex and costly telemetering facility is being employed.

Illustrations of five different metering options that may be appropriate in various circumstances are shown in Figures C-6 through C-9. Estimated 1980 costs for each option and costs added because of the cogenerator are shown in the figures. The costs do not include instruments, transformers that might be used for interconnections above 240 volts and 200 amperes, nor do they allow for additional costs of three-phase metering.

## METERING OPTIONS

All metering options (including simultaneous) assume single-phase metering without demand indicators. Estimated 1980 "plug-in" costs are for purposes of "order of magnitude" only. Three-phase metering will probably be more costly. Costs do not include instrument transformers that will be required for interconnections rated above 600 volts or above 200 amps.

Option 2

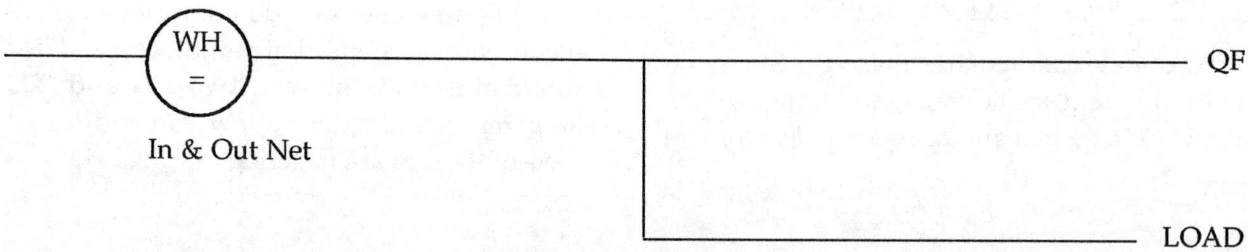


Simplest, assumes rate for billing is the same as the rate for buy back.

| Costs: | | |
|---|---|---|
| | Meter | $ 25 |
| | Labor | 58 |
| | Total | $ 83 |

Option 2

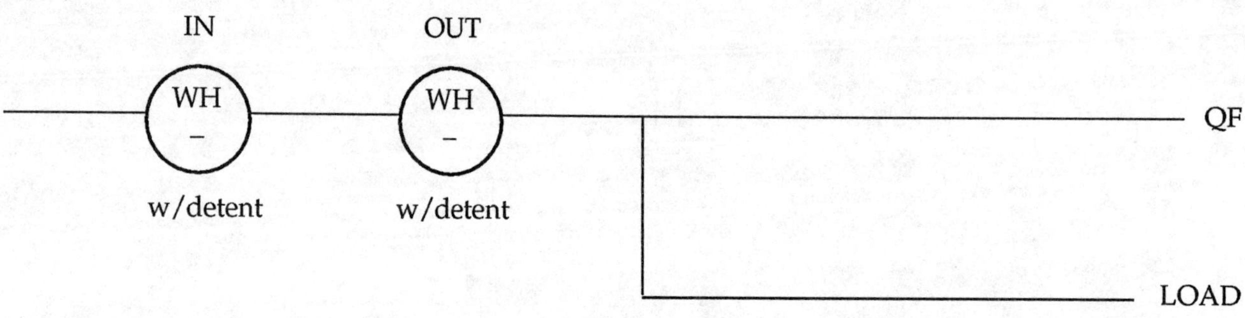


Needed if rate for billing is different than the rate for buy back.

| Costs: | | |
|---|---|---|
| | Meter | $ 72 |
| | Labor | 77 |
| | Total | $149 |

Figure C-6

## METERING OPTIONS

Option 3

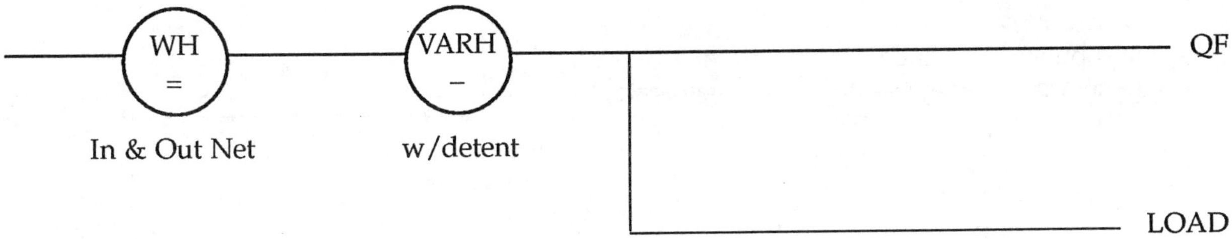


This option is for metering varhours for billing.

| Costs: | | |
|---|---|---|
| | Meters | $ 62 |
| | Phase Shifter | 280 |
| | Labor | 160 |
| | Total | $502 |

Option 4

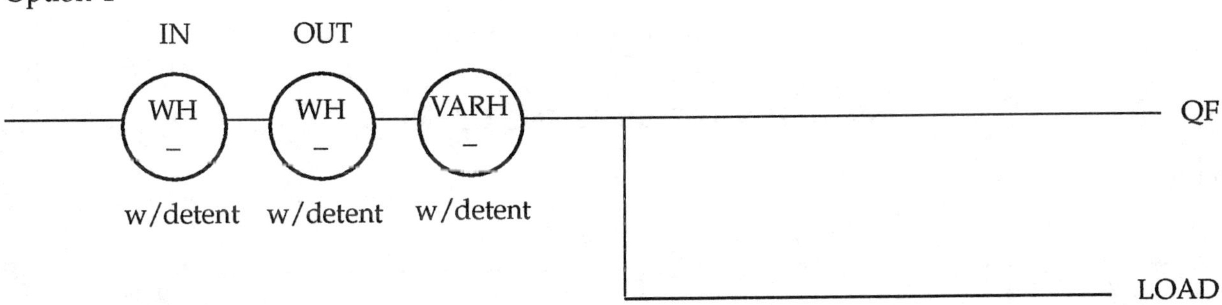


For metering varhours, or determining average power factor, and also gives option of different rates for billing and buy back.

| Costs: | | |
|---|---|---|
| | Meter | $111 |
| | Phase Shifter | 280 |
| | Labor | 200 |
| | Total | $591 |

**Figure C-7**

## METERING OPTIONS

Option 5

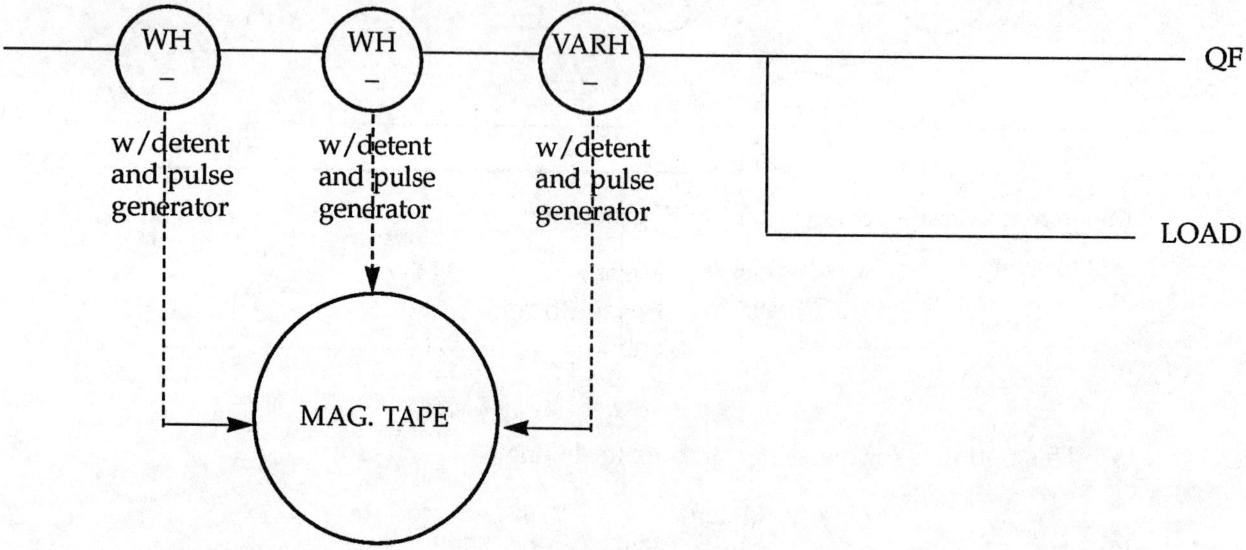


Provides the same metering as Option 4 but has time of day capabilities

| Costs: | | |
|---|---|---|
| | Meters | $ 135 |
| | Phase Shifter | 280 |
| | Magnetic Tape | 750 |
| | Labor | 450 |
| | Total | $1615 |

## SIMULTANEOUS METERING OPTIONS

Option A

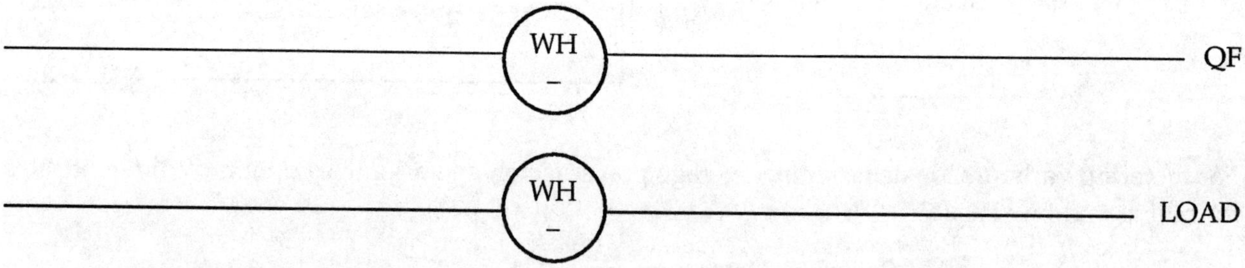


Assumes QF load and QF generation are on separate service laterals.

| Costs: | | |
|---|---|---|
| | Meters | $ 50 |
| | Labor | 77 |
| | Total | $127 |

Figure C-8

Option A

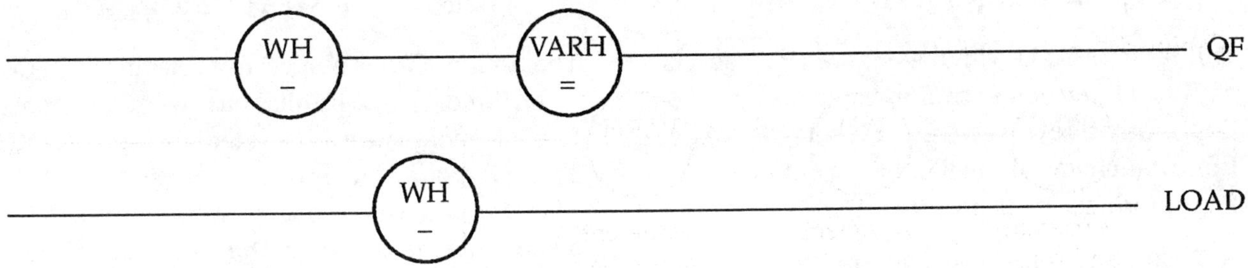


Meters net average power factor of QF generation

| Costs: | | |
|---|---|---|
| | Meters | $ 75 |
| | Phase Shifter | 280 |
| | Labor | 200 |
| | Total | $555 |

Option C

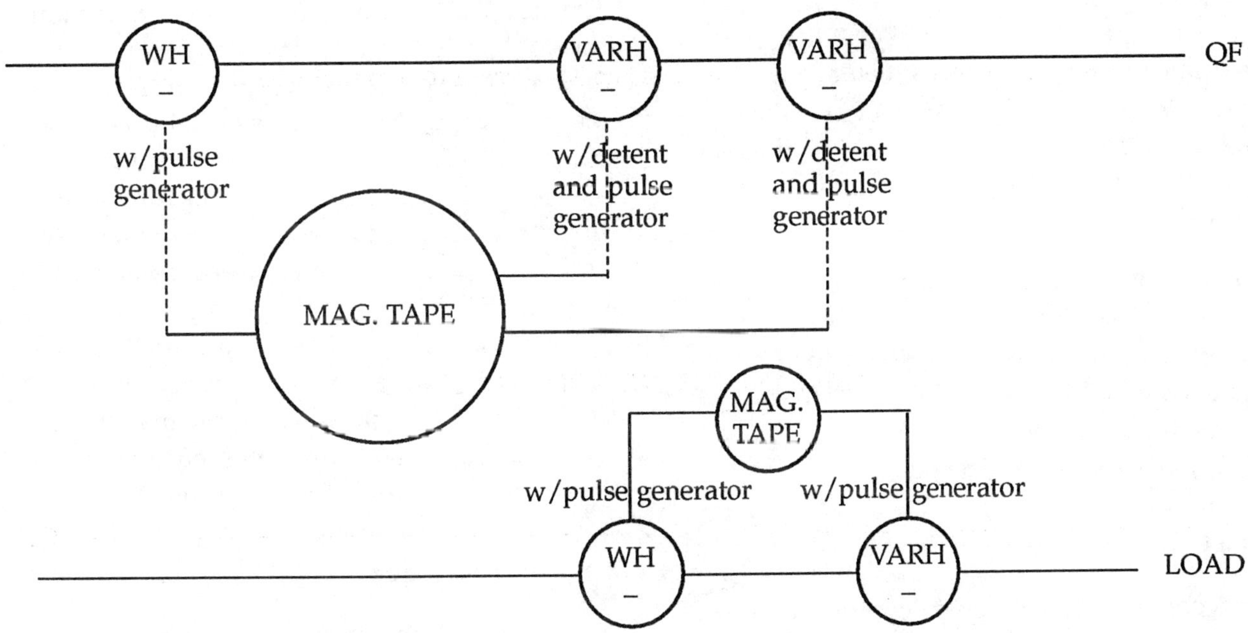


Time-of-day metering. Permits determination of both VARS generated and those required by QF generator. Also, separately meters power factor of load.

| Costs: | | | |
|---|---|---|---|
| | | Meters | $ 192 |
| | (2) | Phase Shifters | 560 |
| | (2) | Magnetic Tapes | 1500 |
| | | Labor | 630 |
| | | Total | $2882 |

**Figure C-9**

## OPTIMUM COGENERATION SYSTEMS ENGINEERING AND DESIGN

This section deals with the engineering aspects of five of the cogeneration systems options. Representative cases were chosen, one each from the chemical, textile, paper, refining and food products industries.

A preliminary design and cost analysis are presented for each of the five optimum system configurations. The analysis includes system schematics, energy balance schematics, system and site layouts (where possible), and preliminary specifications for all major components. The analyses in this study were performed at the preliminary engineering level, prior to undertaking a detailed engineering and economic analysis, to arrive at closer estimates of the feasibility and economic viability of the selected systems. A more detailed analysis would require consideration of the thermal and electrical load coincidences as well as the rate structure of the particular utility serving the site.

For engineers seeking a more detailed and extensive series of studies, The Energy Cogeneration Handbook: Criteria for Central Plant Design, by George Polimeros (Industrial Press Inc., 200 Madison Avenue, NY.NY. 10157) is highly recommended

### The Chemical Industry

The chemical industry is competitive and dynamic. Product lifetimes are generally in the 15 to 20 year range, and as the mix of products changes in a plant so do the thermal and power demands. However, recognizing that substantial gains can be achieved through reducing the use of energy, most large companies have instituted comprehensive energy accounting procedures and housekeeping measures. One of the most visible of these is Union Carbide.

## UNION CARBIDE (CHEMICALS) INSTITUTE, WEST VIRGINIA

The Union Carbide specialty chemicals plant in Institute, West Virginia, has the general pattern of energy use illustrated in figure C-10. The cogeneration option proposed for this plant would involve the interposition of a steam turbine between the 400 psig header and the 75 psig header, replacing the pressure reducing valves. The single automatic extraction noncondensing steam turbine configuration shown in figure C-11 was derived after consulting with Union Carbide on the steam-load demand pattern.

### Heat Balance

The heat balance for the proposed extraction turbine-generator is shown in Figure C-11. Steam is supplied to the turbine from the plant's existing 400 psig steam main. The values shown on the balance were obtained from a U.S. turbine/generator manufacturer using the throttle conditions, extraction pressure, and back pressure which were determined during the initial technical assessment of cogeneration alternatives. There are various minor leakages from valve stems and the steam sealing system that are not shown. These leakages and generator efficiency are included in the 7,700 kW generation.

### System Description

The turbine/generator system would be designed for 400 psig, 580°, 325,000 lb/hr inlet, with a single automatic extraction at 200 psig and a 75 psig exhaust. The turbine would be noncondensing, with the 200 psig and 75 psig flows going to meet process needs. Speed would be governed at 3,600 rpm, and turbine would be direct coupled to an air cooled generator. To minimize additional load on the existing ventilation system, the generator would be totally enclosed, with heat removed from the air by an auxiliary cooling water system with a water-to-air heat exchanger.

The turbine would be a single flow, multistage unit with springbacked; labyrinth type sealing, split-sleeve type bearings, and a mechanical/hydraulic control system. The set would be mounted on a concrete pedestal with extraction and exhaust piping coming from below. It would require a separate lube oil system with reservoir, pumps, and coolers.

### Layout Drawing

The equipment layout drawing for the turbine generator set is shown in Figure C-12. The turbine is pedestal mounted with auxiliaries located on the mat beneath it. Their functions are labeled on the drawing.

### Major Component Specifications

Following is a list of specifications for the turbine/generator:

Turbine/Generator

- 7.7 MW rated output
- 400 psig, 580°F throttle conditions
- Multistage, single casing, single ended
- 325,000 lbs/hr throttle flow, 125,000 lbs/hr automatic extraction at 200 psig
- 75 psig back pressure
- Pedestal mounted (concrete)
- Downward exhaust and extraction piping
- 3,600 rpm
- Shaft packing spring-backed labyrinth type, with steam sealing
- Split-sleeve radial bearings
- Thrust bearing required
- Speed governing system, with 110 percent emergency trip
- Complete with gland-steam condenser, lube-oil reservoir, pumps, and coolers
- Generator enclosed, air cooled, with water-to-air heat exchangers
- Turning gear system required
- Local control panel plus control from a centralized location

### Cost Data

Cost data for the proposed steam turbine/generator are as follows:

| | |
|---|---|
| Turbine/generator base price | $1,765,000 |
| Erection | 250,000 |
| Pedestal and mat | 250,000 |
| Miscellaneous mechanical (piping) and electrical components | 50,000 |
| Total direct costs: | $2,815,000 |
| Indirect Construction Cost (10%), | |
| Contingency (10%), and Engineering (2%) | $500,000 |
| $/kW (net additional plant cogen cap.) | 366 |

Since the unit is to be installed inside an existing structure, no building or site preparation is required. Also, a general contractor may not be needed because the turbine manufacturer could erect the turbine, and the foundation work could be subcontracted. However, piping would have to be installed from the turbine to the existing process steam network. Operations and maintenance costs are estimated to be .20¢/kWh, exclusive of fuel.

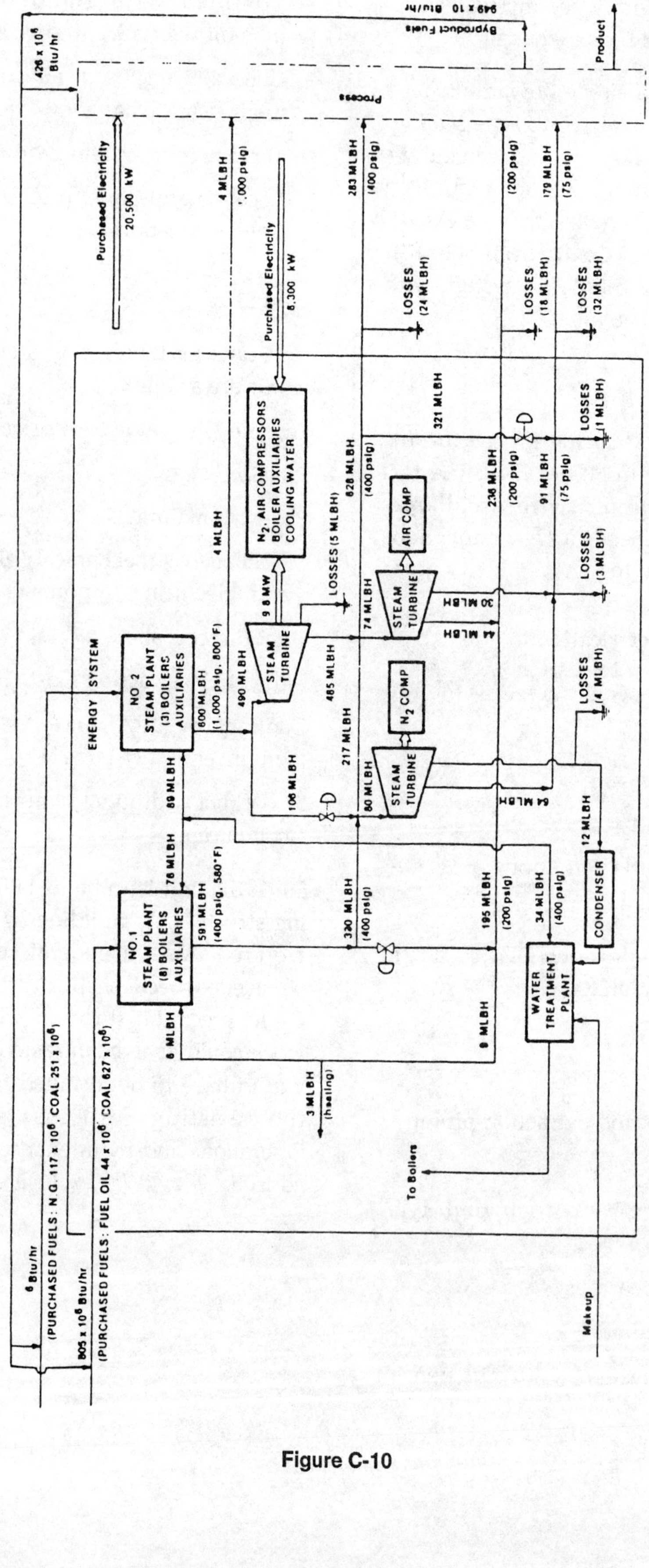


Site-specific Chemicals Plant - Institute, West Virginia (Base Case with Existing Cogeneration)

**Figure C-10**

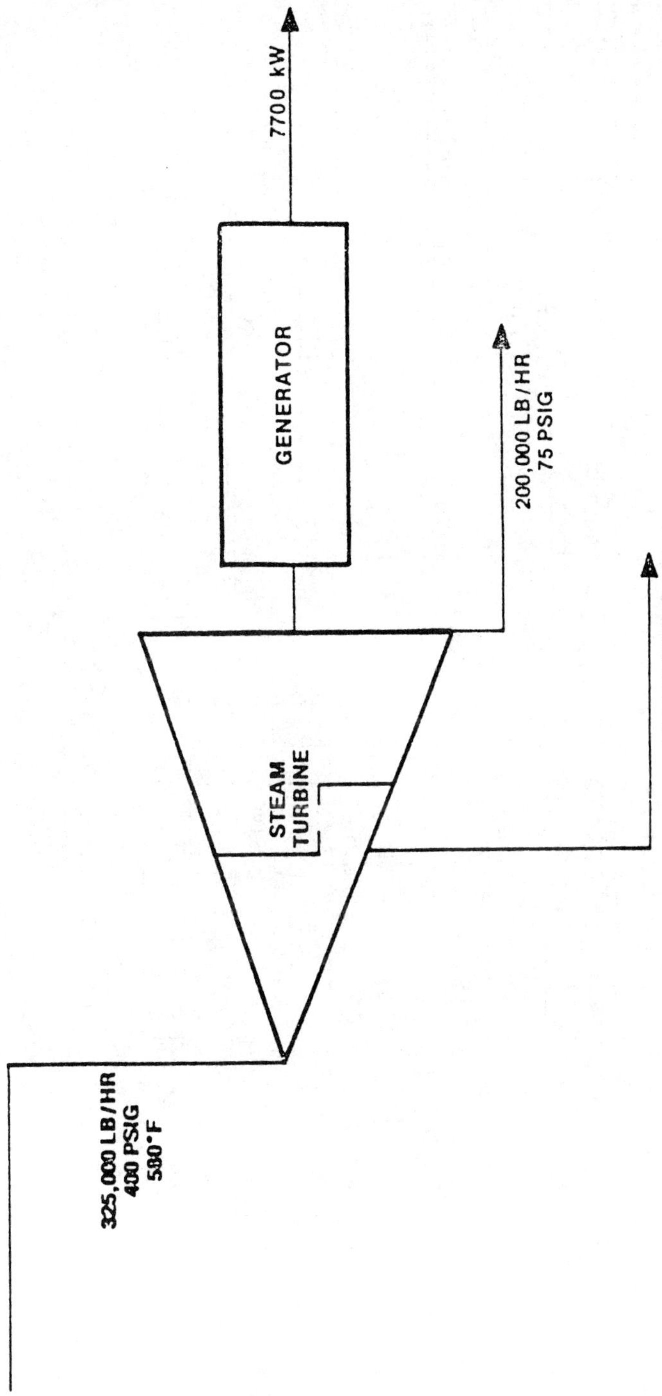


Union Carbide Turbine/Generator Heat Balance

Figure C-11

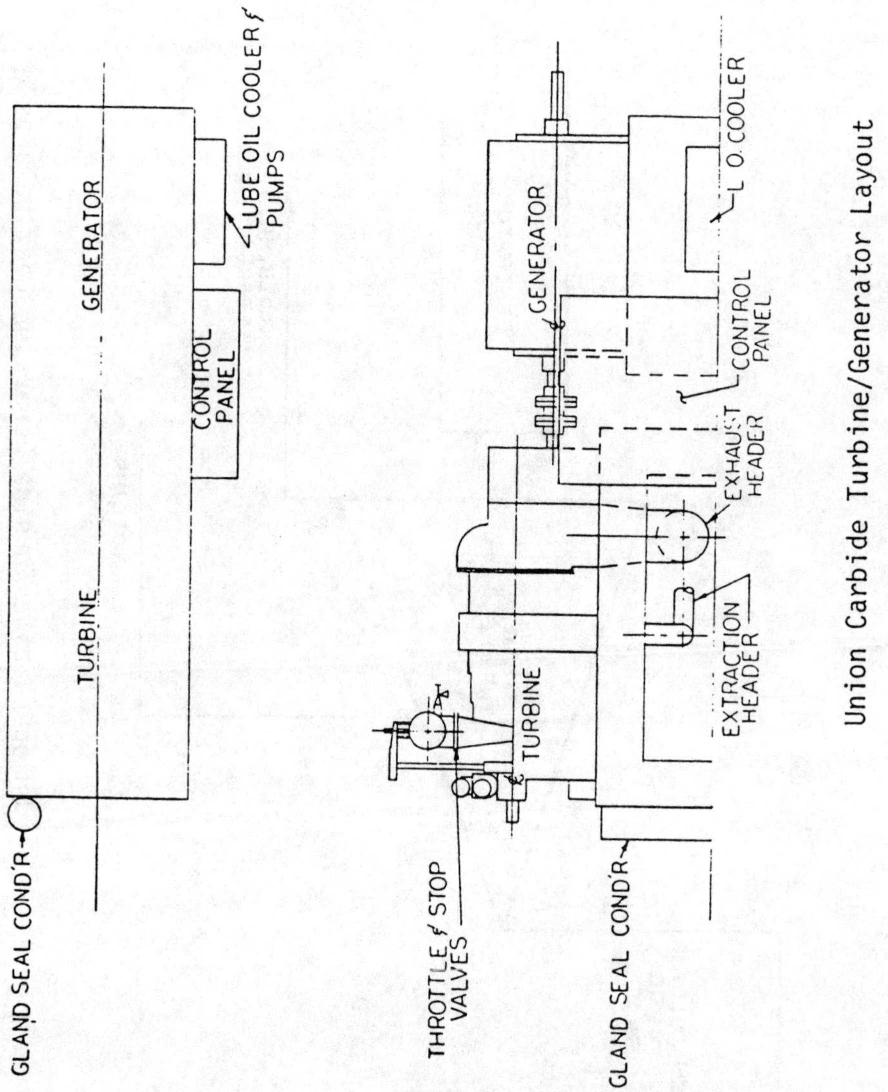
GLAND SEAL COND'R
TURBINE
GENERATOR
CONTROL PANEL
LUBE OIL COOLER & PUMPS
THROTTLE & STOP VALVES
GLAND SEAL COND'R
TURBINE
EXTRACTION HEADER
EXHAUST HEADER
GENERATOR
CONTROL PANEL
L.O. COOLER
Union Carbide Turbine/Generator Layout

Figure C-12

## J.P. STEVENS (TEXTILES), CHERAW, SOUTH CAROLINA

The Cheraw complex consists of two separate finishing plants, called Delta 2 and Delta 3 where "greige" (or "grey") cloth obtained from several Stevens' mills is treated, dyed and/or printed. The products of Delta 2 are knits and woven goods in natural and man-made fibers. Delta 3 produces unprinted "yard" goods from cotton and cotton/polyester fabrics in one color only. The complex includes a waste treatment facility and boiler house which are used by both finishing plants. The complex occupies an area of 500,000 to 600,000 square feet and employs approximately 1,100. Finishing operations often run 3 shifts per day, 5 to 6 days per week.

Delta 2 is a batch process operation used for smaller orders. All types of cotton polyester woven and knit blends are processed into checks and plaids of several colors for men's and women's wear. Initially, the grey cloth is desized, bleached and washed on a continuous range. It is then stored wet in batches before being sent to the batch dying operation. The dye-wash machines are large cylindrical tanks similar to pressure cookers. Next, the cloth undergoes several wash-due cycles, depending on the type of cloth and end result desired, after which it is dried, mechanically shaved, brushed, pressed, shined and preshrunk by a mechanical sanforizing process. A permanent press finish is achieved by applying resins and heat setting the impregnated cloth.

The processes in Delta 3 are similar to those of Delta 2 except that they are all continuous. For this reason, the plant is used mostly to finish large orders.

The finishing process uses more thermal energy than other textile processes. The Cheraw complex has four boilers which generate the required steam at 125 psig. Two are dual fuel, burning either gas or residual oil. Oil is burned most of the time except when gas is available. The other two boilers burn coal, oil or gas. Coal is the usual fuel. These tri-fuel boilers supply the baseload needs of the complex, the other two boilers being operated as needed. About 15 to 20% of the steam generated is used in boiler house auxiliaries. Steam is used either directly in the process or to make hot water. Pressure reducing valves are used to decrease the supply pressure to 100 psi or less for most operations. One exception is the caustic evaporation process which uses 120 psig steam to reclaim caustic recovered from the desizing operation. This is one of the largest single uses of steam in the plant. Steam is also used for drying by heat dry cans. The batch dying in Delta 2 uses steam-in-shell and tube heat exchangers for dye-wash water heat. The continuous bleaching and dying ranges of Delta 3 use steam directly and indirectly in heat exchangers to produce hot water. Water temperatures range from 140 to 180°F. Only 10% of the Delta 3 condensate is returned to the boiler house; all of the Delta 2 condensate is lost. However, condensate and hot water are recycled within the plants to reclaim heat. All of the raw water intake, except the boiler make-up, is preheated in heat exchangers by the water discharged from the plant. Although total energy usages are known, data for the various steam uses are not available. Steam usage pressures and hot water temperatures may be estimated, but quantities used in each operation are unknown.

Natural gas is in short supply in Cheraw. The plants are supplied by the TRANSCO pipeline in North Carolina. The boiler gas contract is preferred interruptible. Process gas requirements are supplied through a firm contract. This gas is used for infrared preheating, singing, and drying by heating air in heat exchangers. Electricity is supplied by Carolina Power and Light. Electric motors throughout the plants are responsible for most of the electric demand. One short-order range of Delta 2 uses 1200 kW of electricity for infrared preheating instead of natural gas to facilitate fast startup. Although the power consumption is minimal from the standpoint of energy use, it is costly because of its high billing demand requirement.

### Heat Balance

The heat balance for the proposed facility is shown in Figure C-13. The turbine/generator balance was obtained from a manufacturer using the throttle conditions and back pressure determined previously. Minor leakages from main steam valves and the sealing system are not shown. The leakages, turbine efficiency, and generator losses are included in the gross generation value of 5.75 MW. Boiler efficiency is based on typical efficiencies of existing stoker-fired boilers of this size and output. Superheater outlet pressure is 20 psig greater than the turbine throttle pressure because of a pressure drop in the main steam line. Currently, condensate is not returned to the boiler house so it was assumed that the textile process use destroys water quality. Thus, hot water production through a closed shell and tube heat exchanger will allow a portion of the boiler water to be saved, minimizing makeup requirements. Regenerative feedwater heating would not be justified for such a small system.

Electrical auxiliaries are required for precipitators, building lighting, and motors for equipment such as pumps, fans, boiler grate, crushers, coal conveyers, coal feeders, ash handling systems, etc.

### System Description

The proposed cogeneration system will use a 132,860 lb/hr coal-fueled, stoker-fired boiler. Steam conditions will be 920 psig, 850°F. The turbine/generator will produce 4.25 MW of power. The exhaust from the turbine will be at 125 psig. This exhaust will supply all the thermal process needs for the plant. The steam needs, 42,000 lbs/hr of 125 psig saturated steam will be met by 39,380 lbs/hr of the exhaust steam which is desuperheated. The 840,000 lbs/hr of 155° hot water is created by condensing 88,7000 lbs/hr of exhaust steam in a closed heat exchanger. This condensate may be saved. The steam that goes directly to the process is nonrecoverable as clean condensate. This results in a normal continuous makeup requirement of 39,380 lbs/hr.

The coal would be supplied by rail, as is presently being done for the existing boiler plant. Unloading and yard storage facilities would have to be expanded to accommodate the increased fuel needs. Bottom hopper coal cars would continue to unload on a trestle. Coal will fall into a new hopper from which it could be routed directly into a new storage silo at the cogeneration plant or, alternatively, to the yard for storage. Coal is presently purchased crushed, sized, and washed. If the existing contracts were continued, the coal could go directly to the stokers. However, a crusher is provided in case run-of-mine coal is purchased. Belt conveyers will transfer the coal at the plant site.

Ash will be periodically removed from the precipitator and boiler and transported by truck to an off-site landfill.

Flue gas cleanup will consist of cyclone-separators and electro-static precipitators. Scrubbers are not presently required but could become necessary as a result of future regulatory requirements.

The present boiler house would remain as a standby. A new building would be required for the cogeneration facility. The existing steam and hot water distribution system in the plant would be used. The boiler would be constructed in a semiopen configuration with the turbine plant completely enclosed.

### Layout Drawing

The layout drawings for the cogeneration plant are shown in Figures C-14 and C-15. The turbine and auxiliaries are contained in a separate building from the boiler, which is built in a semiopen configuration.

Figure C-16 shows the layout on a site of the J.P. Stevens' Delta 2/3 finishing plants. The cogeneration facility is located near the existing boilers so that they can connect to the existing feeder network. The present rail line and coal storage facilities are still used as previously mentioned, but will need to be expanded.

### Major Component Specifications

Major components and their specifications are as follows:

- Boiler
  - 132,860 lbs/hr
  - 920 psig, 850°F superheater outlet
  - Coal-fueled, spreader stoker
  - Balanced draft (forced and induced draft fans)
  - Traveling grate
  - Lungstrom air heater, economizer
  - Steam soot blowing
- Turbine/Generator
  - 5.250 MW rated output
  - 900 psig, 850°F throttle conditions, 132,860 lbs/hour throttle flow
  - Multistage, single casing, single ended
  - 125 psig backpressure
  - Pedestal mounted (concrete)
  - 3,600 rpm
  - Downward exhaust and extraction piping
  - Shaft packing spring-backed labyrinth type, with steam sealing.
  - Split-sleeve radial bearings
  - Thrust bearing required
  - Speed governing system with 110 percent emergency trip
  - Complete with gland steam condenser, lube oil reservoir, pumps, and coolers
  - Generator enclosed, air cooled, with water to air heat exchangers
  - Turning gear system required
  - Local control panel with provision for central control

### Cost Data

Cost data for the coal-fired steam turbine cogeneration facility are as follows:

| | |
|---|---|
| Boiler, 133,000 lbs/hr stoker fired coal (installed) | $2,250,000 |
| Turbine/generator (installed) | 1,500,000 |
| Coal unloading and handling | 750,000 |
| Pollution control, precipitators | 1,000,000 |
| Miscellaneous mechanical, including deaerator, piping, water treatment | 1,000,000 |
| Miscellaneous civil, including site preparation and foundations | 2,015,000 |
| Miscellaneous electrical | 300,000 |
| Total Direct Costs | $8,815,000 |
| Indirect construction costs | $1,000,000 |
| Contingency (10%) | 750,000 |
| Engineering and client charges (10%) | 750,000 |
| TOTAL COST | $11,315,000 |
| $/kW (net additional plant cogen capacity) | 2,660 |

A new building will be required next to the existing boiler house. The existing boilers will be used as standby so no costs are allowed for dismantling of existing facilities. A new building will be required and existing coal unloading and storage facilities will have to be extended.

Site preparation costs should be minimal because the existing site is cleared and relatively level.

The boiler will be semiopen construction. Components will be shop assembled, but much of the erection will be in the field.

Cost estimates are based on quotations and costs of similar systems which have been previously constructed.

The construction will require a general contractor and will be of sufficient scope that indirect construction costs, approximately equal to 40 percent of direct labor costs, would be included. O&M costs are estimated to be .35¢/kWh, based on previous experience with similar facilities.

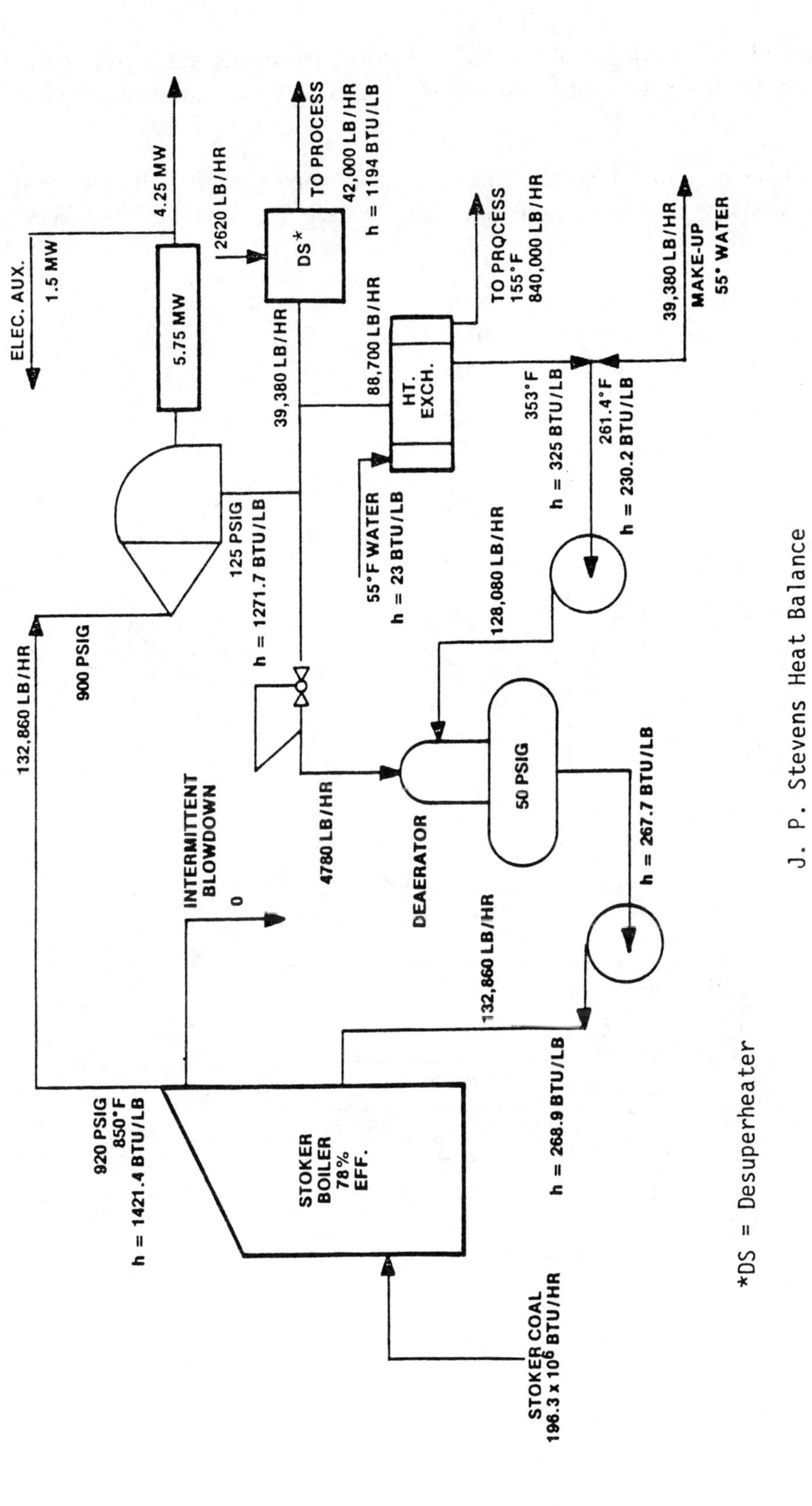
ELEC. AUX.
1.5 MW
4.25 MW
5.75 MW
2620 LB/HR
DS*
TO PROCESS
42,000 LB/HR
h = 1194 BTU/LB
39,380 LB/HR
88,700 LB/HR
HT. EXCH.
TO PROCESS
155°F
840,000 LB/HR
39,380 LB/HR
MAKE-UP
55° WATER
353°F
h = 325 BTU/LB
261.4°F
h = 230.2 BTU/LB
55°F WATER
h = 23 BTU/LB
125 PSIG
h = 1271.7 BTU/LB
128,080 LB/HR
132,860 LB/HR
900 PSIG
INTERMITTENT
BLOWDOWN
0
4780 LB/HR
DEAERATOR
50 PSIG
h = 267.7 BTU/LB
132,860 LB/HR
920 PSIG
850°F
h = 1421.4 BTU/LB
STOKER
BOILER
78%
EFF.
h = 268.9 BTU/LB
STOKER COAL
196.3 x 10^6 BTU/HR
J. P. Stevens Heat Balance
*DS = Desuperheater

Figure C-13

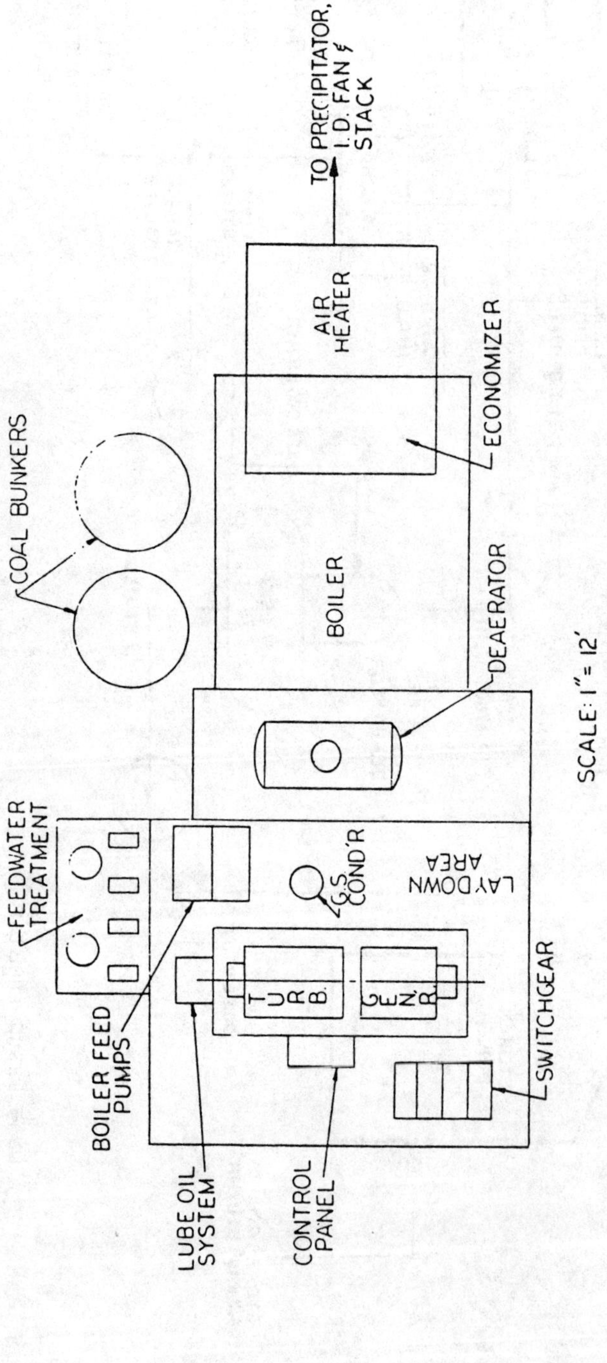


Plan of J. P. Stevens Cogeneration Plant

Figure C-14

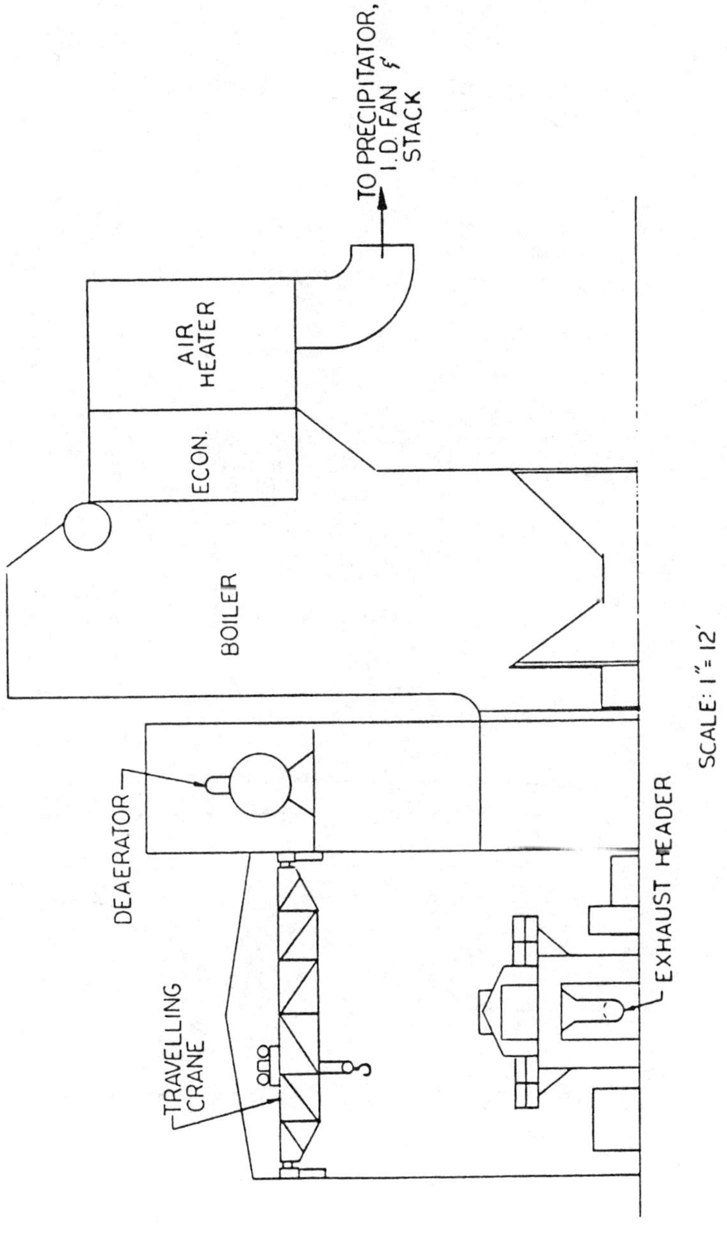


Elevation of J. P. Stevens Cogeneration Plant

**Figure C-15**

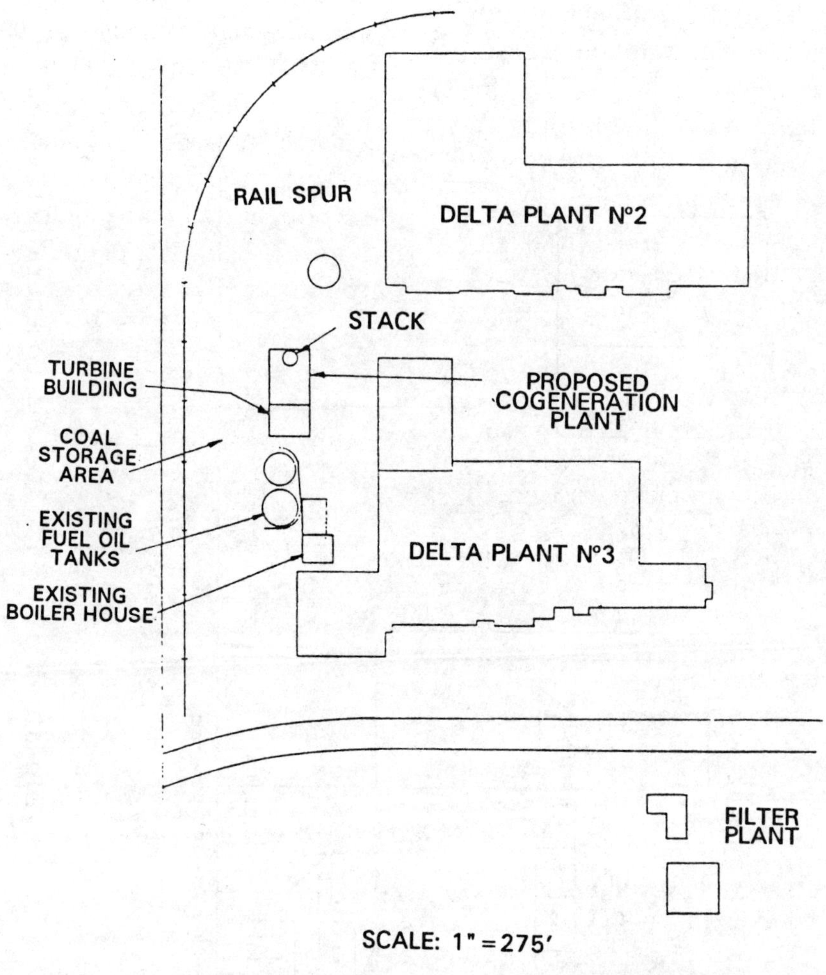


Site Plan of J. P. Stevens Cogeneration Plant

**Figure C-16**

## SCOTT PAPER COMPANY (PAPER AND PULP), MOBILE, ALABAMA

The Scott Paper Mill in Mobile, Alabama, is an integrated, bleached Kraft paper and pulp mill producing 800 tons per day of tissue paper and 600 tons per day of writing paper. Typical fuel usage is given in Table C-2 which shows that about 43% of the required energy is supplied by process byproducts (black liquor, wood and bark). Winter consumption of steam by unit process is given in Table C-3. This table shows that about 23% of the steam is utilized in pulping operations (digester, bleaching and pulp drying), 35% in paper drying (tissue and printing and converting machines), 10% in energy and material recovery (evaporation), 16% in feedwater heating, and 7% in turbogenerators. As shown in Table C-2, 110 x 106 Btu/hr of No. 2 oil or natural gas is utilized in paper machine hood dryers in addition to the utilization of 650 x $10^6$ Btu/hr of steam. Electrical consumption is given in Table C-4. About 63% or 38,000 kW of the required 58,900 kW is generated in plant by means of multiple extraction back-pressure turbines in the wintertime. In the summer, because of the reduction in steam, the electrical output is about 27,000 kW. Table C-5 lists the types and capabilities of the power boilers, recovery boilers, kilns, paper machines, hoods and turbogenerators.

Figure C-17 shows a winter steam balance and fuel inputs to the process. A summer steam and fuel balance would be similar, except that steam requirements are about 85% of the winter demands and in plant electrical generation is about 27,000 kW, with purchased electricity of 32,000 kW making up the additional deficit. Work-to-heat ratios for the summer and winter are given in table C-6. The summer work-to-heat ratios reflect the reduced summer electrical generation and steam requirements. Figures C-18 and C-19 give the electrical and steam demand curves. The diurnal variation of both electricity and steam is flat, varying normally only 2 to 3% with the time of day. The steam and electrical demand will, however, change is a paper machine (of which there are nine in the plant) is taken off line for repair or to reduce production. Electrical demand as shown in Table C-4 and Figure C-18 is fairly flat, varying from 56,000 kW with a demand constant of 59,000 to 60,000 kW for about 4,000 hours. Demand is above 58,000 kW and 52,000 kW for 7,000 and 8,000 hours, respectively. Minimum demand, representative of vacation and holiday periods, is about 30,000 kW. The steam demand curve (Figure C-19) peaks at 1.4 x 106 lb/hr and varies linearly from 1.3 x $10^6$ to 1.0 x 6 lb/hr. The period after 8,000 hours is representative of vacation and holiday periods. The steam demand curve slopes slightly because steam demand is a function of the make-up water temperature which varies seasonally. It is the difference between make-up water temperature in the summer and the winter which accounts for the 15% difference in steam requirements between the two periods.

### Heat Balance

The heat balance for the coal/wood cogeneration plant is shown in Figure C-20. The feedwater is deaerated by the existing system, so a deaerator is not needed. However, a feedwater pump is still needed to pump the feedwater to the high-pressure boiler.

Boiler efficiency is based on 85 percent wood, 15 percent coal firing. The efficiency of wood-fired boilers is lower than for boilers using other fuels because of the high moisture content of wood fuel. Efficiencies based on the higher heating value include the heat that is lost in evaporating the moisture contained in the wood. The wood used for this plant is assumed to contain 50 percent moisture. The coal and wood are burned together on the same grate, and the efficiency represents the composite of both fuels.

Superheater outlet pressure is 25 psi greater than turbine throttle pressure because of main steam line pressure drop.

Boiler blowdown is continuous. A nominal average value of 2 percent has been assumed. Actual requirements would depend upon an analysis of water quality. Sootblowing is intermittent and the average amount of steam used is small.

Electrical auxiliary power is required for precipitators, building lighting, pump drives, fans, wood feeders, blowers, wood hoggers, coal crushers, chain and belt conveyors, ash handling systems, etc.

### System Description

The proposed cogeneration system will use a new 475,000 lb/hr combination coal/wood firing boiler. Steam conditions will be 1,275 psig, 900°F. The turbine/generator will produce 32.3 MW of power. The 50 psig exhaust from the turbine will go into the existing process.

Both wood and coal may be burned in a single boiler simultaneously. However, separate fuel handling facilities would be needed for each fuel. The fuels are burned on a traveling grate which dumps ash off of its end. There should be no problem with combining the ashes if the ash handling system is designed to handle the maximum. Ash from wood is nominally 1 to 2 percent, but it occupies a large volume because it is not very dense. Also, this percentage of ash can be higher and can contain sand and dirt. Coal ash, 4 to 20 percent, is denser but might come off the grate in larger chunks. Coal would be fed into the boiler by a mechanical spreader stoker which throws the coal across the grate surface. Wood would be pneumatically blown into the furnace (although it can also be mechanically fed). Particulates would be relatively high, and some of the cinders in the flue gas would be collected and reinjected into the furnace. Since wood has little to no sulfur and only 15 percent of the fuel would be coal, it was assumed that scrubbers were not needed. However, centrifugal dust collectors and precipitators would be needed to electrostatically collect the particulates. Because of the large amount of moisture in the wood, the flue gas volumes are greater than for coal alone. Thus, precipitators, fans, furnaces, etc., must be designed for the larger flows.

Coal and wood fuel may both enter the plant by rail. The system must be designed so that the expected variation in fuel can be accommodated. Coal will arrive in bottom-hoppered cars which will unload on a trestle. From the unloading area, the coal may be conveyed to the boiler house, crushers, or a dead storage pile for future use. Wood will also be unloaded on the trestles, and may similarly be conveyed to the boiler house or storage piles. However, it may also need to have iron and large chunks removed or be hogged (cut into burnable size pieces). Wood may also arrive by truck, and after unloading the fuel must be able to follow the same routing.

Upon entering the boiler house, the respective fuels are stored in live storage facilities for immediate use. Separate coal silos and wood bins will be needed. Wood flow from bins is a major consideration, so the bins will be designed with sides sloping outward toward the bottom. Screw conveyors on the entire bottom will remove the wood as needed. Conveying within the boiler will be drag chain and belt conveyors.

Belts will also handle the ash, conveying it from the precipitators to a central location from which it may be loaded on trucks for removal.

The condensate is supplied from the existing feedwater treatment area where it is deaerated. However, since the boiler is at a higher pressure than the existing boiler, additional chemicals would have to be added.

In addition to the new system proposed, it will be necessary to modify an existing boiler from wood/oil to wood/coal. The boiler is presently designed to burn 300,000 lb/hr of wood or 400,000 lb/hr of oil. Combinations of wood and oil are also permissible. The modification would require that nominally 80 percent wood, 20 percent coal be burned (on a Btu basis).

Modifications of this traveling grate boiler could be extensive, and a determination of the exact requirements is beyond the scope of this study. However, some basics are appropriate for discussion. The total output, with coal, is still to be maintained at 300,000 lb/hr of steam. Thus, the total ash to be removed from the boiler will be increased because of the use of coal. Furnace exit gas temperatures will have to be checked to ensure that coal ash does not deposit on tube banks or superheaters. Boiler efficiency will improve over wood firing alone, but will be reduced from oil firing. Fans will be adequate. Coal spreader stokers, bunkers, and conveyors would be required at the boiler house. Additionally, conveyors from fuel storage piles would be required. Unloading, crushing,a nd dead storage facilities might be shared with the new facility. Wed scrubbers would probably not be needed for the relatively small amount of coal burned, but the existing particulate control system would have to be examined to see if it was capable of handling the coal fly ash.

### Layout Drawing

The layout drawings for the cogeneration plant are shown in Figures C-21 and C-22. The boiler is built in an open configuration but the steel work envelope is shown. Many auxiliaries are needed because of the dual fuel requirement. The turbine building, however, is a conventional configuration. Feedwater comes from the existing plant.

Site drawings for the facility were not available, so a nominal site configuration is shown in Figure C-23. Unloading facilities are required as well as separate storage piles, reclamation areas, and conveyors for the wood and the coal.

### Major Component Specifications

Major components and their specifications are as follows:

- Boiler
  - 475,000 lb/hr steam
  - 1,275 psig, 900°F superheater
  - 85 percent wood fuel, 15 percent coal (Btu basis)
  - Wood misc. chips and bark, 50 percent moisture
  - Outdoor open construction
  - Field erected boiler
  - 358.5°F feedwater temperature
  - Economizer
  - Tubular air preheater, steam coil air preheating
  - Steam soot blowing
  - Balanced draft forced and induced draft fan, air admission under grate
  - Overfire air fan required, air introduced in rear wall
  - Traveling grate
  - Coal fed by spreader stoker, wood pneumatically injected
  - Cyclone steam separators in steam drum
  - Cinder recovery system reinjection by air stream
  - Dry ash handling system
- Pollution Control
  - Centrifugal dust collection

- Electrostatic precipitators

- Turbine/Generator
  - 32.3 MW rated output
  - 1,250 psig, 900°F throttle conditions
  - Multistage, single casing
  - 50 psig back pressure
  - Pedestal mounted (concrete)
  - Downward exhaust and extraction piping
  - 3,600 rpm
  - Shaft packing, spring backed labyrinth type, with steam sealing
  - Split sleeve radial bearings
  - Thrust bearing required
  - Speed governing system, with 110 percent emergency trip
  - Complete with gland steam condenser, lube oil reservoir, pumps, and coolers
  - Generator enclosed, hydrogen cooled, with water to hydrogen coolers
  - Turning gear system required
  - Local control panel with provision for central control
- Fuel Handling System
  - Rail unloading facilities for wood and coal, truck unloading for wood
  - Drag chain and conveyor belts for wood waste, belt conveyors for coal
  - Unloading facilities with conveyor access to boiler, crushers (coal) hoggers (wood), or storage piles
  - Live storage
    - -- Coal (bunkers at boilerhouse)
    - -- Wood (bins at boilerhouse with screw conveyors)

**Cost Data**

Cost data for the proposed power generation system are shown in Table C-7. Since the new additions are substantial, it is assumed that a new building is required to house the turbine plant. The boiler will be open construction.

Boiler modifications are difficult to estimate without a detailed study of the existing boiler, which is beyond the scope of this study. Additionally, boiler specifications and drawings were not available. Thus, a nominal estimate of 35 percent of new boiler cost has been assumed for this expense.

Site drawings were not available for this facility, so civil and site preparation costs were assumed for a nominal plant of this type.

The original study considered two separate boilers to supply the required 475,000 lb/hr of steam. Although a single boiler of this size is among the largest currently built with wood burning capability, it was believed more economical to provide a single boiler. This is reflected in the costing.

The basis for these costs are manufacturers quotations, data from previous and similar jobs, and engineering estimates. Indirect construction costs are taken to be 40 percent of direct labor costs.

Operations and maintenance costs are estimated to be .30¢/kWh. This is based on costs experienced by similar plants.

## TYPICAL FUEL INPUT

| | $10^6$ Btu/hr | Percent of Total Plant |
|---|---|---|
| Boilers | | |
| Natural Gas | 410 | 15.6 |
| No. 6 Oil | 640 | 24.3 |
| TOTAL PURCHASED FUEL | 1,050 | 39.9 |
| Black Liquor | 930 | 35.4 |
| Wood Bark | 200 | 7.6 |
| TOTAL BY PRODUCT FUEL | 1,130 | 43.0 |
| TOTAL BOILER FUEL | 2,180 | 82.9 |
| Paper Machine Hood Dryers | | |
| No. 2 Oil & Natural Gas | 110 | 4.2 |
| Lime Kiln | | |
| High S No. 6 Oil | 140 | 5.3 |
| Purchased Electricity* | 200 | 7.6 |
| TOTAL PLANT FUEL | 2,630 | 100.0 |

* 1 kWh – 10,000 Btu

Table C-2

## WINTER STEAM CONSUMPTION

| | $10^6$ Btu/hr | Percent of Plant Total |
|---|---|---|
| Turbogenerators | 135 | 7.2 |
| Digesters | 200 | 10.7 |
| Bleach Plants | 175 | 9.3 |
| Evaporator | 190 | 10.2 |
| Tissue Machine | 260 | 13.9 |
| Printing and Converting | 440 | 23.5 |
| Feedwater Heating | 310 | 16.5 |
| Pulp Drying | 70 | 3.7 |
| Unaccounted | 85 | 4.5 |
| TOTAL | 1,865 | 99.5 |
| Make-up | 10 | 0.5 |
| TOTAL STEAM | 1,875 | 100.0 |

**Table C-3**

## ELECTRICAL LOADS – kW

| | Maximum Period | Average | Minimum Period |
|---|---|---|---|
| Wood Preparation | 1,860 | 1,750 | 1,642 |
| Continuous Pulping | 2,068 | 1,990 | 1,789 |
| Batch Pulping | 6,154 | 4,935 | 3,899 |
| Bleaching | 6,503 | 6,060 | 5,515 |
| P&C Stock Prep. | 4,702 | 4,735 | 3,787 |
| P&C Machines | 13,214 | 12,110 | 11,265 |
| T/M Stock Prep. | 2,626 | 2,385 | 2,202 |
| T/M Machines | 14,497 | 13,190 | 11,774 |
| T/M Finishing | 1,905 | 1,760 | 1,548 |
| Pulp Dryers | 1,190 | 970 | 692 |
| Evaporators | 759 | 695 | 456 |
| Recovery | 2,406 | 2,295 | 2,157 |
| Causticizing | 1,085 | 855 | 962 |
| Water | 1,976 | 1,845 | 1,752 |
| Power Boilers | 1,761 | 1,560 | 1,385 |
| Waste Treatment | 1,205 | 765 | 333 |
| Miscellaneous | 1,384 | 1,340 | 1,258 |
| TOTAL HOUR | 61,305 | 58,880 | 56,017 |
| DAY | 1,471,320 | 1,413,120 | 1,334,410 |

Table C-4

## LIST OF POWER BOILERS, KILNS, PAPER MACHINES, HOODS, AND TURBOGENERATORS

| User | Fuel | Capacity | Stack Outlet Temperature |
|---|---|---|---|
| #1 Boiler | Resid. Oil 1/2%S, Nat. Gas | 90,000#/hr | 320°F |
| #2 Boiler | Resid. Oil 1/2%S, Nat. Gas | 90,000#/hr | 320°F |
| #3 Boiler | Resid. Oil 1/2%S, Nat. Gas | 90,000#/hr | 320°F |
| #4 Boiler | Resid. Oil 1/2%S, Wood Waste | 160,000#/hr | 350°F |
| #5 Boiler | Resid. Oil 1/2%S, Nat. Gas | 200,000#/hr | 350°F |
| #6 Boiler | Resid. Oil 1/2%S, Wood Waste | 300,000#/hr | 150°F (Scrubber) |
| #3 Recovery | Black Liquor Resid. Oil | 85,000#/hr | 320°F |
| #4 Recovery | Black Liquor Resid. Oil | 130,000#/hr | 320°F |
| #5 Recovery | Black Liquor Resid. Oil | 130,000#/hr | 320°F |
| #6 Recovery | Black Liquor Resid. Oil | 204,000#/hr | 350°F |
| #1 Kiln | Resid. Oil Nat. Gas | 65 T/D Lime | 180°F (Scrubber) |
| #1 Kiln | Resid. Oil Nat. Gas | 65 T/D Lime | 180°F (Scrubber) |
| #2 Kiln | Resid. Oil Nat. Gas | 65 T/D Lime | 180°F (Scrubber) |
| #3 Kiln | Resid. Oil Nat. Gas | 130 T/D Lime | 180°F (Scrubber) |
| #4 Kiln | Resid. Oil Nat. Gas | 130 T/D Lime | 180°F (Scrubber) |
| Machine Hoods | Distillate #2, Nat. Gas | – | 650°F |

Steam Conditions: 575 psig, 740°F, Feedwater Temperature 355°F

| | | |
|---|---|---|
| Turbogenerators | 1 | 7,500 kW |
| | 2 | 7,500 kW |
| | 3 | 13,500 kW |
| | 4 | 19,500 kW |
| TOTAL: | | 48,000 kW |

Table C-5

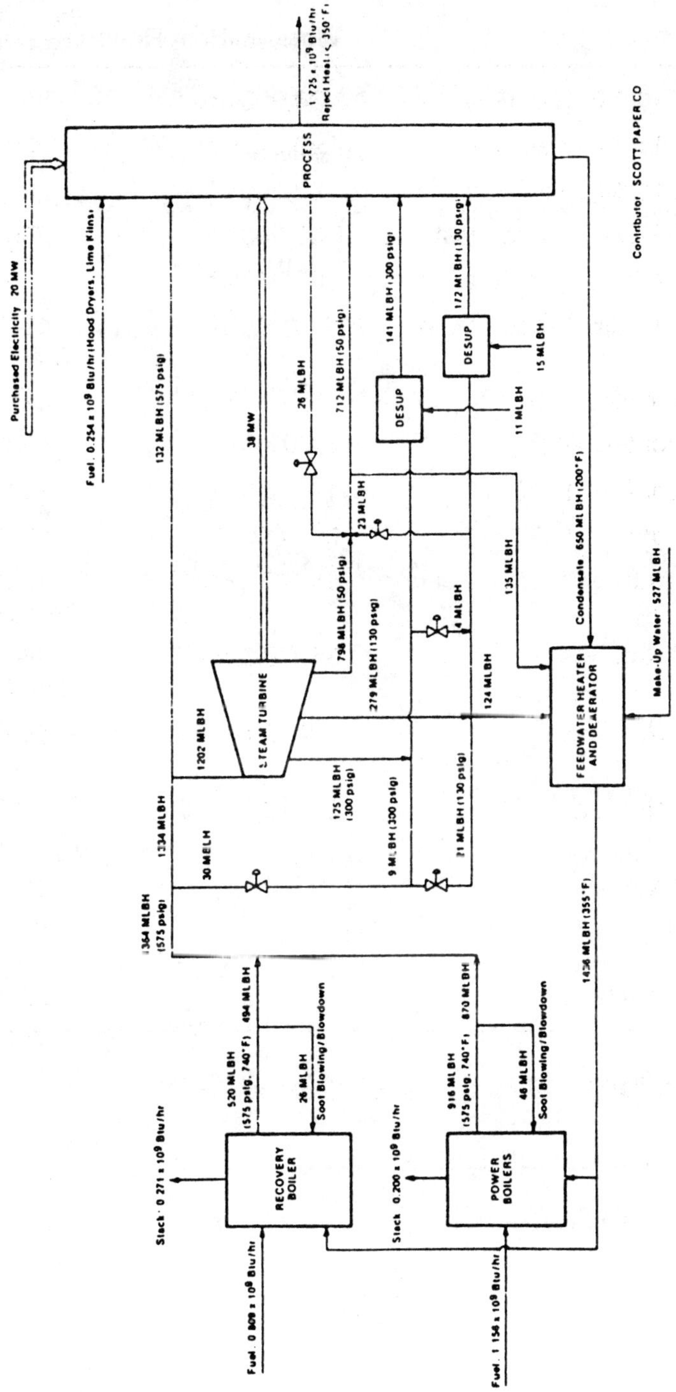


Pulp and Paper Industry - Site-Specific Integrated Mill - Mobile, Alabama (Current Average Winter Energy Use Pattern)

**Figure C-17**

## WORK-TO-HEAT RATIOS

| Winter | Work-to-Heat Ratios |
|---|---|
| Total System Work/Steam | 58,000 kWh/1.364 x $10^6$ lb |
| | 42.5 kWh/$10^3$ lb |
| Existing Cogeneration System Work/Steam | 38,000 kWh/1.364 x $10^6$ lb |
| | 27.9 kWh/$10^3$ lb |
| Potential Work/Hood Dryer Process Heat | 20,000 kWh/(6.8 x 114 x $10^6$ Btu) |
| | 220 kWh/$10^6$ Btu |
| Potential Work/Lime Kiln Process | 20,000 kWh/(0.2 x 140 x $10^6$ Btu) |
| | 714 kWh/$10^6$ Btu |
| Potential Work/Total Hood Dryer | 170 kWh/$10^6$ Btu |
| **Summer** | |
| Total System Work/Steam | 50 kWh/$10^3$ lb |
| Existing Cogeneration System Work/Steam | 23.3 kWh/$10^3$ lb |
| Potential Work/Hood Dryer Process Heat | 32,000 kWh/(0.8 x 114 x $10^6$ Btu) |
| | 350 kWh/$10^6$ Btu |
| Potential Work/Lime Kiln Process Heat* | 32,000 kWh/(0.2 x 140 x $10^6$ Btu) |
| | 1,140 kWh/$10^6$ Btu |
| Potential Work/(Total Hood Dryer + Lime Kiln*) | 224 kWh/$10^6$ Btu |

* Assumes preheat of lime kiln combustion air.

**Table C-6**

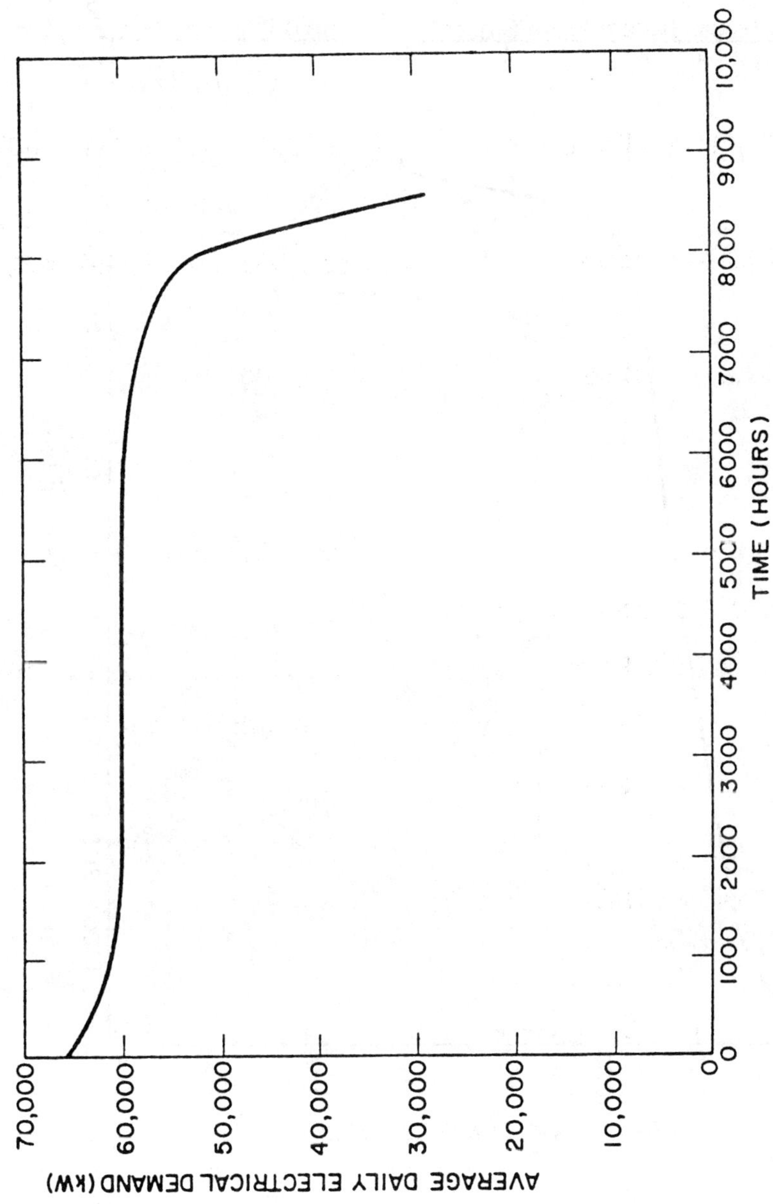


Electrical Demand Curve - Scott Paper Company - Integrated Mill - Mobile, Alabama

Figure C-18

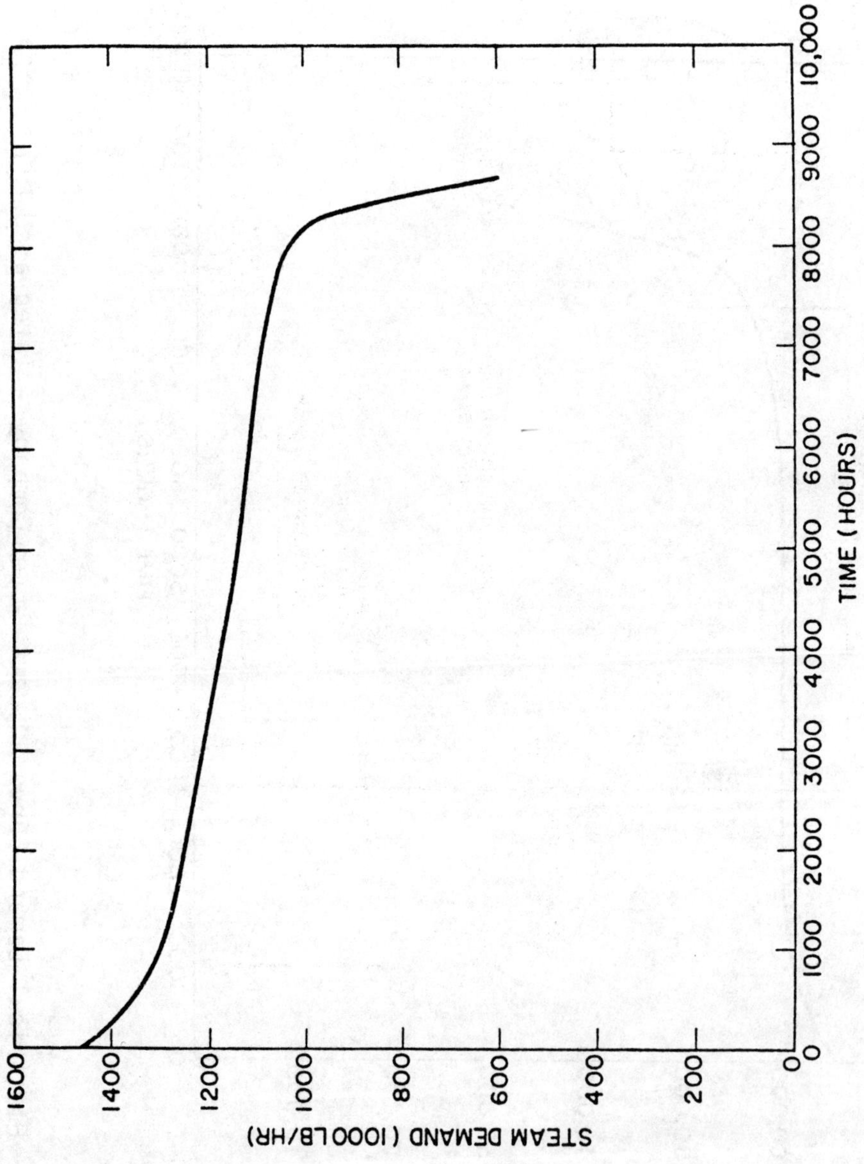


Steam Demand Curve - Scott Paper Company - Integrated Mill - Mobile, Alabama

**Figure C-19**

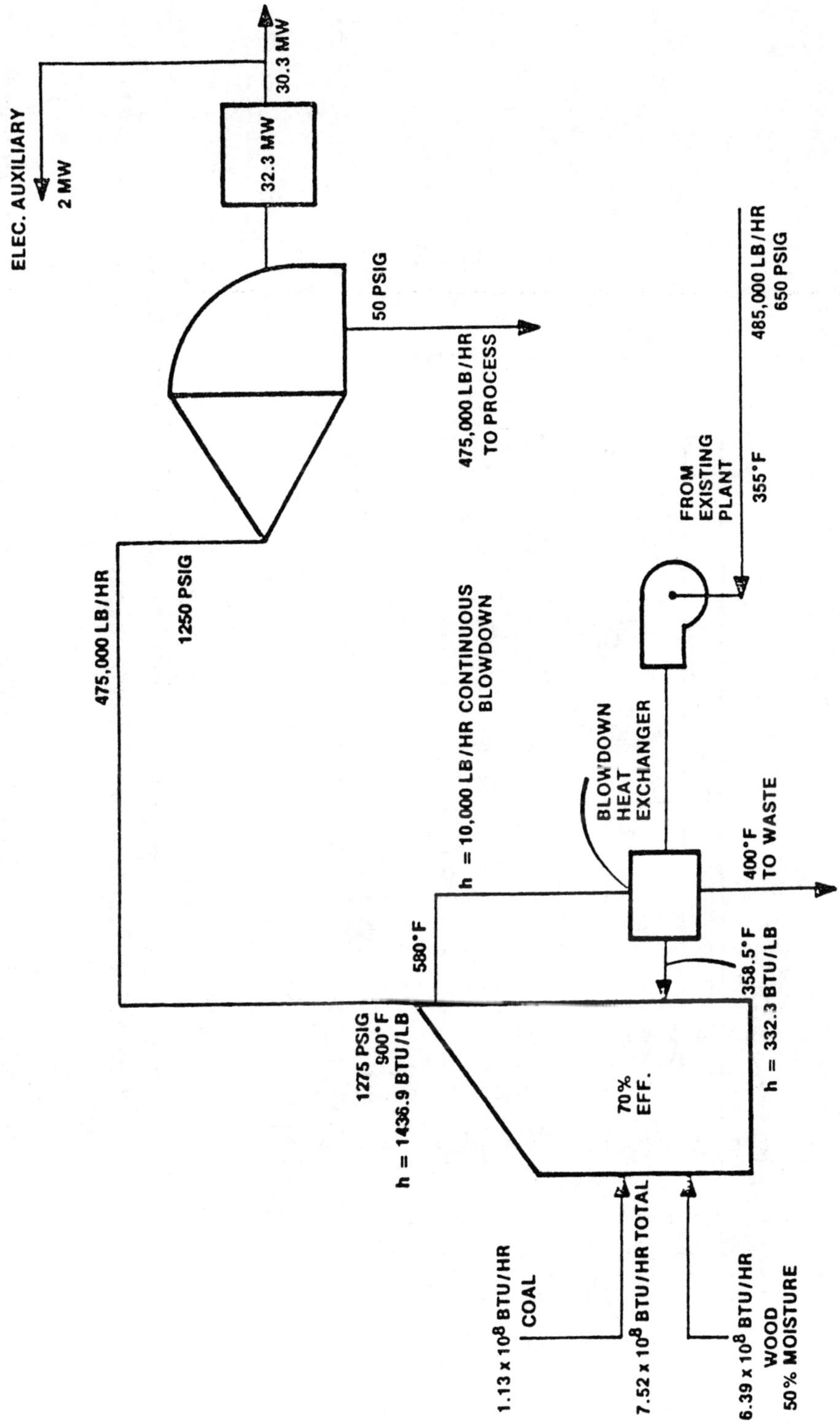


Scott Paper Company Heat Balance

**Figure C-20**

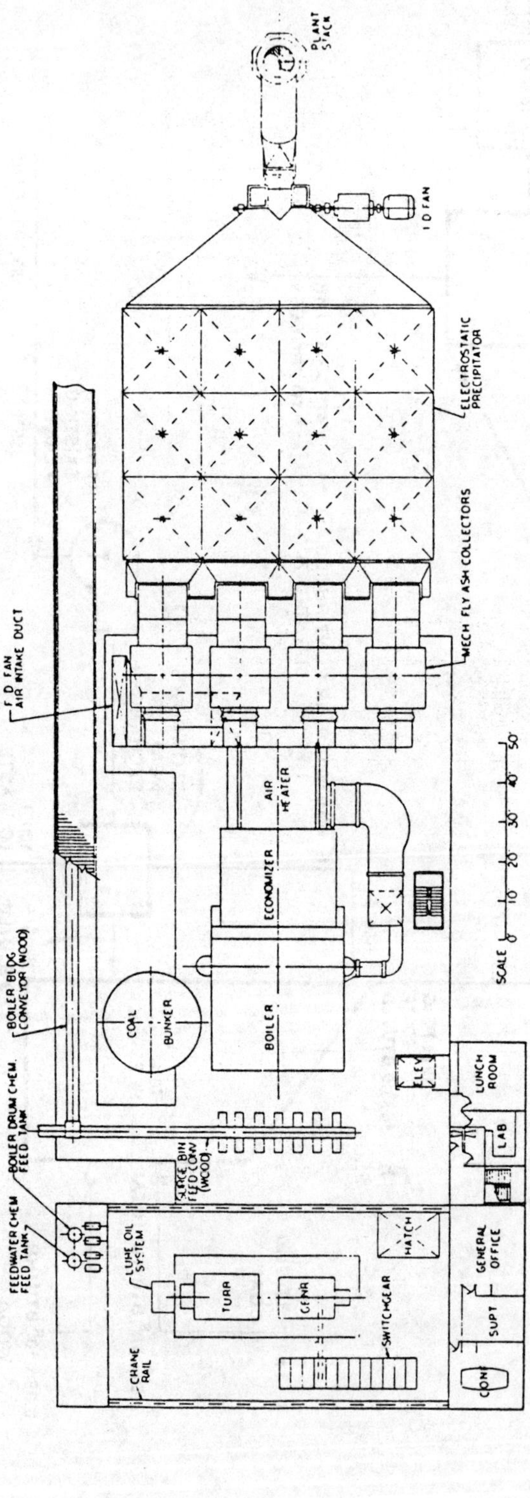


Composite Plant View of Scott Paper Company Cogeneration

Figure C-21

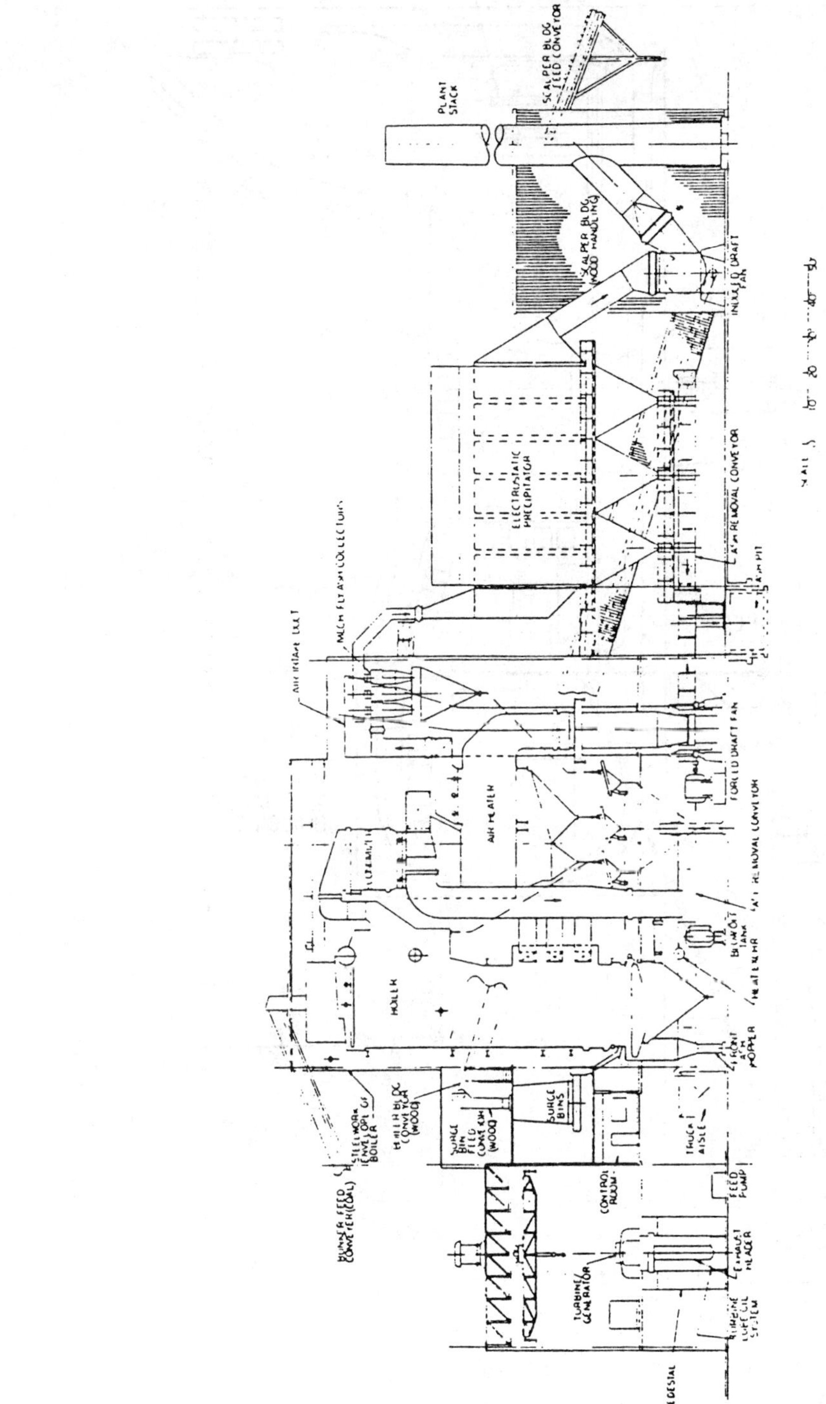


Composite Elevation of Scott Paper Company Cogeneration Plant

Figure C-22

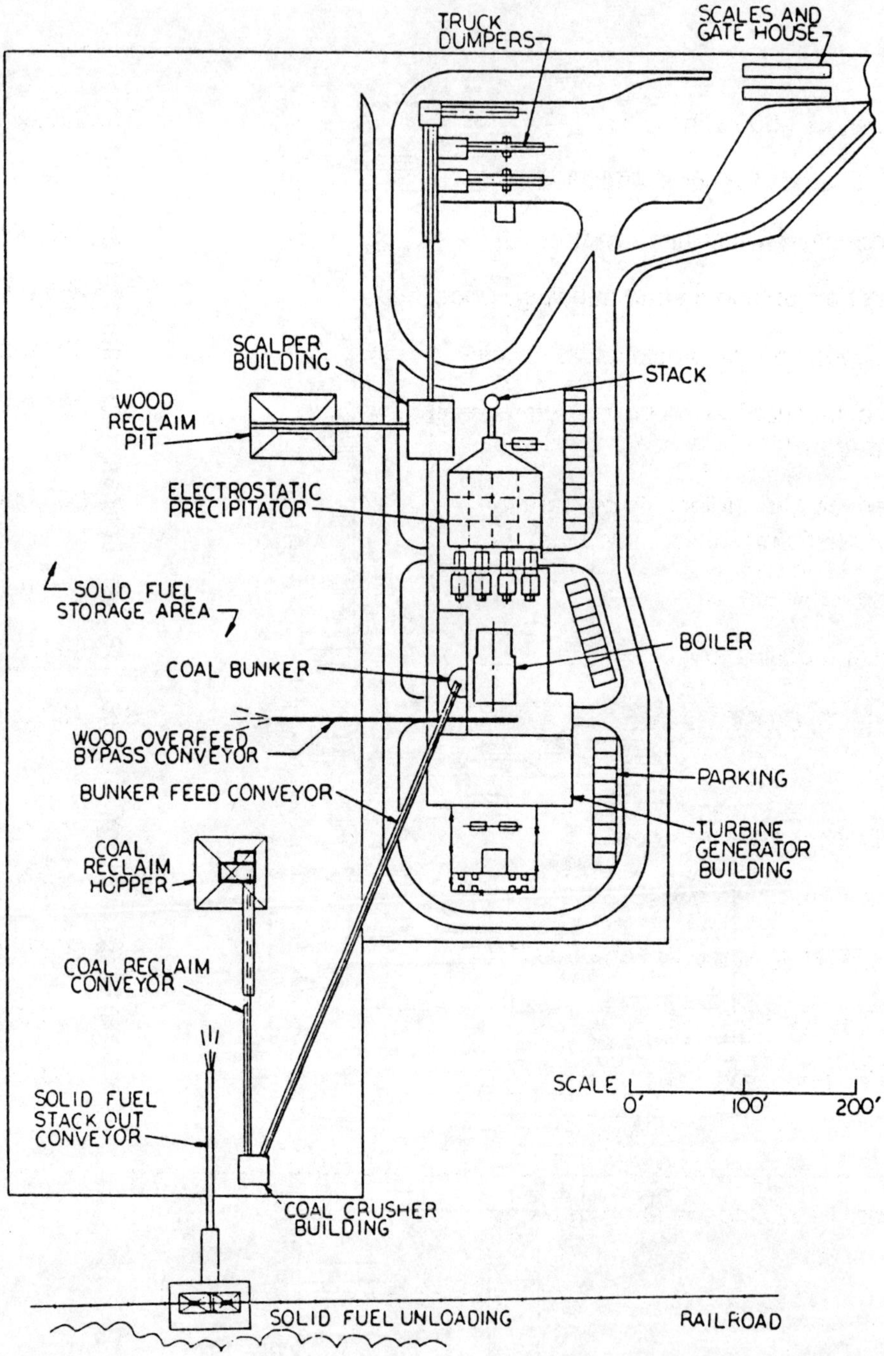


Site Drawing of New Scott Cogeneration Plant

**Figure C-23**

## SCOTT PAPER COMPANY COST DATA

| | |
|---|---|
| Boiler, 475,000 lb/hr | $10,000,000 |
| 85% wood, 15% coal (installed) | |
| Turbine/generator (installed) | $3,700,000 |
| Materials unloading and handling, wood, coal | $3,250,000 |
| Pollution control, precipitators | $3,000,000 |
| Misc. mechanical, including pumps, piping, water treatment | $2,750,000 |
| Misc. civil, including site preparation, foundations building, and stack | $4,000,000 |
| Misc. electrical equipment | $2,500,000 |
| Boiler modifications | $5,500,000 |
| DIRECT COST TOTAL | $34,700,000 |
| Indirect construction costs | $4,800,000 |
| Contingency (10%) | $3,500,000 |
| Engineering and client charges (10%) | $3,500,000 |
| TOTAL COST | $46,500,000 |
| $/kW (net)* | $3,345 |

* Net Additional Plant Cogeneration Capacity

Table C-7

## LARGE REFINER, NORCO, LOUISIANA

The distribution of refinery capacity in the U.S. is such that about 60% of the nation's refining capacity is in 8% of the plants whose industrial capacities exceed 100,000 bpcd (barrels per calendar day). Plants of this size are medium-to-high complexity and are generally engaged in the manufacture of chemicals. cogeneration is generally practiced in some form, either within the plant boundaries or via the local utility. The Exxon Baytown Refinery, for example, shares a gas turbine cogeneration plant with the neighboring chemical facility in the same complex. The Exxon Linden, New Jersey plant obtains all the power and steam generated by the local utility in exchange for fuel supplies. Although some refineries (e.g., the Exxon Baton Rouge facility) are considering coal as a primary fuel, the fuels used to generate process steam in direct fired boilers, in most cases, are generated internally in the refinery. The situation is often complicated by the interchange of byproduct fuels between the refinery and the chemical plant. Smaller refineries (50,000 bpcd), on the other hand, are much less complex, but they rarely cogenerate, mostly for economic reasons. Although their number exceeds 65% of the total number of refineries, small refineries account for only about 18% of the total refining capacity.

The general view is that refining capacity will remain fairly constant in the United States, with some structural changes in the input and output mix, and the manufacture of petrochemicals will not sensibly rise in relation to basic refining.

Exxon contributed data that are "typical" of "large" and "small" refineries in the U.S. The term "large" refineries was defined as those refineries whose individual capacities exceed 200,000 bpcd. However, the extent to which an energy use pattern is "typical" of such refineries is more difficult to assess because it depends to a large extent on the product structure and the input mix of raw materials. The point is made by Exxon that although the largest number of "large" refineries is located in Texas and Louisiana (PAD 3) the product mix that is nationally most prevalent is in PAD 2. A subsequent review of refinery energy profiles by Exxon, however, indivated that a specific refinery in Norco, Louisiana, is reasonably typical of an average large refinery. This is the case study presented below.

The primary use of steam in refineries is for pumping and compression. The fuels consumed to generate steam vary seasonally. Another factor that contributes to seasonal energy uses is the particular product mix demanded by the market. "Typical" seasonal variations in product demand (such as the increase in gasoline demand during the summer) can be anticipated to some extent based on historical data. However, the fluctuations in current market conditions are becoming increasingly difficult to forecast.

From a technical standpoint, the potential for increased cogeneration in the petroleum refining industry is quite substantial. The structural variety allows for the use of all virtually all technologies, including steam turbines, gas turbines, diesels and fuel cells. Cogeneration is practiced most prevalently in the larger refineries using steam turbines and gas turbines.

The primary barriers to cogeneration are not technical. In the petroleum industry, the physics appear to be subservient to the issues of safety and cost and their corollaries of pollution control, permit availability, and financing.

The patterns of energy use derived from Exxon's report on the typical large and small refinery are shown in figures C-24 and C-25.

### Heat Balance

The heat balance for the pulverized coal cogeneration plant is shown in Figure C-26. The turbine generation is based on a manufacturer's calculations using the given throttle conditions and back pressure. Minor leakages from main steam valves and the steam sealing system are not shown. Sealing steam excess would be condensed in a gland steam condenser, giving off a small amount of heat to the feedwater. All losses are included in the generation of 10.30 MW. Boiler efficiency is based on the performance and losses of typical pulverized coal-fired boilers of this size and output. Superheater outlet pressure is 25 psi greater than turbine throttle pressure because of main steam line pressure drop. Condensate is fed to the boiler from an existing feedwater to the 1,525 psig boiler. Soot blowing and blowdown losses reduce the amount of steam available. The amount of soot blowing required would depend on the type of coal being burned. Since the refinery does not currently burn coal, there are no existing sources. Additionally, an operating plant might purchase coal from a number of suppliers. Blowdown requirements would depend upon an analysis of the feedwater. Soot blowing would be intermittent using steam from the superheater outlet. Blowdown of drum water would probably be continuous. A nominal average value of 20 percent has been assumed for blowdown. Soot blowing has been neglected.

Electrical auxiliary power is required for precipitators, scrubbers, building lighting pump drives, fans, pulverizers, crushers, coal conveyors, the ash handling system, etc.

### System Description

The proposed cogeneration system would use a 450,000 lb/hr pulverized coal boiler. Steam conditions will be 1,525 psig, 950°F. The turbine/generator will produce 10.3 MW of power. The 625 psig exhaust from the turbine will go to the process and an existing turbine.

The coal will arrive at the plant in bottom hopper rail cars which would unload on a new trestle. The coal will then fall into a bin where large chunks will be broken up before being transported by conveyor belts. The coal may go to a crusher house, direct to the boiler house, or to a storage pile. Its destination will depend upon boiler house needs and the condition in which it is received. At the boiler house, there will be "live storage" bunkers which will feed the coal to the pulverizers. The pulverized coal is blown directly into the furnace where combustion takes place. Part of the ash falls to the bottom of the boiler where it is removed by a water sluicing system. The rest passes out with the flue gas where it will be removed by electrostatic precipitators. Wet limestone scrubbers will be required on a boiler of this size. The scrubbers will also aid in particulate removal. The boiler will have a feedwater economizer as well as a regenerative type air preheater. Steam soot blowing will be required using steam from the superheater outlet. Forced and induced draft fans will be required. Air will enter with the pulverized fuel as well as separately through the walls. Ash will be removed from the precipitator and boiler hoppers to a central location by a water sluicing system. Water will then be separated and ash removed offsite.

The turbine would be a single flow multistage unit with spring backed, labyrinth type sealing, split-sleeve type bearings, and a mechanical/hydraulic type control system. The generator would be totally enclosed and air cooled, with heat being removed by water-to-air heat exchangers. The set would be mounted on a concrete pedestal with exhaust piping coming from below. It would require a separate lube oil system with reservoir, pumps, and coolers.

Condensate is supplied from an existing feedwater treatment area where it is deaerated. Thus, no deaerator is needed in the cycle. However, since the boiler is at a higher pressure than the existing boiler, additional chemicals would have to be added.

### Layout Drawing

The layout drawings for the cogeneration plant are shown in Figures C-27 and C-28. The turbine building is separate from the boiler, which is open to the environment but weatherproofed and canopied. Site drawings for the refinery were not available so only a nominal site configuration is shown in Figure C-29. Unloading facilities are shown as well as storage and fuel handling equipment.

### Major Component Specifications

Major components and their specifications are as follows:

- Boiler
  - 450,000 lb/hr
  - 1,525 psig, 950°F superheater outlet conditions
  - Coal fuel, pulverized
  - Outdoor open construction, weatherproofed and canopied
  - Field erected boiler, suspended from above
  - 256.5°F feedwater temperature
  - Economizer
  - Air heater (lungstrom type)
  - Steam soot blowing
  - Balanced draft – forced and induced fan
  - Pulverizers, crushers
  - Ignition with #2 oil, electric spark
  - Cyclone steam separators in steam drum
- Pollution Control
  - Electrostatic precipitators
  - Wet limestone scrubbers with steam coils for reheating flue gas
  - Turbine/Generator
  - 10.3 MW rated output
  - 1,500 psig, 950°F, throttle conditions
  - Multistage, single casing, single ended
  - 625 psig back pressure
  - Pedestal mounted (concrete)
  - Downward exhaust and extraction piping
  - 3,600 rpm
  - Shaft packing spring backed labyrinth type, with steam sealing
  - Split-sleeve radial bearings
  - Thrust bearing required
  - Speed governing system with 110 percent emergency trip
  - Complete with gland steam condenser, lube oil reservoir, pumps, and coolers
  - Generator enclosed, air cooled, with water-to-air heat exchangers
  - Turning gear system required
  - Local control panel with provision for central control

### Cost Data

Cost data for the pulverized coal fired steam turbine cogeneration facility are shown in Table C-8. The new system is extensive enough to warrant a new building for the turbine plant. The boiler will be of open outdoor construction, requiring only weatherproofing and a canopy. However, because of its size, most of the erection will be in the field. Since site drawings were not available, site preparation costs may only be estimated. A general contractor will be required to coordinate the various aspects of the construction. Indirect construction costs are taken to be 40 percent of direct labor costs.

The basis for the cost estimates are manufacturers' quotes, previous jobs, and engineering estimates. O&M costs are estimated to be .30¢/kWh, based on similar experience.

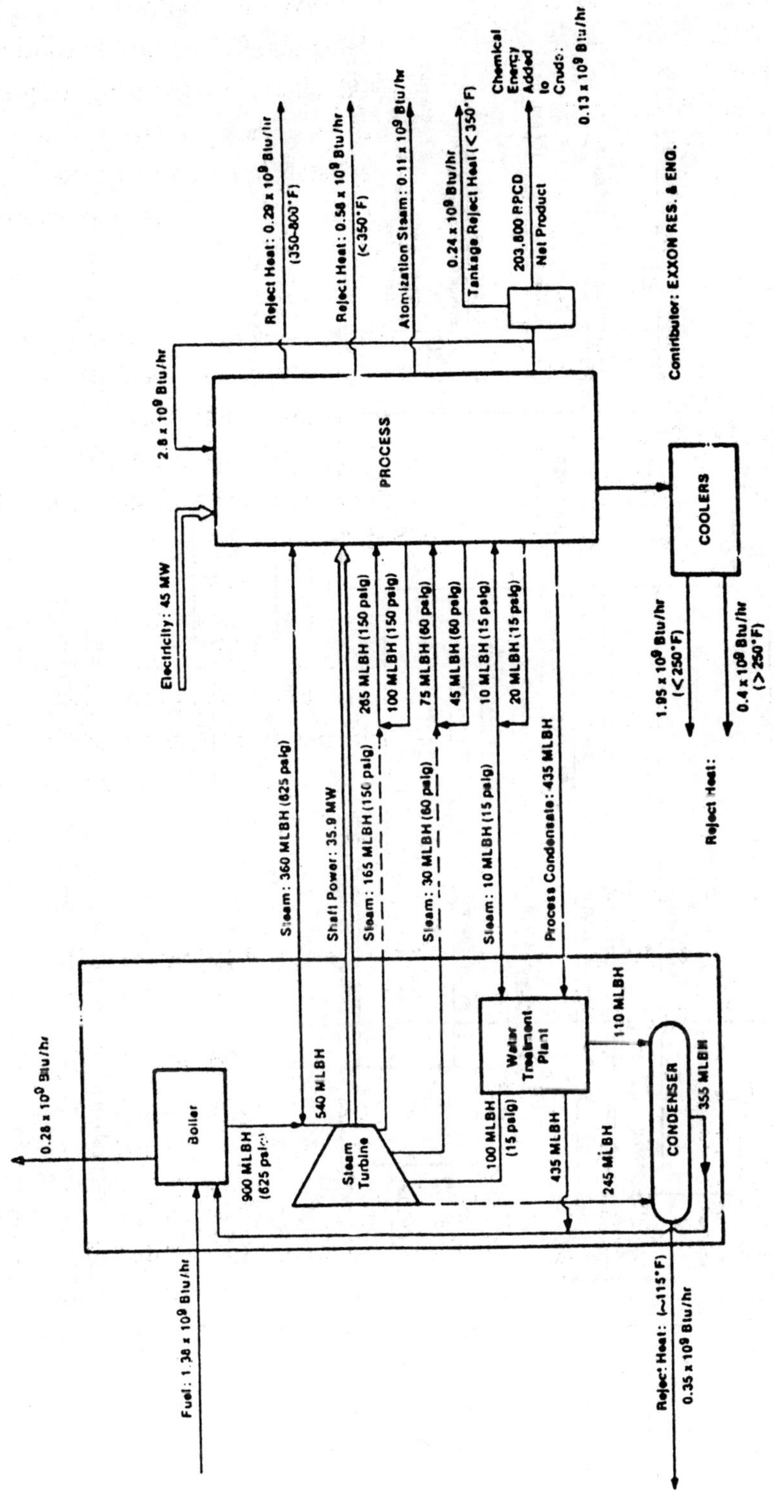


Petroleum Refining Industry - Non-Site-Specific Average Large Refinery - Pad 2
(Average Energy Use Pattern)

Figure C-24

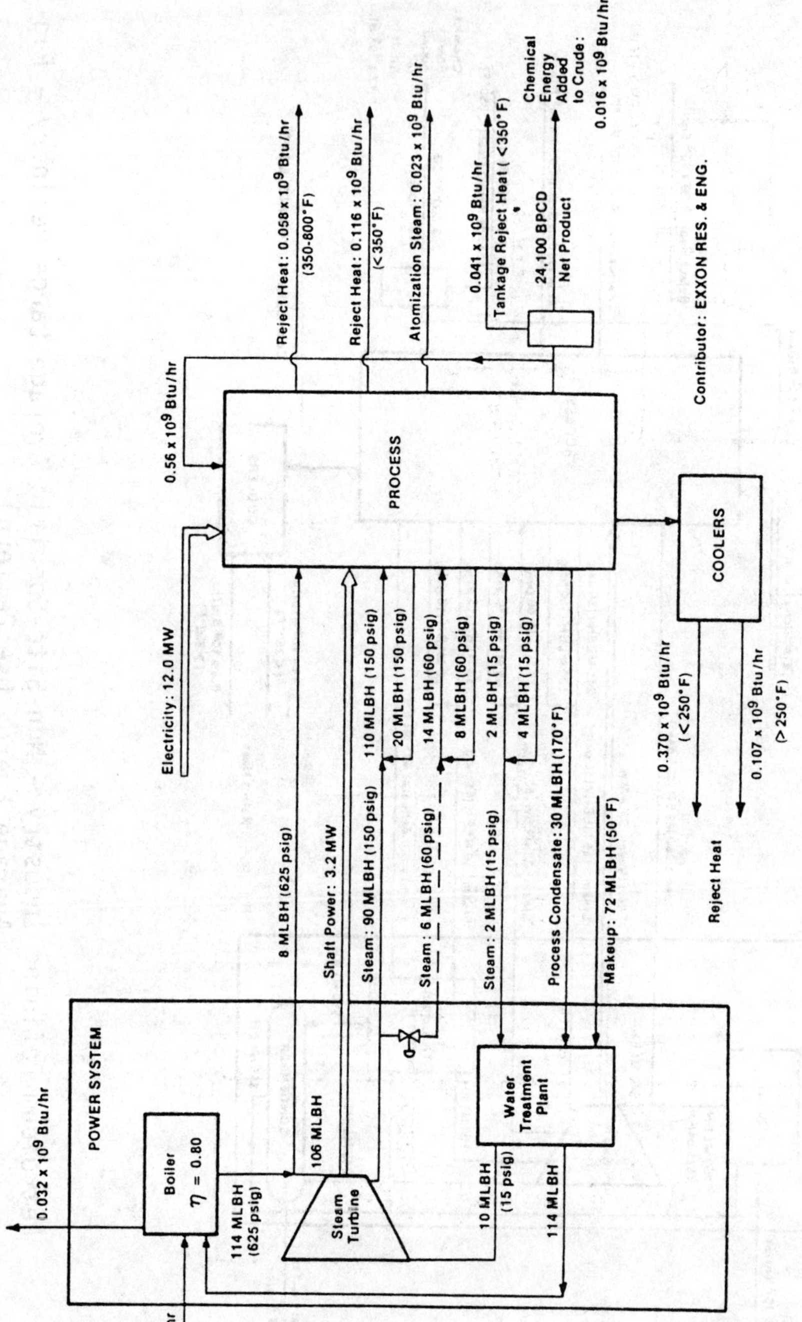


Petroleum Refining Industry - Non-Site-Specific Average Small Refinery (Average Energy Use Pattern)

Figure C-25

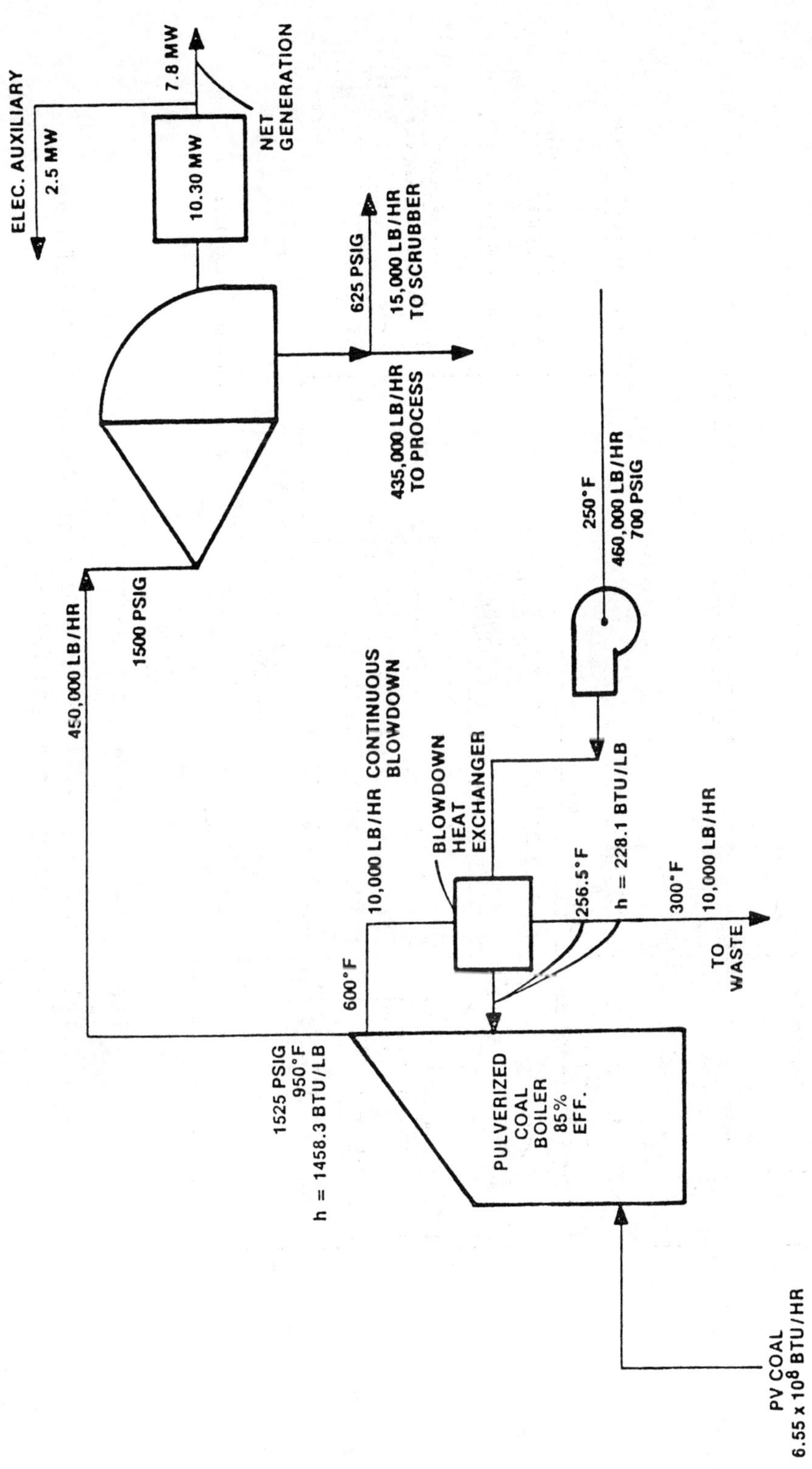


Large Refinery Heat Balance

Figure C-26

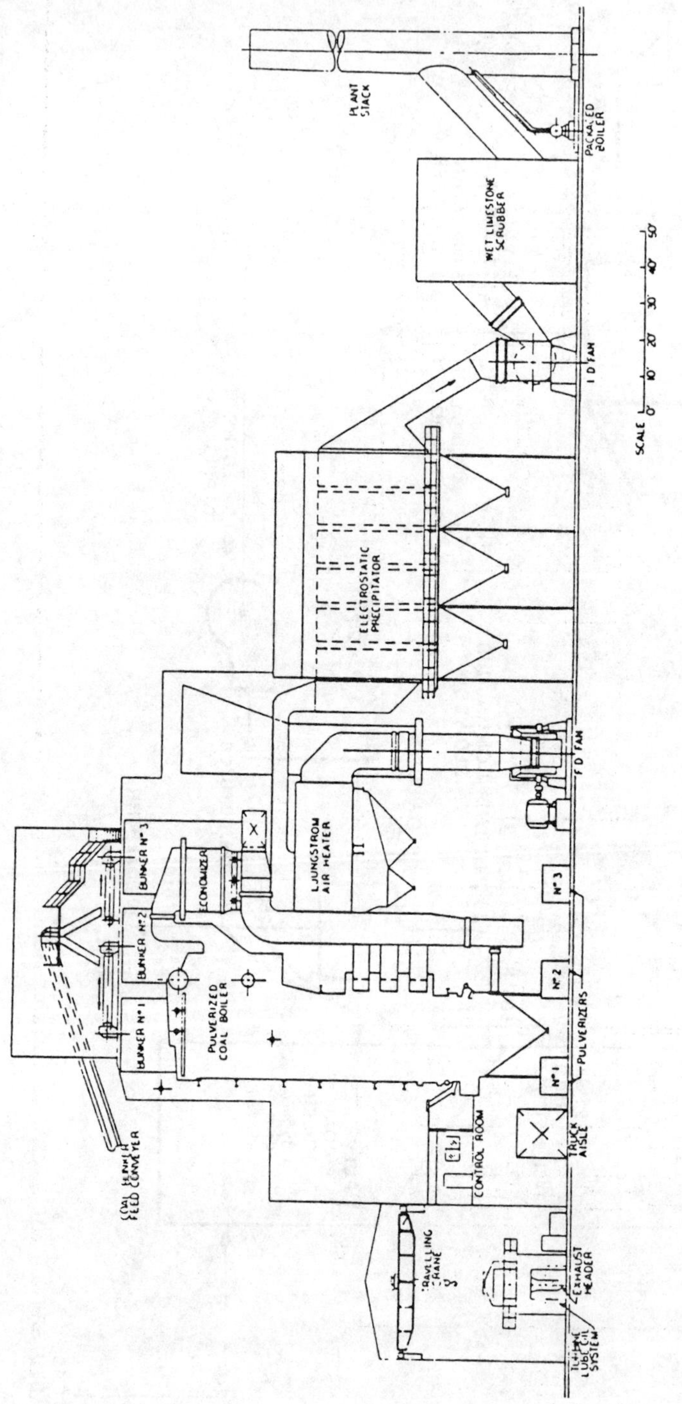


Composite Elevation of Large Refinery Cogeneration Plant

**Figure C-27**

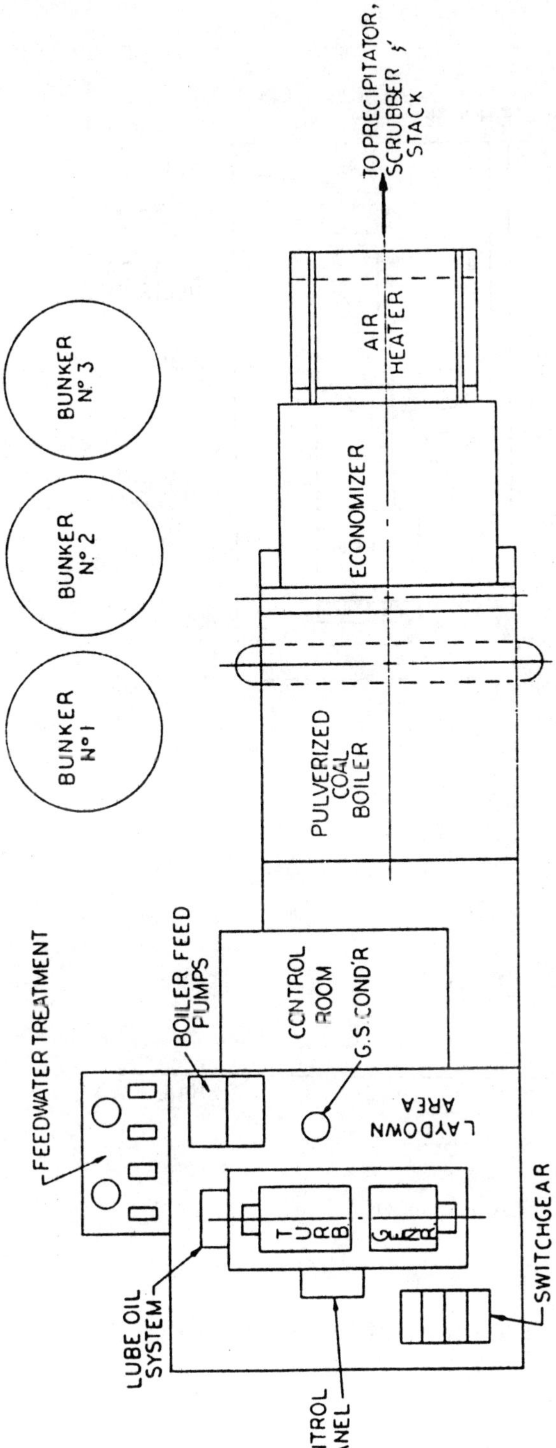


Partial Plan of Large Refinery Cogeneration Plant

**Figure C-28**

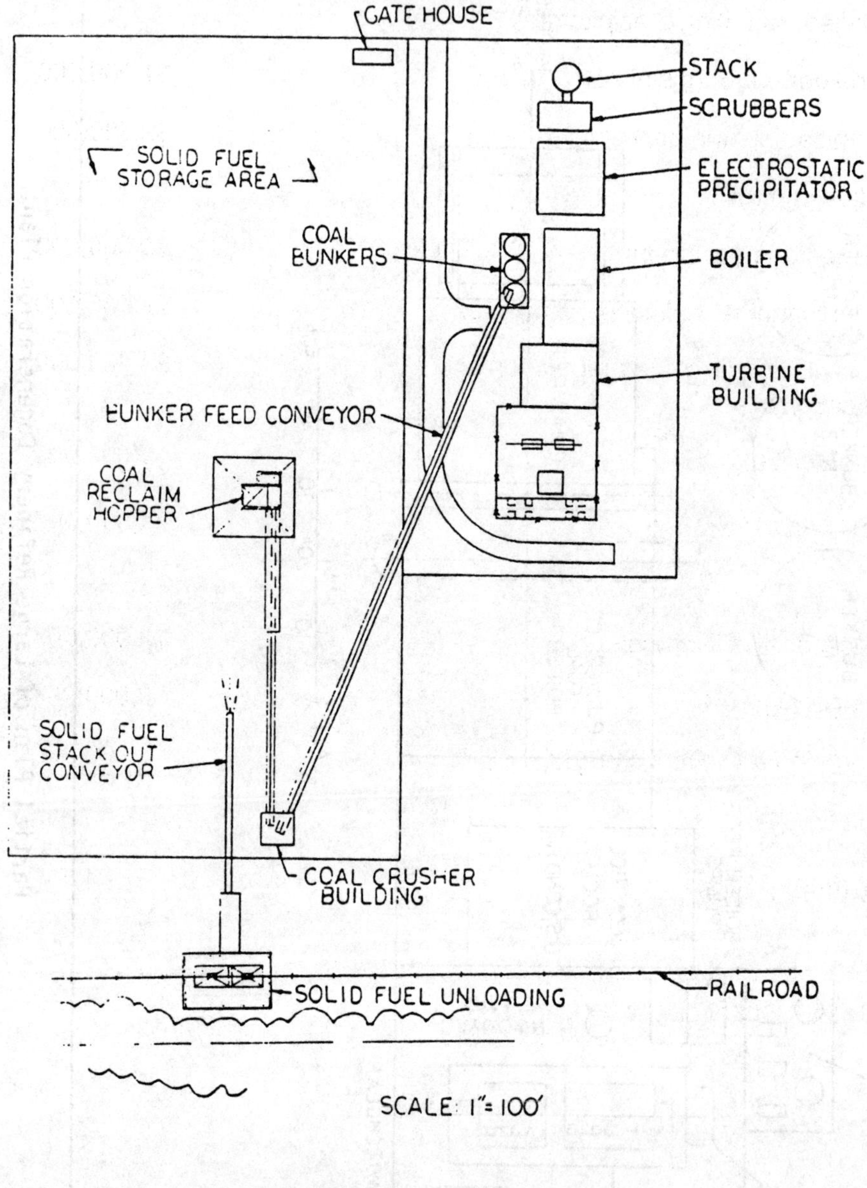


Site Drawing of Large Refinery Pulverized Coal Cogeneration Plant

Figure C-29

## LARGE REFINERY COGENERATION SYSTEM COST DATA

| | |
|---|---|
| Boiler 450,000 lbs/hr pulverized coal fired (installed, including precipitators) | $9,500,000 |
| Turbine/generator (installed) | $1,900,000 |
| Coal unloading and handling | $2,500,000 |
| Pollution control | |
| Wet limestone scrubbers | $6,000,000 |
| Misc. mechanical including piping, etc. | $2,500,000 |
| Misc. civil, including site preparation and foundations | $3,500,000 |
| Misc. electrical | 2,000,000 |
| DIRECT COST TOTAL | $27,900,000 |
| Indirect construction costs | $3,900,000 |
| Contingency (10%) | 2,800,000 |
| Engineering and client charges | 2,800,000 |
| TOTAL COST | $37,400,000 |
| $/kW (net)* | 4,795 |

* Net Additional Plant Cogeneration Capacity

**Table C-8**

## GREYHOUND, KANSAS CITY, MISSOURI

The Armour meat processing plant in Kansas City, Missouri, employs about 350 people and is primarily engaged in the manufacture of sausages and other prepared meat products, including bacon, boneless and smoked ham, hot dogs, luncheon meats, and meat products. The primary energy inputs to the plant are electricity, natural gas, distillate fuel, and propane.

Electricity accounts for approximately 90% of the total specific energy cost. The end-use breakdown of this power is approximately as follows: 75% for refrigeration, 20% for machine motor drives, and 5% for lights and miscellaneous. Natural gas is used with propane for cooking on a high priority supply and for boiler generation of steam on an interruptible supply. Distillate fuel oil (No. 2) is used as a backup fuel for steam generation when the interruptible gas supply is cut off during the winter months.

There are eight batch-operated smokehouses in the plant. Hot dogs are cooled separately in a continuous operation. An additional smokehouse was installed and operated in January 1978 at a time when one of the smokehouses was being serviced. When service was completed, the local utility refused to supply natural gas because of the prohibition against new gas use in any existing plant served by the pipeline. As a result, the company installed an 18,000 gallon propane tank to operate the additional smokehouses.

The steam is generated in two dual fuel fired boilers at 15 psig. Steam is either used directly for space heating and in some unit operations (cooking, sausage blending, hot dog peeling and cryovac packaging) or indirectly to heat city water to about 140°F, mostly for cleaning purposes.

The plant operates almost continuously with three shifts per day, 5 to 7 days a week. Figure C-30 shows the current pattern of energy use and the pattern of energy use that would need to be satisfied by a self-sufficient cogeneration system. This pattern was modified to accommodate the company's plan to increase average production by about 30%.

### Heat Balance

The heat balance for the proposed system is shown in Figure C-31. Performance is scaled from manufacturers' data for a gas turbine generator in the size range being considered. Corrections were made for inlet and exhaust duct losses, including the effect of a heat recovery boiler. Radiation and convection losses from the heat recovery boiler are assumed to be 2 percent. Exhaust gas is only reduced to 275°F, because lower temperatures would promote condensation of corrosive products. Electrical power is required to compress the natural gas fuel to 160 psig before it is fed into the engine. This is necessary because all manufacturers' performance data are based on receipt of compressed fuel gas. Once the gas is compressed, it must be reduced to at least 160°F. The heat can be recovered by the city water in a heat exchanger before it enters the recovery boiler. There are also other lesser electrical auxiliary requirements, such as for the recirculation pump on the boiler, feedpump, etc.

### System Descriptions

The gas turbine cogeneration system will have a 908 kW net output. This size unit would be a skid-mounted gas turbine/generator set. It would be mounted outdoors with an appropriate enclosure for weather protection, and the waste heat recovery boiler would be provided with an enclosure. Control of the unit would be by a remote panel in the existing boilerhouse. Steam from the waste-heat boiler would supply all the plant's process needs. The saturated 15 psig steam would supply process needs directly (6000 lb/hr), and 1,2500 lb/hr would mix with city water in a open heat exchanger to produce 18,000 lb/hr of 140°F water. The hot water heat exchanger would be enclosed with the heat recovery boiler.

The fuel would be natural gas, although the unit would be purchased with a fuel fuel option to burn distillate oil if gas is interrupted. A separate motor-driven gas compressor is necessary to pressurize the natural gas to 140 to 160 psig to feed the engine. The pressurized gas may also be used to start the engine. Connection to existing oil storage facilities must be provided.

The gas turbine will be geared down to 1,800 rpm at the generator. A flexible-drive coupling connects the output shaft of the gearbox to the generator shaft; the generator is water cooled; and the gas turbine is mounted on a concrete mat. An inlet air silencer and filter will be provided.

The feedwater system would include a water tank and feedpump which should supply water to the waste heat boiler at a rate based on water level. The feedpump would be motor driven at constant speed, with a control valve throttling the flow as needed. Feedwater treatment would be handled by chemicals, but an exact treatment program can only be determined after a water analysis is made. Then, compatibility of the proposed treatment program with the process use must be determined. The waste heat boiler would use spiral tubes with economizer and evaporator sections. Boiler and steam separation would take place in a drum. Tubes would be carbon steel. A recirculation pump would recirculate water between the drum and evaporator section.

### Layout Drawing

The layout drawing is shown in Figure C-32. The gas turbine/generator and heat recovery boiler have their own self-contained weatherproof enclosures. A shelter enclosed the open heat exchanger, gas compressor, and auxiliary equipment. Control is remote from the existing boilerhouse. Figure C-33 shows the layout on a plot plan of the Kansas City plant. It is located near the existing boiler (utility) building so that the existing gas line and water supply may be used.

### Major Component Specifications

Major components and their specifications are as follows:

- Gas Turbine/Generator
  - 948 kW @ 50°F ambient, sea level
  - Continuous duty
  - Dual Fuel capability (natural gas and distillate oil)
  - Constant rpm, geared to 1,800 rpm generator
  - Skid-mounted unit with weatherproof enclosure
  - 4,160 volt air-cooled generator
  - Industrial heavy duty
  - Necessary fuel pumps and controls to be mounted on skid unit
  - Compressed gas start
  - Remote control panels
  - Two-bearing, open drip proof generator
  - Inlet air silencer
- Heat Recovery Boiler
  - 7,270 lb/hr of 15 psig saturated steam; gas flow 61.750 lb/hr, 800°F
  - Condensate on the tube side, with gas pass over the tubes
  - Provisions for soot blowing (oil firing)
  - Damper to bypass gas flow

- Provisions for mechanical cleaning of tubes
- Forced recirculation drum type
- Separate evaporator and economizer sections
- Carbon steel tubes in a spiral configuration

**Cost Data**

Cost data for the proposed gas turbine plant are shown in Table C-9. Indirect construction costs are relatively low due to the number of components that are shop assembled. The basis for the gas turbine/generator cost is a manufacturer's estimate. Other costs are estimated from available data and quotations.

Operation and maintenance costs are estimated to be 0.35/kWh. This assumes one major gas turbine/generator overhaul every 3 years and 2 hours of operator/maintenance attention per day. Lube-oil consumption is negligible, unlike a diesel which unavoidably burns oil. Tube cleaning of the waste heat boiler may be done during the overhaul period. Gas side maintenance of the boiler should be minimal if natural gas is the turbine fuel.

If high pressure (125) natural gas service was unavailable to the Greyhound plant, it would be possible to eliminate the fuel compressor from the gas turbine system. This would decrease the system capital cost by $50,000 increase the net output by 40 kW, and increase the ROI to 16.2 percent as compared with the preliminary design estimates. Based on discussions with a gas utility representative, however, it is unlikely that such high pressure service could be made available without significant cost to the customer.

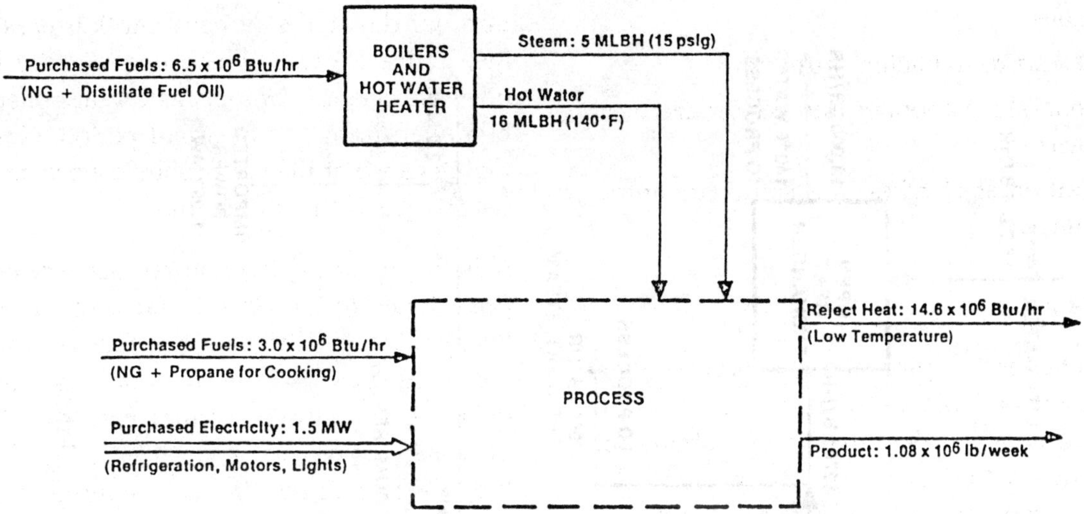


Contributors: GREYHOUND CORP.
ARMOUR & CO.

Food and Kindred Products - Site-Specific Meat Packing Plant - Kansas City, Missouri (Current Average Energy Use Pattern)

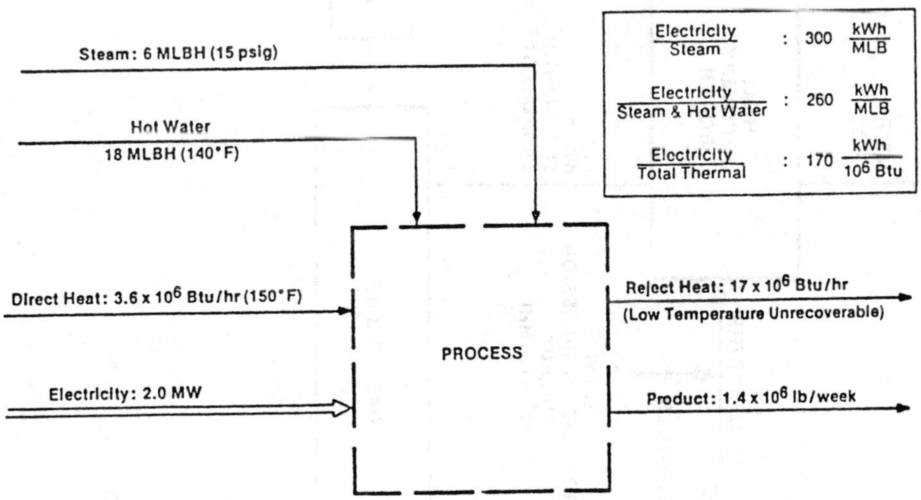


Contributors: GREYHOUND CORP.
ARMOUR & CO.

Food and Kindred Products - Site-Specific Meat Packing Plant - Kansas City, Missouri (Average Process Energy Demands for Design)

**Figure C-30**

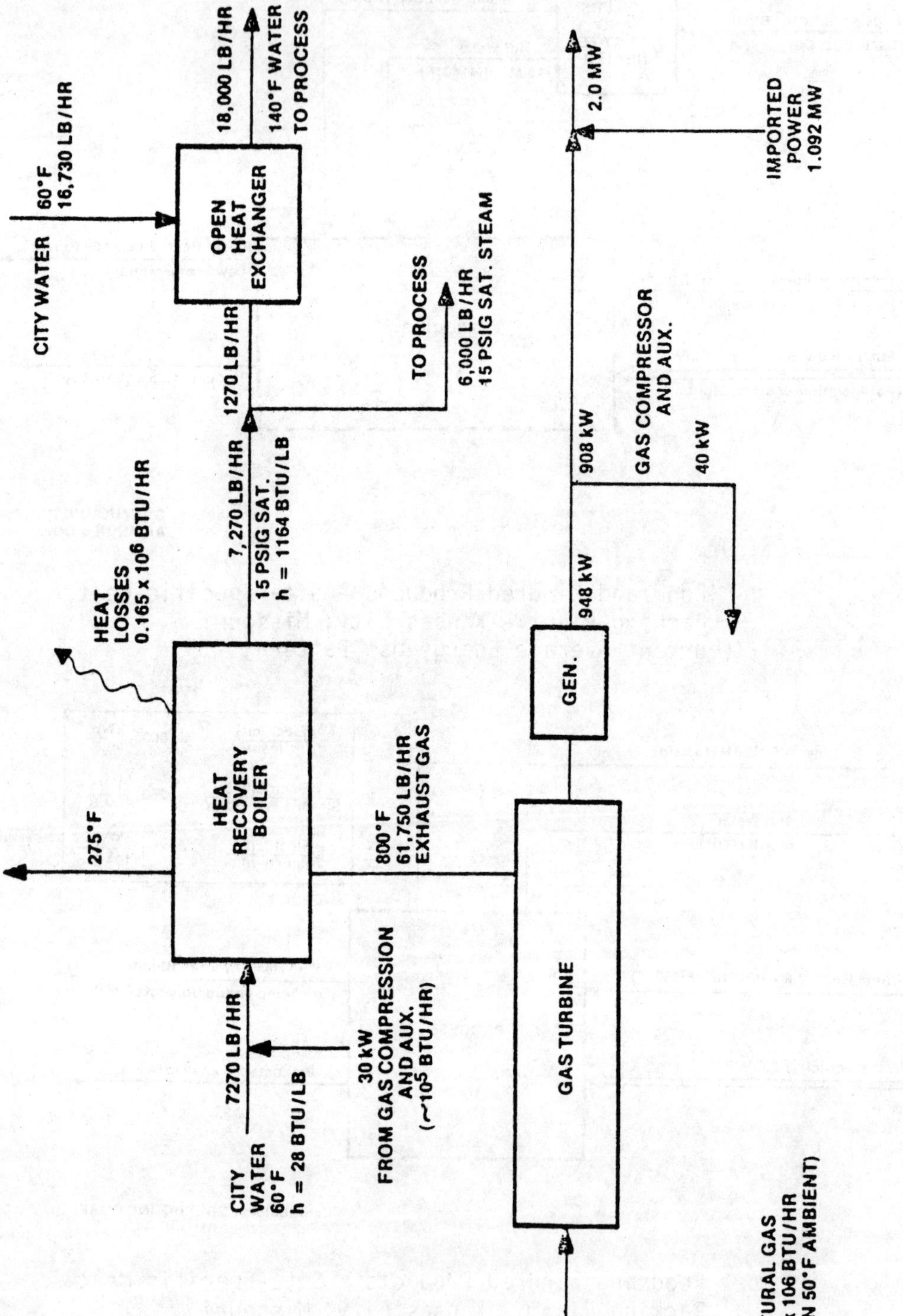


Greyhound Food Processing Plant Heat Balance

**Figure C-31**

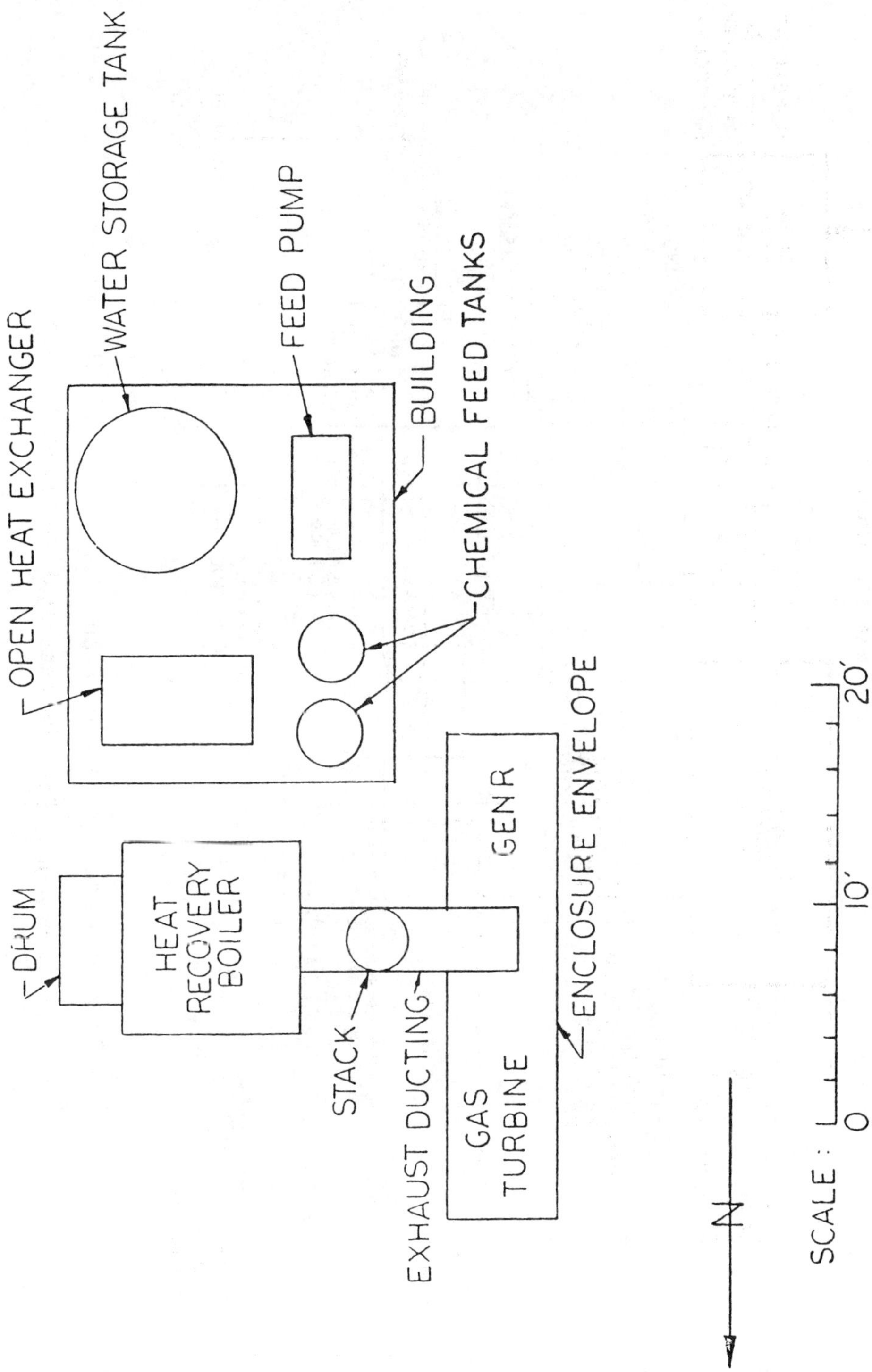


Greyhound Food Processing Plant Gas Turbine Plant Layout

Figure C-32

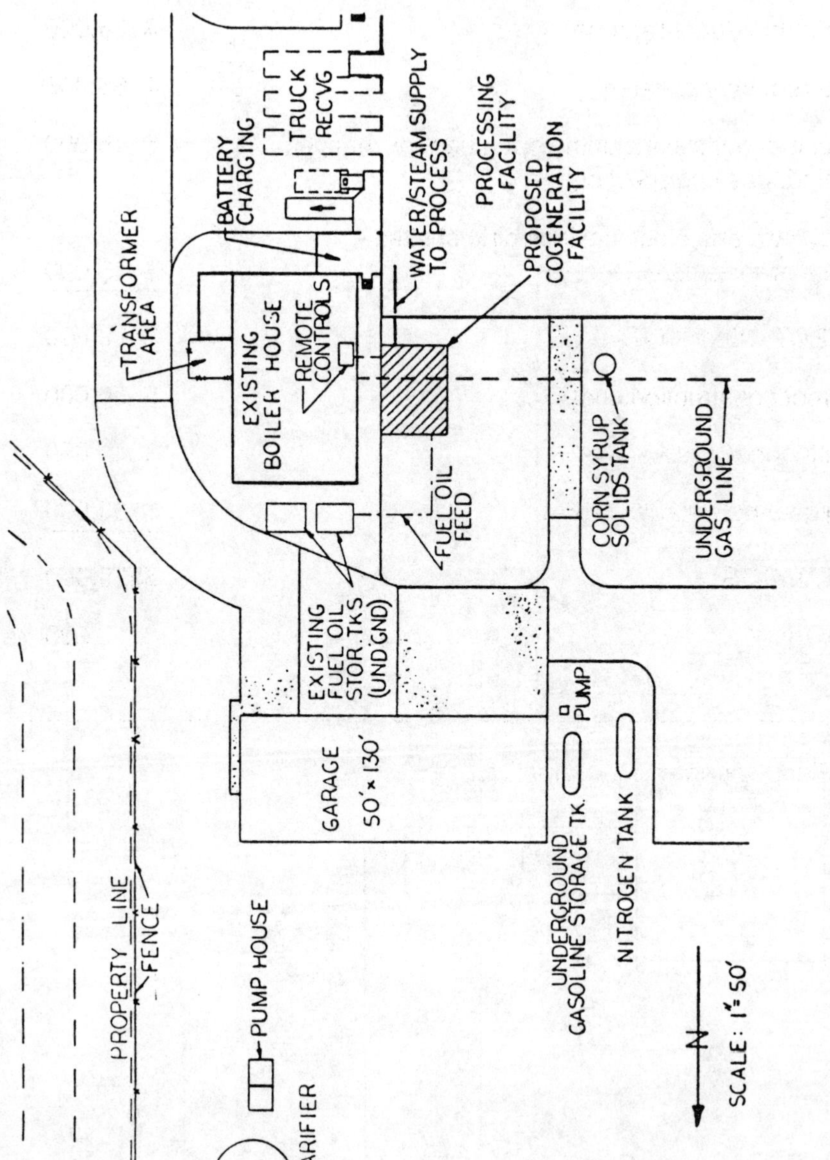


Greyhound Food Processing Plant Gas Turbine Plant Location

Figure C-33

## GREYHOUND COST DATA

| Installed equipment costs | |
|---|---|
| Gas turbine/generator set | $275,000 |
| Heat recovery boiler | $ 65,000 |
| Misc. mechanical including open heat exchanger, piping, gas compressor, etc. | $ 45,000 |
| Misc. civil, and electrical including shelter, site preparation | $ 15,000 |
| DIRECT COST TOTAL | $400,000 |
| Indirect construction cost | $ 50,000 |
| Contingency (10%) | $ 40,000 |
| Engineering services (15%) | $ 60,000 |
| TOTAL COST | $550,000 |
| $kW (net)* | 606 |

* Net Additional Plant Cogeneration Capacity

**Table C-9**

# Section C
## Engineering and Cogeneration

## Appendix

# Configurations

The following are examples of gas turbine and reciprocating engine configurations together with a simple heat balance for a pass-out steam turbine/generation system. There are numerous variations on the theme that can be configured to suit a particular requirement.

When deciding on a particular gas turbine or reciprocating engine configuration and sizing the basic rules of thumb are:

*i)* *Always consider supplementary firing.*
*ii)* *Always consider a combined cycle.*

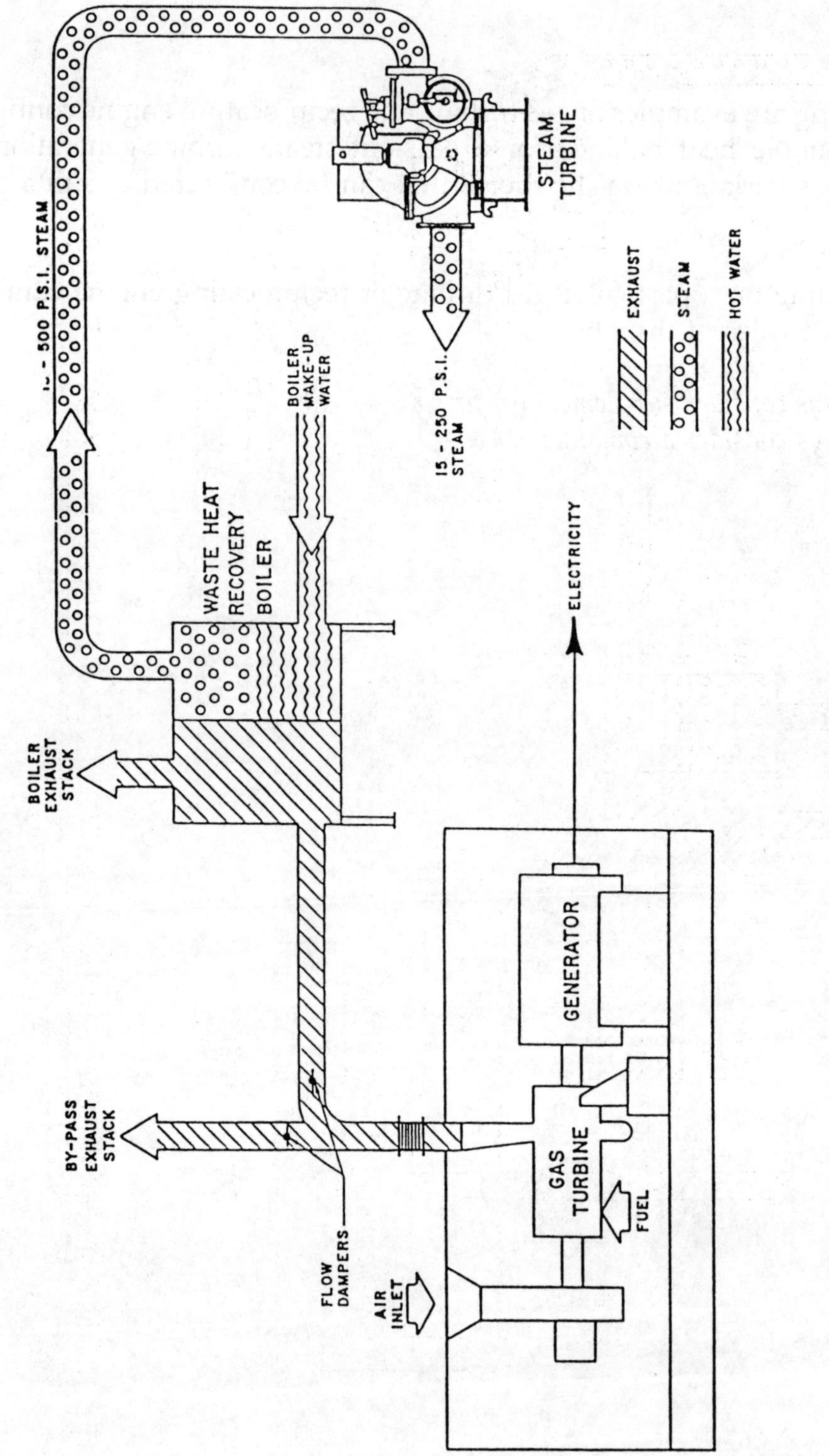


HEAT RECOVERY FROM COMBINED CYCLE GAS TURBINE EXHAUST GAS

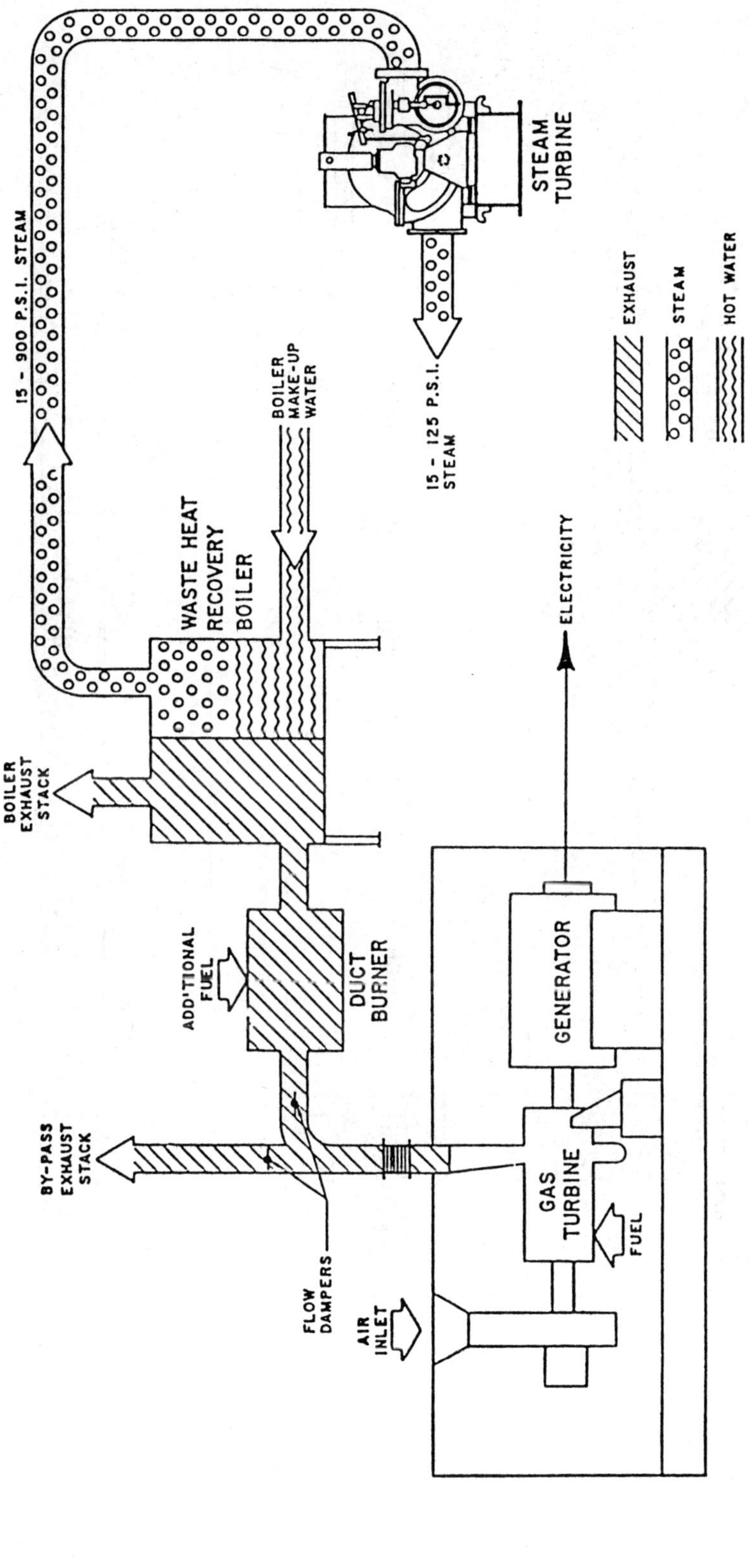


HEAT RECOVERY FROM COMBINED CYCLE GAS TURBINE EXHAUST GAS

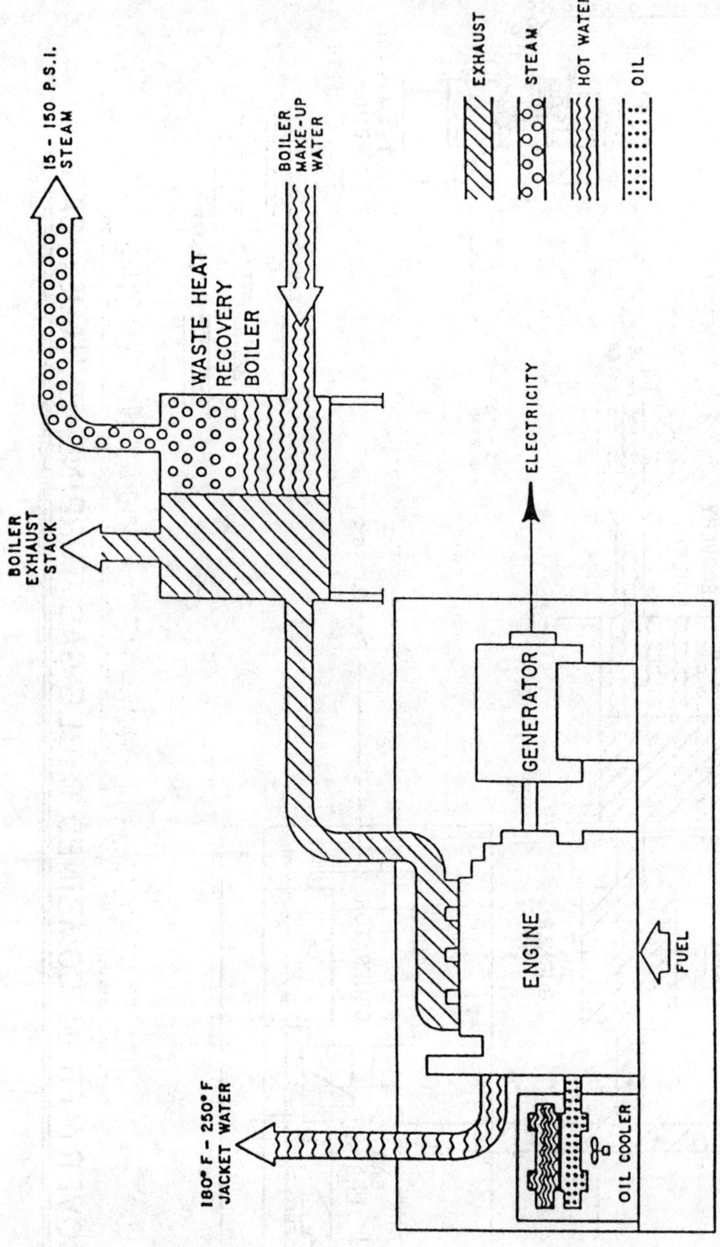


HEAT RECOVERY FROM RECIPROCATING ENGINE

EXHAUST GAS AND JACKET COOLANT

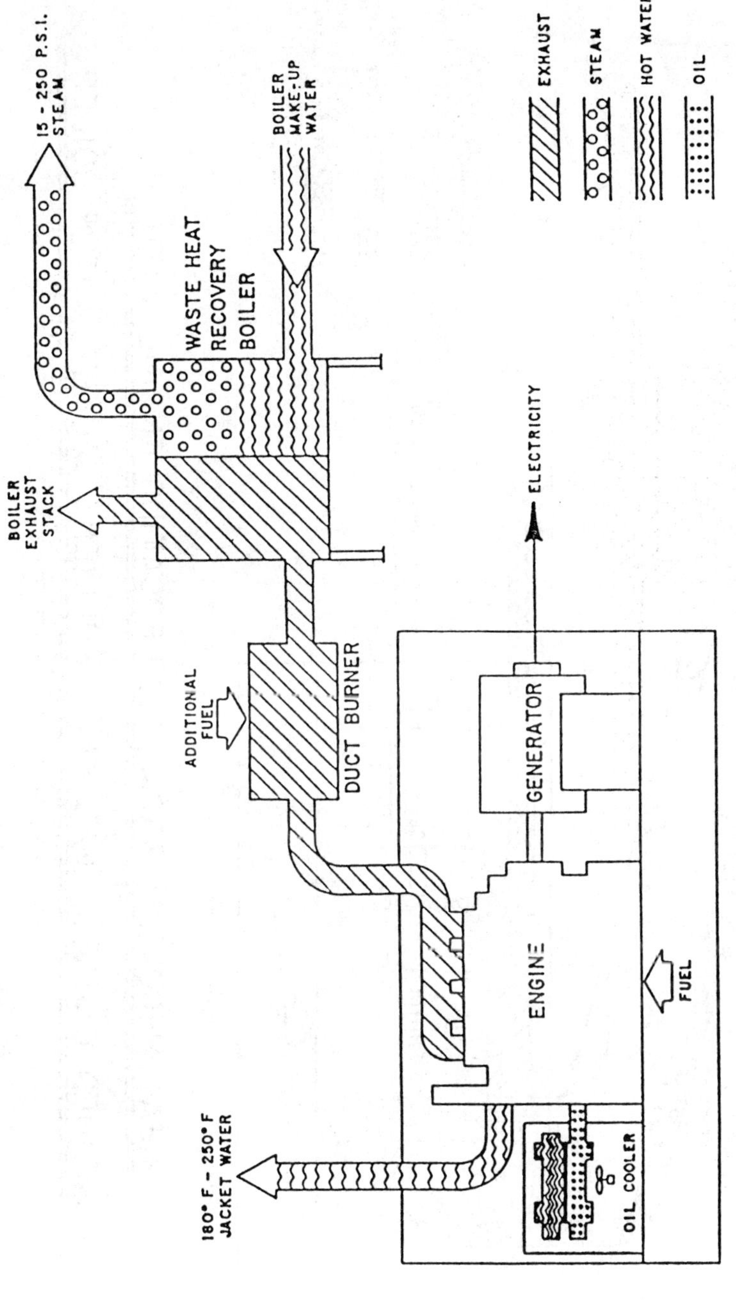


HEAT RECOVERY FROM RECIPROCATING ENGINE

EXHAUST GAS AND JACKET COOLANT

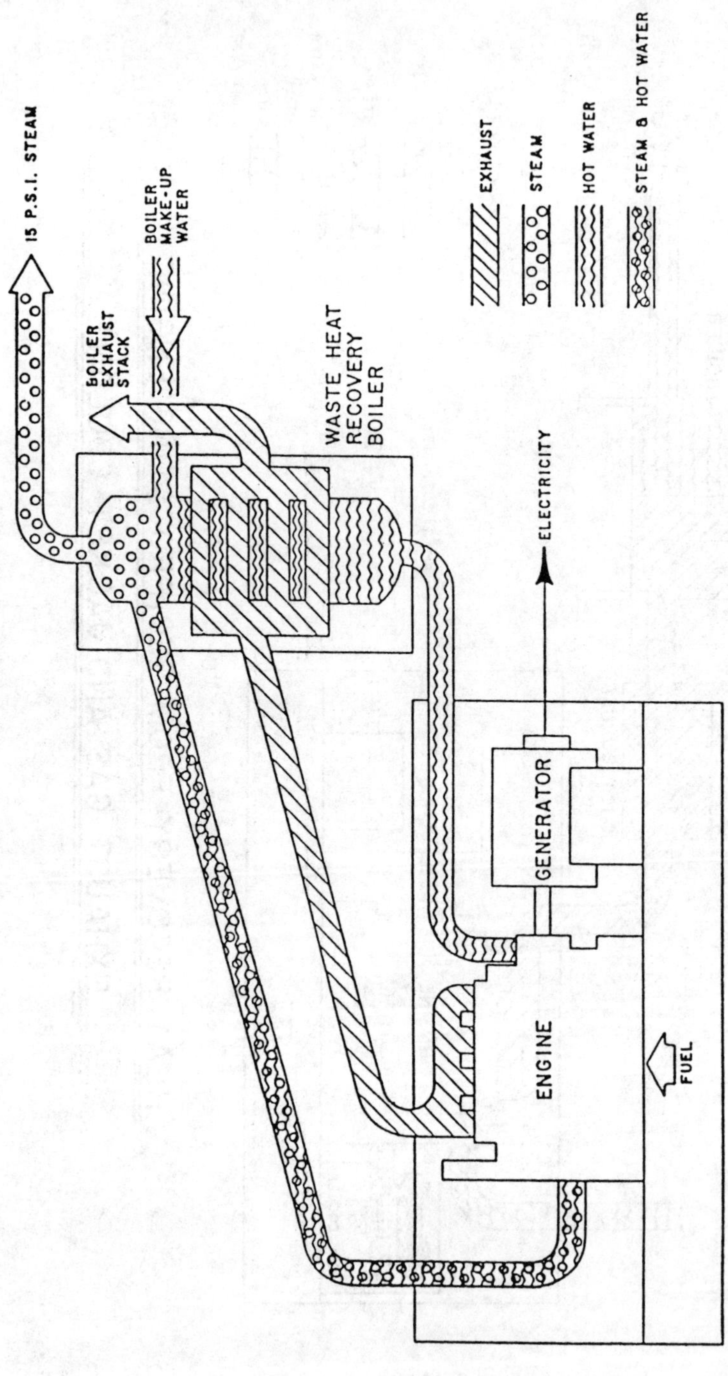


HEAT RECOVERY FROM RECIPROCATING ENGINE WITH

COMBINED EXHAUST GAS AND EBULLIENT JACKET WATER BOILER

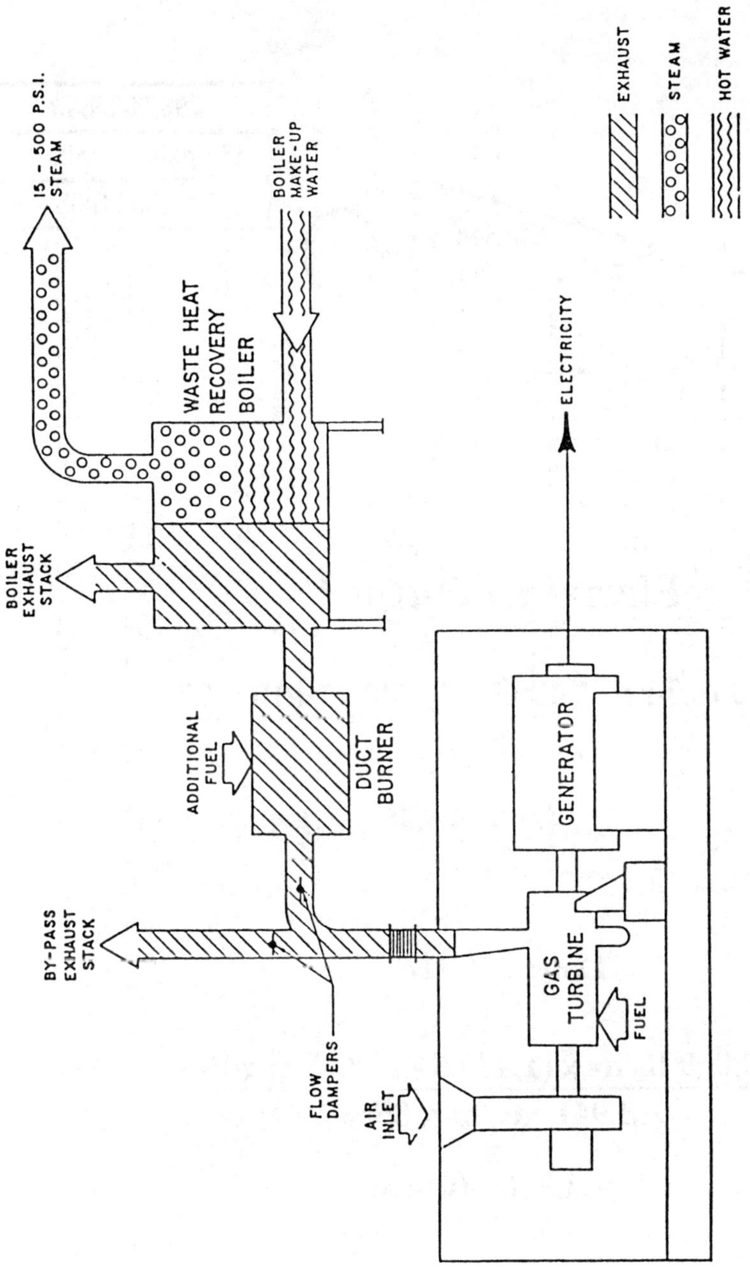


HEAT RECOVERY FROM GAS TURBINE EXHAUST GAS

# Passout Steam Turbine Heat Rate

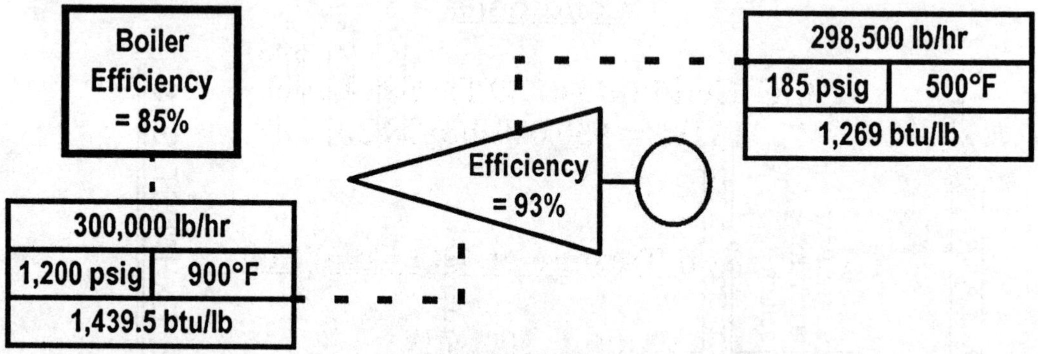


## *Electrical Output*

$$\frac{300{,}000 \text{ lb/hr} \times (1{,}439.5 - 1{,}269) \text{ btu/lb} \times 93\%}{3{,}412 \text{ btu/kw}}$$

$$= 13{,}941 \text{ kw/hr}$$

## *Heat Rate*

$$\frac{300{,}000 \text{ lb/hr} \times (1{,}439.5 - 1{,}269) \text{ btu/lb}}{13{,}941 \text{ kw/hr} \times 85\%}$$

$$= 4{,}316 \text{ btu/kw}$$

# Example of Derating a Gas Turbine

## Conditions

ISO Conditions: 59°F; Sea Level
LHV = 950,000 btu/MCF

## Theoretical Turbine at ISO Conditions

Output: 100 MW
Heat Rate: 10,000 btu/kw
Exhaust Temp: 900°F
Exhaust Flow: 3,000,000 lb/hr
Inlet Loss: 4" W.G.
Outlet Loss: 5" W.G.

## Site Conditions

Elevation: 2,000 ft
Ambient Temperature: 80°F
Total Inlet Losses: 5.5" W.G.
Total Outlet Losses: 13" W.G.

# Rules of Thumb

REDUCE OUTPUT BY:

| | |
|---|---|
| 3% | For every 1,000 ft above sea level |
| 4% | For every 10°F above 59°F |
| 0.5% | For every 1.5" W.G. of increased inlet loss |
| 0.5% | For every 4" W.G. of increased outlet loss |

INCREASE HEAT RATE BY:

| | |
|---|---|
| 0% | For elevation |
| 2% | For every 10°F above 59°F |
| 0.5% | For every 3" W.G. of increased inlet loss |
| 0.5% | For every 3" W.G. of increased outlet loss |

INCREASE EXHAUST TEMPERATURE BY:

| | |
|---|---|
| 0°F | For elevation |
| 1°F | For every 1°F rise in inlet temperature |
| 1°F | For every 1" W.G. of increased inlet loss |
| 1°F | For every 1" W.G. of increased outlet loss |

REDUCE EXHAUST FLOW BY:

| | |
|---|---|
| 3% | For every 1,000 ft above sea level |
| 2% | For every 10°F above 59°F |
| 0.5% | For every 2" W.G. of increased inlet loss |
| 0% | For exhaust loss |

IPPI: Example of Derating a Gas Turbine

# Calculations

## NET POWER OUTPUT:

100 MW x .94 x .92 x .995 x .99 = 85.19 MW
Percentage output reduction = 12.03%

## NET INCREASE IN HEAT RATE:

10,000 Btu/hr x 1 x 1.04 x 1.0025 x 1.0133 = 10,556
Percentage increase in heat rate = 5.56%

## NET INCREASE IN EXHAUST TEMPERATURE:

900°F + 0 + 21 + 1.5 + 8= 930.5°F
Increase in exhaust temperature = 30.5°F

## NET LOSS IN EXHAUST FLOW:

3,000,000 lb/hr x .94 x .96 x .996 = 2,696,371 lb/hr
Loss: 303,629 lb/hr
Percentage: 10.12%

# HHV Fuel Consumption

**ISO FUEL CONSUMPTION:**

Net heat output x Heat Rate / LHV
85,190 KW x 10,556 btu/kwh / 950,000 btu/MCF
= 946.6 MCF/hr or 90.00 KWh/MCF

**FUEL PURCHASED AT THE HIGHER HEATING VALUE:**

HHV = LLV x 1.11
Fuel purchased = 946.6 MCF/hr x 1.11
= 1,050.7 MCF/hr or 81.08 KWh/MCF

IPPI: Example of Derating a Gas Turbine

# Project Development

There are two major questions which are asked at the outset when considering a cogeneration project.

i) How much does it cost to install a plant, say in terms of $/kw?.
ii) How long does it take to permit a plant?.

Unfortunately, it is not possible to answer these fundamental questions at the outset.

*The installation of a cogeneration plant is site specific. Two identical configurations utilizing identical equipment can vary in the final cost of installation by as much as 50%.*

*The permitting process is site specific. Two identical configurations utilizing identical equipment can vary in the time taken to permit from as little as 6 months to never.*

These questions can only be answered through the carrying out of a process which may entail a detailed analysis and actual submittal of permit applications. This process can be both costly and time consuming.

The following Gantt charts lay out the schedule of a theoretical project which takes just under 3 years of elapsed time to complete from the Initial Feasibility Study to operation. Obviously, this time frame will vary from project to project.

The charts are based upon a methodology that is used by Independent Power Producers, Inc. which can be applied to both large and small projects and is designed to minimize costs and avoid unnecessary errors. The methodology breaks the project down into four basic phases. These are:

- Feasibility Study Phase
- Permitting and Negotiation Phase
- Procurement and Design Phase
- Installation Phase

Milestones and decision points are associated with each phase. However, it should be noted that the permitting process and the obtaining of permits in a timely manner are critical to the whole decision making process and there is no definitive method of estimating the time it may take to obtain all the necessary permits.

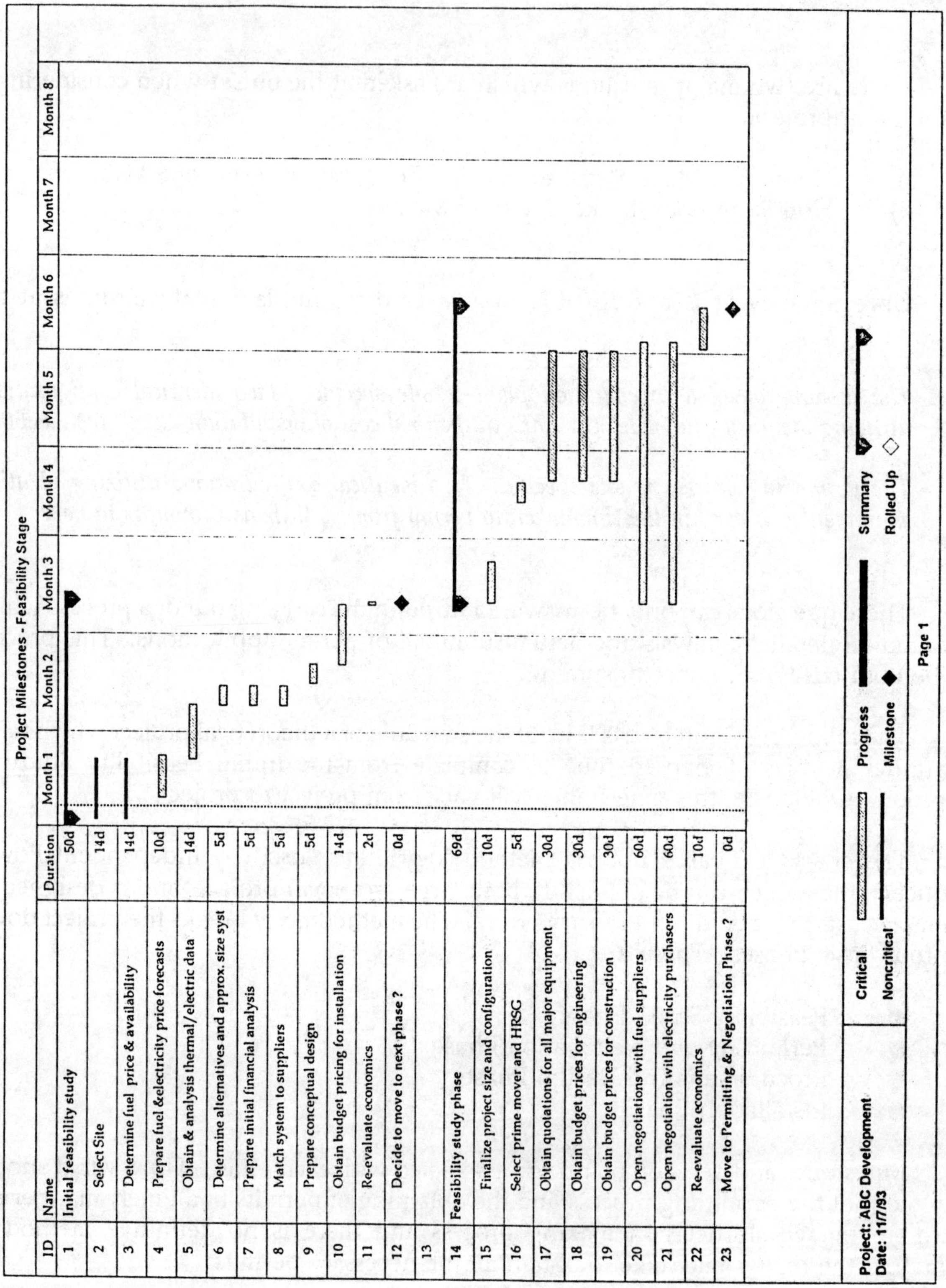

Project Milestones - Feasibility Stage
ID
Name
Duration
Month 1
Month 2
Month 3
Month 4
Month 5
Month 6
Month 7
Month 8
1 Initial feasibility study 50d
2 Select Site 14d
3 Determine fuel price & availability 14d
4 Prepare fuel &electricity price forecasts 10d
5 Obtain & analysis thermal/electric data 14d
6 Determine alternatives and approx. size syst 5d
7 Prepare initial financial analysis 5d
8 Match system to suppliers 5d
9 Prepare conceptual design 5d
10 Obtain budget pricing for installation 14d
11 Re-evaluate economics 2d
12 Decide to move to next phase ? 0d
13
14 Feasibility study phase 69d
15 Finalize project size and configuration 10d
16 Select prime mover and HRSG 5d
17 Obtain quotations for all major equipment 30d
18 Obtain budget prices for engineering 30d
19 Obtain budget prices for construction 30d
20 Open negotiations with fuel suppliers 60d
21 Open negotiation with electricity purchasers 60d
22 Re-evaluate economics 10d
23 Move to Permitting & Negotiation Phase ? 0d
Project: ABC Development
Date: 11/7/93
Critical
Noncritical
Progress
Milestone
Summary
Rolled Up
Page 1

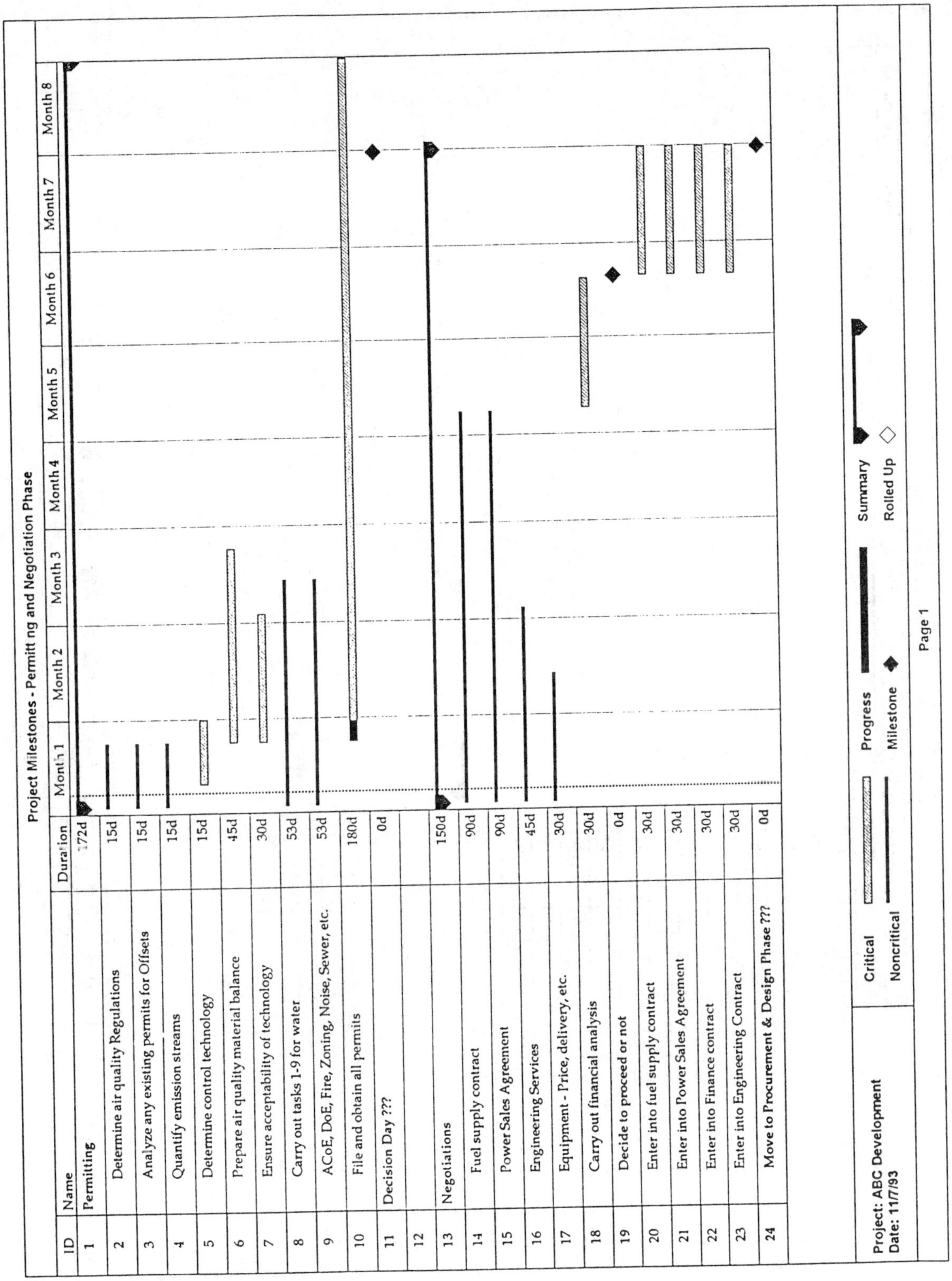
Project Milestones - Permitting and Negotiation Phase
ID
Name
Duration
Month 1
Month 2
Month 3
Month 4
Month 5
Month 6
Month 7
Month 8
1 Permitting 172d
2 Determine air quality Regulations 15d
3 Analyze any existing permits for Offsets 15d
4 Quantify emission streams 15d
5 Determine control technology 15d
6 Prepare air quality material balance 45d
7 Ensure acceptability of technology 30d
8 Carry out tasks 1-9 for water 53d
9 ACoE, DoE, Fire, Zoning, Noise, Sewer, etc. 53d
10 File and obtain all permits 180d
11 Decision Day ??? 0d
12
13 Negotiations 150d
14 Fuel supply contract 90d
15 Power Sales Agreement 90d
16 Engineering Services 45d
17 Equipment - Price, delivery, etc. 30d
18 Carry out financial analysis 30d
19 Decide to proceed or not 0d
20 Enter into fuel supply contract 30d
21 Enter into Power Sales Agreement 30d
22 Enter into Finance contract 30d
23 Enter into Engineering Contract 30d
24 Move to Procurement & Design Phase ??? 0d
Project: ABC Development
Date: 11/7/93
Critical
Noncritical
Progress
Milestone
Summary
Rolled Up
Page 1

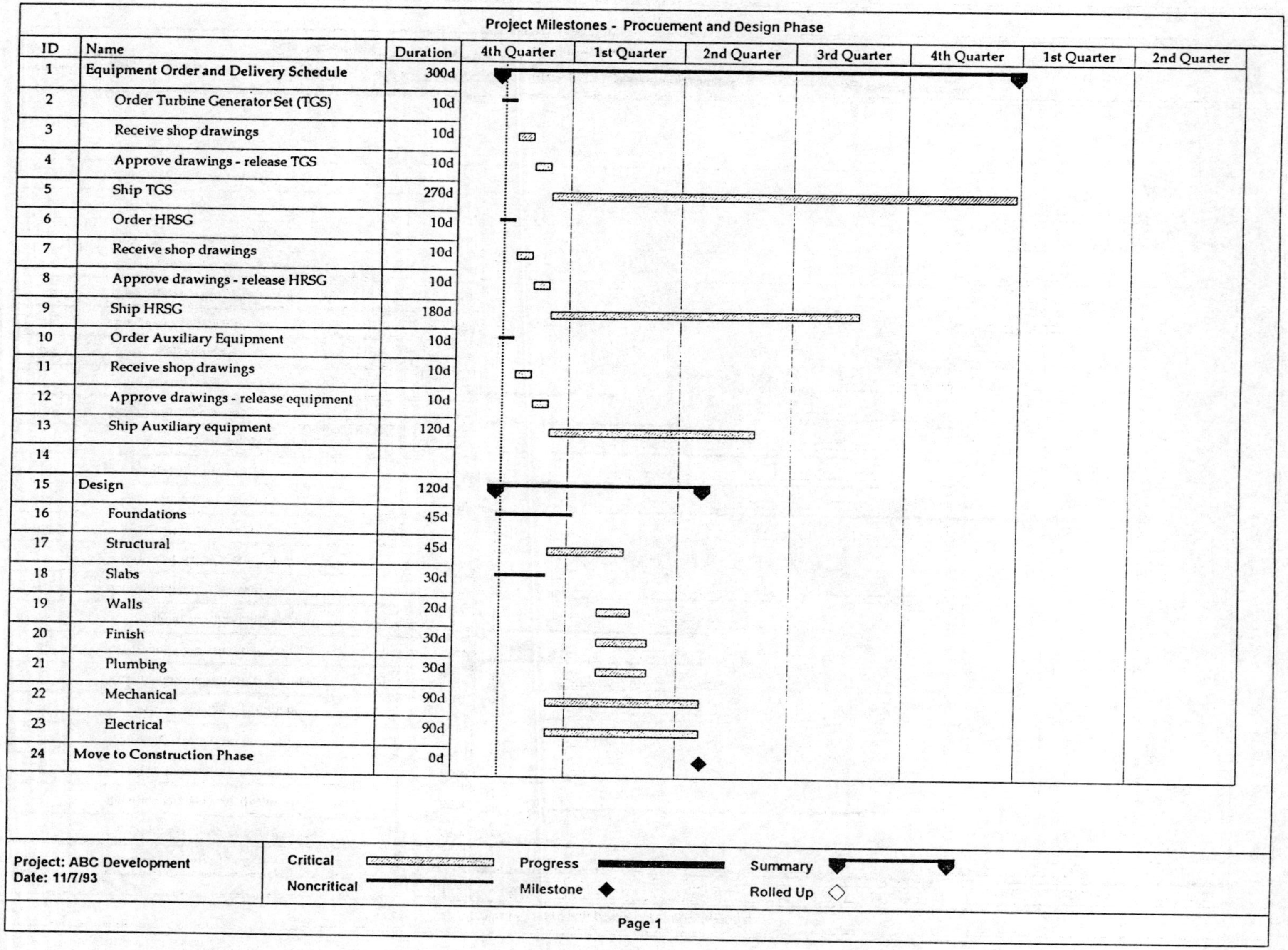
Project Milestones - Procuement and Design Phase
ID
Name
Duration
4th Quarter
1st Quarter
2nd Quarter
3rd Quarter
4th Quarter
1st Quarter
2nd Quarter
1 Equipment Order and Delivery Schedule 300d
2 Order Turbine Generator Set (TGS) 10d
3 Receive shop drawings 10d
4 Approve drawings - release TGS 10d
5 Ship TGS 270d
6 Order HRSG 10d
7 Receive shop drawings 10d
8 Approve drawings - release HRSG 10d
9 Ship HRSG 180d
10 Order Auxiliary Equipment 10d
11 Receive shop drawings 10d
12 Approve drawings - release equipment 10d
13 Ship Auxiliary equipment 120d
14
15 Design 120d
16 Foundations 45d
17 Structural 45d
18 Slabs 30d
19 Walls 20d
20 Finish 30d
21 Plumbing 30d
22 Mechanical 90d
23 Electrical 90d
24 Move to Construction Phase 0d
Project: ABC Development
Date: 11/7/93
Critical
Noncritical
Progress
Milestone
Summary
Rolled Up
Page 1

Project Milestones - Turbine Installation

| ID | Name | Duration |
|---|---|---|
| 1 | Gas Turbine - Installation Schedule | 139d |
| 2 | Foundations | 4w |
| 3 | Receive & install equipment | 77d |
| 4 | Install Turbine | 3w |
| 5 | Install Generator | 2w |
| 6 | Install Starting Equipment | 2w |
| 7 | Install Mechanical Equipment | 8w |
| 8 | Install inlet manifold | 4w |
| 9 | Align rotational equipment | 4w |
| 10 | Install Lube oil coolers | 3w |
| 11 | Install Exhaust equipment | 5w |
| 12 | Install inlet system | 4w |
| 13 | Install electrical equipment | 6w |
| 14 | Pipework installation | 12w |
| 15 | Cable and wiring installation | 12w |
| 16 | Installation inspection | 10w |
| 17 | Training program | 6w |
| 18 | Synchronization | 0d |
| 19 | Testing & Operations | 6w |
| 20 | Operational | 0d |

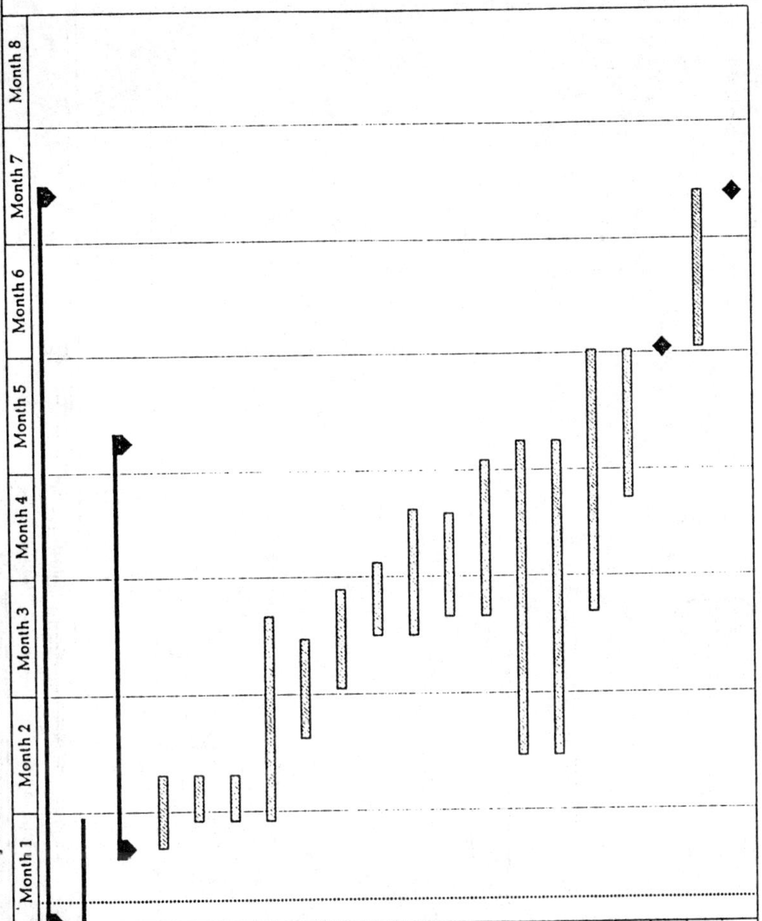


Project: ABC Development
Date: 11/7/93

Critical — Progress — Summary
Noncritical — Milestone — Rolled Up

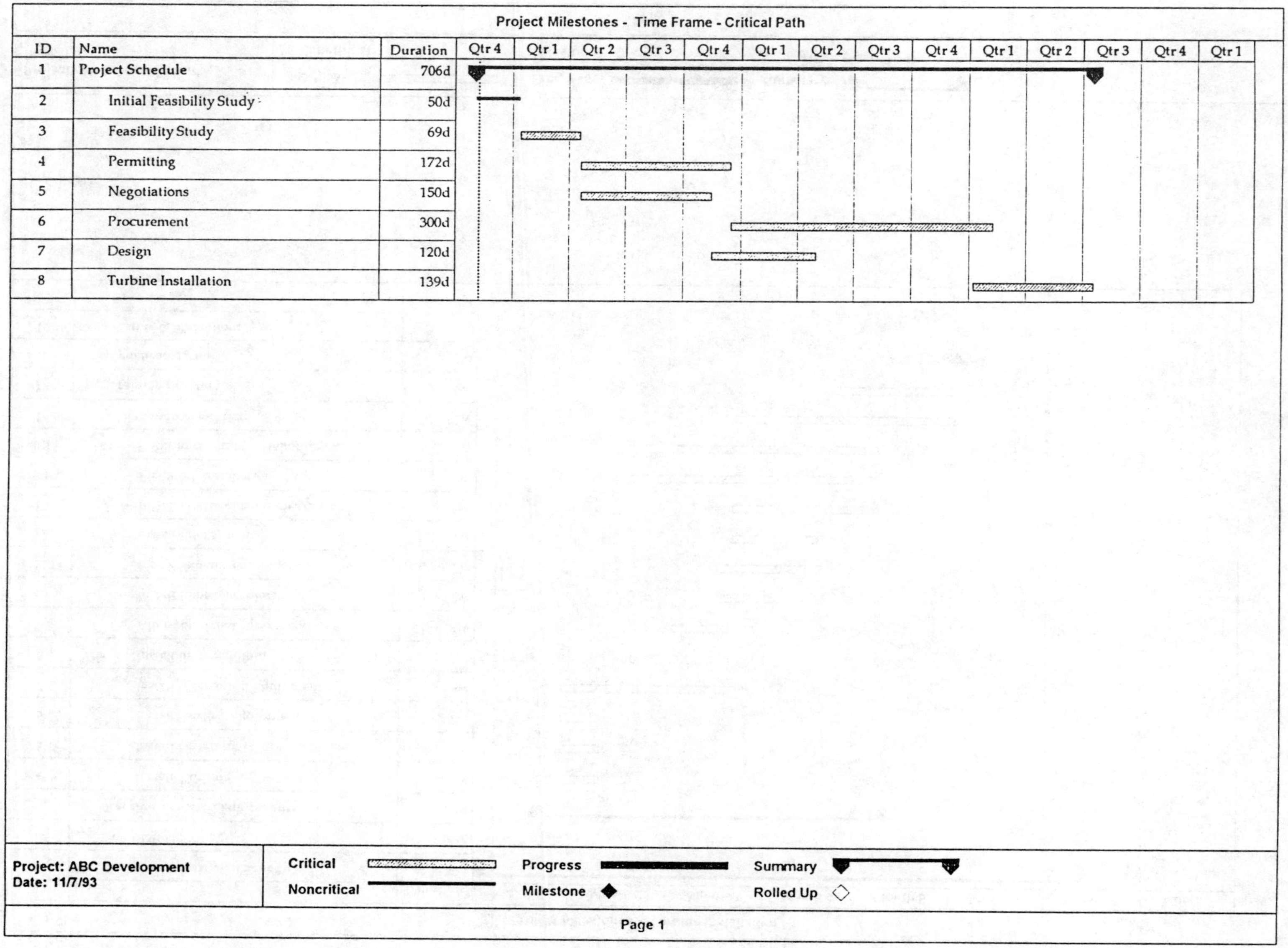
Project Milestones - Time Frame - Critical Path
ID
Name
Duration
Qtr 4
Qtr 1
Qtr 2
Qtr 3
Qtr 4
Qtr 1
Qtr 2
Qtr 3
Qtr 4
Qtr 1
Qtr 2
Qtr 3
Qtr 4
Qtr 1
1 Project Schedule 706d
2 Initial Feasibility Study 50d
3 Feasibility Study 69d
4 Permitting 172d
5 Negotiations 150d
6 Procurement 300d
7 Design 120d
8 Turbine Installation 139d
Project: ABC Development
Date: 11/7/93
Critical
Noncritical
Progress
Milestone
Summary
Rolled Up
Page 1

# Equipment Manufacturers

The following section contains the names of equipment manufacturers, particularly those associated with the production of prime movers. It is not intended that this list is all encompassing, or that it endorses any particular manufacturer.

Also, as no individual manufacturer's product range covers the entire range of size or types of products, each manufacturer should be contacted to ascertain his range of products, heat rates, prices and other specific information as required.

# MANUFACTURERS LIST

Below you will find a partial list of manufacturers of cogeneration equipment. This list is meant to be a good starting point for your selection of equipment.

*Gas Turbines*

Asea Brown Boveri
Turbine Power Division
1460 Livingston Avenue
North Brunswick, NJ 08902
908 932 6069
908 932 6194 (F)

General Electric Company
Industrial and Power Systems
One River Road
Schenectady, NY 12345
518 385 2323
518 385 7850 (F)

Westinghouse Electric Corp.
Power Generation Business Unit
4400 Alafaya Trail
Orlando, FL 32826
407 281 2000
407 281 3303 (F)

Siemens
Gas Turbine Division
1040 South 70 Street
Milwaukee, WI 53214

Solar Turbines Incorporated
P.O. Box 85376
San Diego, CA 92186-5376
619 544 5000
619 544 5875 (F)

Allison Gas Turbines
Industrial Engine Sales
P.O. Box 420, Speed Code U-6
Indianapolis, IN 46206-0420
317 230 4151
317 230 2900 (F)

Dresser-Rand Company
Turbo Products Division
P.O. Box 560
Olean, NY 14760
716 375 3000
716 375 3178 (F)

Pratt & Whitney Canada Inc.
1000 Marie Victorin
Longueuil, Quebec J4G 1A1
514 677 9411
514 442 7298 (F)

Stewart & Stevenson, Inc.
P.O. Box 1637 (77251-1637)
2707 North Loop West
Houston, TX 77008
713 868 7700
713 868 7692 (F)

*Heat Recovery Steam Generators*

Vogt Machine Company
P.O. Box 1918
Louisville, KY 40201
502 634 1500
502 634 0549 (F)

ABB/Combustion Engineering
Combustion Engineering, Inc.
1000 Prospect Hill Road
Windsor, CT 06095-0500
203 285 3824
203 285 4950 (F)

Foster Wheeler Energy Corporation
8 Peach Tree Hill Road
Livingston, NJ 07039
201 535 2400

Zurn/NEPCO
18578 N.E. 67 Court
P.O. Box 747
Redmond, WA 98073-0747
206 869 3000

Vaporphase
600 South Holmes
St. Louis, MO 63122
314 821 7900

## *Reciprocating Engines*

Cooper Energy Services
North Sandusky Street
Mount Vernon, OH 43050
614 393 8200
614 393 8373 (F)

Fairbanks-Morse
Engine Division
701 Lawton Avenue
Beloit, WI 53511-5492
608 364 4411
608 364 0382 (F)

Caterpillar Inc.
Engine Division
P.O. Box 610
Mossville, IL 61629
309 675 1000
309 578 6466 (F)

Detroit Diesel Corp.
13400 Outer Drive
Detroit, MI 48239-4001
313 592 5000
313 592 7288 (F)

Waukesha Engine Division
1000 West ST. Paul Avenue
Waukesha, WI 53188
414 547 3311
414 549 2795 (F)

Onan Corporation
Industrial Business Group
1400 73RD Avenue N.E.
Minneapolis, MN 55432
612 574 5000
612 574 5548 (F)

Yanmar Diesel America Corp.
901 Corporate Grove Drive
Buffalo Grove, IL 60089-4533
708 541 1900
708 541 2161 (F)

F.G. Wilson Inc.
10125 N.W. 116TH Way
Suite 6.
Medley, FL 33178
305 883 8100
305 883 9956 (F)

# Section D

## Finance and Cogeneration

# Section D
# Finance and Cogeneration

## FINANCIAL FEASIBILITY OF COGENERATION

With the following information – records of past operating levels and expenses, the efficiency of the proposed equipment, the level of expected boiler operation, the demand for electricity, and the expected price of fuel – it should be possible to predict fairly closely the savings in electrical costs that would result from substituting cogenerated electricity for utility-generated electricity.

To evaluate the desirability of an investment, measures of costs are needed to compare with the benefits. Table D-1 shows the type of costs which may arise in connection with cogeneration.

As may be seen, costs may begin before the cogeneration system is installed and extend throughout the period of continued plant operation. In most cases, the major cost item is likely to be the acquisition and installation of the power equipment, and should be relatively easy to estimate. (See listing of cogeneration equipment manufacturers, their offerings and prices, in the Appendix to Section C: ENGINEERING, above).

It is important that only those costs and benefits which are attributable to an investment be included in the analysis of that investment. For example, costs of equipment replacement or repair not necessitated by the addition of the cogeneration system should not be incorporated into the evaluation, although it may be undertaken jointly for convenience.

## PARTIAL METHODS OF EVALUATION

The simplest procedures which are used by firms to try to evaluate alternative kinds and

**Table D-1**

**Potential costs to consider in investing in cogeneration**

1. Pre-engineering and planning costs
2. Acquisition costs of cogeneration equipment
3. Acquisition costs of necessary additions to existing equipment
4. Replacement costs
5. Costs of modification and repair of existing equipment
6. Space costs
7. Costs of production downtime during installation
8. Costs of adjustments (debugging)
9. Maintenance costs of new equipment
10. Property and/or equipment taxes of cogeneration equipment
11. Change in insurance or hazard costs.

amounts of investments are visual inspection, payback period, and return on investment – approaches which are termed "partial" here because they do not fully assess the economic desirability of alternatives. (There are some investments whose desirability is apparent merely by inspection, and which do not require further economic analysis. An example is an investment characterized by negligible or low costs and a highly certain return. But actions which require significant initial investment and yield benefits over time – as cogeneration is typically characterized – usually require more extensive analysis than visual inspection). These partial methods may be contrasted with the more complete techniques, discussed below, which take into account factors such as timing of cash flows, risk, and taxation effects – factors which are required for full economic assessment of investments.

Despite their shortcomings, the partial techniques of analysis may serve a useful purpose. They can provide a first level measure of profitability which is, relatively speaking, quick, simple, and inexpensive to calculate. They may therefore be useful as initial screening devices for eliminating the more obviously uneconomical investments. These partial techniques (particularly the payback method) may also provide needed information concerning certain sensitive features of an investment. But where partial methods are used, the more comprehensive techniques may also be needed to verify the outcome of the evaluations, and to rank alternative projects as to their relative efficiency.

Following are descriptions, examples, and limitations of the payback method and the return on investment method, two of the more popular partial methods.

## PAYBACK METHOD

The payback (also known as the payout or payoff) method determines the number of years required for the invested capital to be offset by resulting benefits. The required number of years is termed the payback, recovery, or break-even period.

The measure is popularly calculated on a before-tax basis and without discounting, i.e., neglecting the opportunity cost of capital. (The opportunity cost of capital is the return which could be earned by using resources for the next best available investment purpose (e.g., from an investor's standpoint, this might be the earning of interest on savings accounts), rather than for the purpose at hand. It represents an extra return, beyond merely covering other costs, which is necessary in order to make an investment competitive with other opportunities. Investment costs are usually defined as first costs, often neglecting salvage value. Benefits are usually defined as the resulting net change in incoming cash flow, or, in the case of a cost-reducing investment like cogeneration, as the reduction in net outgoing cash flow.

The payback period is calculated as follows:

$$\text{Payback Periods (PP)} = \text{First Cost} / (\text{Yearly Benefits} - \text{Yearly Costs})$$

(This method has the implicit assumption that the expected proceeds from an investment are constant from year-to-year. If expected yearly proceeds are not equal, the customary approach is either (1) to average yearly benefits and costs and use the above formula, or (2) to add the proceeds in successive years until their total equals the first cost).

The disadvantages of the payback method which recommend against its use as a sole criterion for investment decisions may be summarized as follows:

(a) The method does not give consideration to cash flows beyond the payback period, and thus does not measure the efficiency of an investment over its entire life.

Consider, for example, the two alternative investments A and B, presented in Table D-2. Using the undiscounted payback method, a firm would prefer Investment A, which has a payback period of 1.7 years, to Investment B, which has a payback of 2.2 years. Yet, depending upon the true opportunity cost of capital (i.e., the discount rate), Investment B, which continues to yield benefits beyond Investment A, may be a more profitable choice. (For example, with an opportunity cost of 10 percent, Investment A would yield $20,832 in total benefits, and Investment B, $22,383 in total benefits in present value terms).

(b) The neglect of the opportunity cost of capital, that is, failing to discount costs occurring at different times to a common base for comparison, results in the use of inaccurate measures of benefits and costs to calculate the payback period, and, hence, determination of an incorrect payback period. This problem is illustrated by the example of two alternative investments shown in Table D-3. Payback using undiscounted values would result in indifference between Investments C and D. They both have a payback of 2 years, and yield total benefits, undiscounted of $25,000. But because Investment D yields more benefits toward the beginning than Investment C, and thereby allows the investor to realize a larger return on earnings, Investment D would be the preferred choice. In present value terms, with an opportunity cost of 10 percent, Investment C would yield total benefits of $20,697, and Investment D, $21,524.

In short, the payback method gives attention to only one attribute of an investment, i.e., the number of years to recover costs, and, as often calculated, does not even provide an accurate measure of this. It is a measure which many firms appear to overemphasize, tending toward shorter and shorter payback requirements. Firms' preference for very short payback to enable them to reinvest in other investment opportunities may in fact lead to a succession of less efficient, short-lived projects.

**Table D-2 Illustration that payback analysis does not take into account cash flows beyond the payback period**

| Investment Year 1 ($20,000) | Year 2 (Yearly Benefits) | Year 3 Period | Payback Value Benefits | | Total Present |
|---|---|---|---|---|---|
| Investment A | $12,000 | $12,000 | $0 | 1.7 Years | $20,832 |
| Investment B | $ 9,000 | $ 9,000 | $9,000 | 2.2 Years | $22,383 |

**Table D-3 Illustration that the undiscounted payback method can result in inaccurate measures**

| Investment Year 1 ($20,000) | Year 2 (Yearly Benefits) | Year 3 Period | Payback Value Benefits | | Total Present |
|---|---|---|---|---|---|
| Investment A | $ 5,000 | $15,000 | $5,000 | 2 Years | $20,697 |
| Investment B | $15,000 | $ 5,000 | $5,000 | 2 Years | $21,524 |

In both tables, calculations of total present value benefits are for a discount rate of 10%, compounded annually.

Despite its limitations, the payback period has advantages in that it may provide useful information for evaluating an investment. There are several situations in which the payback method might be particularly appropriate:

(a) A rapid payback may be a prime criterion for judging an investment when financial resources are available to the investor for only a short period of time.

(b) The speculative investor who has a very limited time horizon will usually desire rapid recovery of the initial investment.

(c) Where the expected life of the assets is highly uncertain, determination of the break-even life, i.e., payback period, is helpful in assessing the likelihood of achieving a successful investment. (This use of the payback method is discussed later in the section on risk, uncertainty, and sensitivity analysis.)

The shortcomings that result from failure to discount costs and the omission of important cost items can be overcome simply by using a more accurate calculation of payback. Essentially what is desired is to find the number of years, R, for which the value of the following expression is equal to zero:

$$C = \sum_{j=1}^{R} \frac{B_j - P_j}{(1+i)^j}$$

where

C = Initial investment cost
Bj = Benefits in year j
Pj = Costs in year j
R = Break-even number of years
i = Discount rate

Where yearly net benefits are uneven, an iterative process can be used to determine the solution. If, on the other hand, yearly net benefits are expected to be about uniform, the following formula can be used to facilitate the calculation:

$$R = (-\log(1-(iC/M))) / (\log(1+i))$$

where
R = Break-even number years
M= Yearly net benefits
C = Initial investment cost
i = Discount rate

## RETURN ON INVESTMENT METHOD

The return on investment (ROI) or return on assets method calculates average annual benefits, net of yearly costs such as depreciation, as a percentage of the original book value of the investment.

The calculation is as follows:

Return on Investment (ROI) =
(Average Annual Net Benefits / Original Book Value) x 100.

As an example, the calculation of the ROI for an investment in a gas turbine is as follows:

Original Book Value - $1,500,000

Expected Life = 10 years

Annual Depreciation,

using a straight-line method = $1,500,000 / 10 = $150,000

Yearly Operation Maintenance and Repair Costs = $20,000

Expected Savings / Income = $500,000

ROI = ($5,000,000 – ($150,000 + $20,000) / $1,500,000) x 100 = 0.22 x 100 = 22%

The return on investment method is subject to the following principal disadvantages, and therefore, is not recommended as a sole criterion for investment decisions:

(a) Like the payback method, this method does not take into consideration the timing of cash flows, and thereby may incorrectly state the economic efficiency of projects.

(b) The calculation is based on an accounting concept, original book value, which is subject to the peculiarities of the firm's accounting practices, and which generally does not include all costs. The method, therefore, results in only a rough approximation of an investment's value.

The advantages of the return on investment method are that it is simple to compute and a familiar concept in the business community. (A variation of this method is the return on average investment method, in which the average yearly benefit, before taxes and net of depreciation, is divided by the average book value over the life of the project, defined as the original book value plus the final book value, divided by 2. Thus, with no remaining salvage value at the end of 10 years, the return on average investment in the example would be 44 percent).

## COMPREHENSIVE METHODS FOR EVALUATING INVESTMENT ALTERNATIVES

There are additional methods of financial analysis which avoid the problems of the partial methods by taking into account total costs and benefits over the life of the investment and the timing of cash flows by discounting. The methods of this type discussed here are the present value of net benefits method, the annual net benefits method, the benefit/cost ratio method, and the internal rate of return method. Before defining each of these, let us consider an element common to them, the discounting of costs.

## DISCOUNTING OF COSTS

Investment in cogeneration systems, like many capital investments, will generally require a number of expenditures spread over a period of time and will result in cost savings (or revenue receipts) also spread over time. To evaluate correctly the profitability of such investments, it is necessary to convert the various expenditures and receipts to a common basis, because dollars spent or received at different times are not of equal value. If, for instance, the firm's earning opportunity is 8 percent/year compounded annually, $1.00 received today would grow to $1.47 in 5 years and would about double in 9 years. Deferral of the receipt of $1.00 for 5 years would mean that the earnings on it over the interim 5 years would not be realized, and it would be worth less than $1.00 received today. It would, in fact, be equivalent to receiving only $0.68 today. Therefore, other things equal, a firm will prefer early income and deferred expenditures. A real opportunity cost of capital exists even when there is no inflation or deflation and even when equity (nonborrowed) funds are used, as long as alternative productive investments are possible.

Cash flows occurring at different times can be converted to a common basis, i.e., discounted, by means of discounting formulas (also known as interest formulas). The six basic discounting formulas are as follows:

Where:

P = a present sum of money

F = a future sum of money, equivalent to P at the end of N periods of time at i interest rate.

i = a discount or interest rate to reflect the opportunity cost of capital

N = number of interest periods

A = an end-of-period payment (or receipt) in a uniform series of payments (or receipts) over N periods at i interest rate, usually annually;

**Single compound amount formula**
$F = P (l+i)n$
(SCA) Use when given P, to find F

**Single present worth formula**
$P = F (l/(l+i)n)$
(SPW) Use when given F, to find P

**Uniform compound amount formula**
$F = A ((l+i)n-l)/i$
(UCA) Use when given A, to find F

**Uniform sinking fund formula**
$A + F (i/((l+i)n-l)$
(USF) Use when given F, to find A

**Uniform capital recovery formula**
$A = P ((i(l+i)n/((l+i)n-1))$
(UCR) Use when given P, to find A

**Uniform present worth formula**
$P = A (((l+i)n-l) / (i(l+i)n))$
(UPW) Use when given A, to find P.

The discounting of costs can be greatly simplified by the use of tables of discount factors, such as those at the end of this chapter. The discount factors are calculated from the discount formulas for various time periods and opportunity costs (expressed as a rate of discount or interest). In the sample tables provided, factors are calculated for 50 to 100 interest periods (which could be years, months, or any other time interval desired), and for selected discount rates ranging from 10 to 25 percent per period.

To find the comparable value of a given sum of money if it were received (or disbursed) at a different time, the appropriate formula may be used, or the corresponding discount factor may be multiplied by the given sum. For example, if a firm's opportunity earning potential is 10 percent annually, \$500 to be received 5 years from now would be equivalent to receiving \$310 today, (i.e., Present value = \$500, SPW, i = 10%, N = 5 years. = \$500 x 1 / (1 + 0.10)5 = \$500 (0.6209) = \$310).

Alternatively, receiving \$500, five years from now, would be equivalent to receiving equal annual payments of \$82 in each of the five years. I.e., Annual Payment Equivalent, under the uniform sinking fund formula, where i = 10% and N = 5 years, equals –

\$500 (.10 / ((l+.10)5-1)) = \$500 (0.1638) = \$82.

Table D-4 illustrates the conversion to a present equivalent and an annual equivalent of various costs and benefits representative of an investment in cogeneration. The examples assume that the investing firm has an opportunity cost of 15 percent and a 10 years time horizon. The first column gives the kind of cost or benefit incurred. The second column uses a cash flow diagram to describe the pattern of cash outflow or inflow associated with the cost or benefit. The horizontal line of the cash-flow diagram is a time scale, where P indicates the present, the progression of time moves from left to right, and the numbers between the points represent time intervals (in this case, years). The dollar figures below the line represent expenditures (cash outflows), and the dollar figures above the line represent receipts (cash inflows), as viewed from the standpoint of the firm.

Once all costs or benefits are discounted to the present, they may be combined in order to find the total. Thus, in this simplified example, total costs of the investment amount to \$3,067,037 in present value terms, and total present value benefits amount to \$3,513,300 over the 10 years (found by aggregating all the costs and all the benefits in the Present Equivalent Column of Table D-4). This is equivalent to incurring cost of \$618,500 annually and realizing benefits of \$700,000 annually for an annual benefit of \$81,500 in each of the 10 years (found by aggregating the costs and the benefits in the Annual Equivalent Column).

**Table D-4. Illustrative discounting of representative costs and benefits**

| Kind of cost or benefit | Cash flow pattern | Present equivalent | Annual equivalent |
|---|---|---|---|
| Planning for cogeneration | −2 −IP I 2 10<br>X<br>$200,000 | P=$200,000 (SCA 15%, 2 yr.)<br>=$200,000 (1.322)<br>=$264,500 | A=$200,000 (SCA, 15%, 2 years)<br>(UCR, 15% 10 years)<br>=$200,000 (1.32) (0.1993)<br>=$52,700 |
| Purchase and Installation of gas turbine etc | P I 2 10<br>X<br>$2,000,000 | P=$2,000,000 (I)<br>=$2,000,000 | A=$2,000,000 (UCR, 15%, 10 years)<br>=$2,000,000 (0.1993)<br>=$398,600 |
| Cost of production downtime | P I 2 10<br>X<br>$500,000 | P=$500,000 (SPW, 15% 1 yr.)<br>=$500,000 (0.8696)<br>=$400,348 | A=$500,000 (SPW, 15% 1 year)<br>(UCR, 15%, 10 years)<br>=$500,000 (0.8696) (0.1993) =<br>$86,700 |
| Net Increase in Annual operating and maintenance cost due to cogeneration | P I 2 10<br>X X X X<br>$$10^5$ $$10^5$ $$10^5$ $$10^5$ | P=$100,000 (UPW, 15%, 10 yr.)<br>=$100,000 (5.019)<br>=$500,019 | A=$100,000 (I) |
| Replacement of parts | P I 2 7 10<br>X<br>$200,000 | P=$200,000 (SPW, 15%, 7 yr)<br>=$200,000 (0.3759)<br>=$75,200 | P=$200,000 (SPW, 15%, 7 years)<br>(UCR, 15%, 10 years)<br>=$200,000 (.3759)(0.1993)=$15000 |
| Salvage value of equipment at end of ten years | $7(10)^5$<br>X<br>P I 2 10 | P=$700,000 (SPW, 15%, 10 years)<br>=$700,000 (0.2472) =$173,030 | P=$700,000 (USF, 15%, 10 years)<br>=$700,000 (0.0493) = $345000 |
| Annual saving in electricity bills (assuming no change in price per kWh) | $7(10)^5$ $7(10)^5$ $7(10)^5$ $7(10)^5$<br>X X ... X X<br>P I 2 10 | P=$700,000 (UPW, 15%, 10 years)<br>=$700,000 (5.019)<br>=$3,513,300 | A=$700,000 (I) |

## NET PRESENT VALUE (NET BENEFITS) METHOD

This method calculates the difference between the present value of the benefits and the costs resulting from an investment. The difference between benefits and costs is the net present value of the investment. A positive net present value means that the financial position of the investor will be improved by undertaking the investment; a negative net present value means that the investment will result in a financial loss. (The net benefits of a given project will be greater at that level where marginal benefits (the additional benefits from the last increment of the investment) are just equal to the cost investment at the margin, i.e., where the marginal benefit cost ratio is 1.)

To use this method in the evaluation of cogeneration investments, the benefits would be defined as positive cash flows (as, for example, would result from sales or surplus electricity), and/or reductions in cash outflows (as would result from generation on site substituting for utility generation). (The definition of cost reductions as benefits is a useful means of converting the data for an investment for which cash flows are all negative into a format to which benefit-cost methods can be applied).

The formula for calculating net present value, or net benefits, is as follows:

$$NPV = \sum_{j+I}^{n} \frac{(S_j + R_j) - (I_j - V_j + M_j))}{(I + i)^j}$$

where:

$NPV$ = Net present value

$n$ = Number of time intervals over which the investment is analyzed

$S_j$ = Energy cost savings in year j

$R_j$ = Revenue from sale of excess energy received in year j

$I_j$ Investment costs in year j

$V_j$ = Salvage value in year j

$M_j$ = Maintenance and repair costs in year j, and

$i / (I + i)^j$ = Single present value discount formula

The acceptance criteria of a project, as evaluated with the net present value method, are that (1) only those investments having positive net benefits will be accepted (unless the project is mandatory), and (2) when selecting among mutually exclusive investments, the one with the highest positive net benefits will be chosen (or the one with the lowest negative net benefits if none of the alternatives has positive net benefits and the project is mandatory).

In using the net present value method to compare alternative investments, it is important to evaluate the costs and benefits of each alternative over an equal number of years. This may be done in any of the following ways, depending upon the nature of the investment. (1) The costs and benefits can be measured over a time period that is a common multiple of the economic lives of the alternatives. For example, to compare turbine A with a life of 5 years against turbine B with a life of 10 years, alternative A could be evaluated on the basis of one replacement, and alternative B, on the basis of no replacements, such that benefits and costs of both systems would be computed for 10 years. (2) Alternatively, benefits and costs of each alternative can be calculated in annual cost terms, based on 5 years for ! and 10 years for B, and the annual benefits and costs can then be used to calculate the present value of benefits and costs for the desired number of years of service (for example, 12 years service for both systems). (Since annual cost in this procedure is based on the economic life of the system, the procedure assumes that if the point of termination of use is different from the economic life of the system, the full value of the remaining life is realizable in terms of salvage value). This avoids the need to find a common multiple of system life. (3) If either system will be used for only a limited period of time, which is less than a common multiple of the economic lives of the alternatives, the estimated cash flows associated with each system over the period of analysis can be discounted to present value, making sure to take into account the expected remaining value of each system at the time of terminated use. For example, the problem might be to choose between turbine A, with its economic life of 5 years, and turbine B, with its economic life of 10 years, where intended use of either would be only 7 years due to expected closure of the plant at the beginning of the eighth year. System A would require one replacement because its expected life is only 5 years and it is needed 7 years. The remaining value of the replacement after 2 years of use would be discounted to present value and deducted from present value cost of the system. Similarly, an estimate would be required of the value of the remaining 3 years of life of system B. If removal costs are prohibitive, or if there is not a good resale market for the equipment, the remailing value of both systems at the end of 7 years should be evaluated at zero dollars. This holds true even though the equipment could provide additional years of service if the existing operation were continued. (4) An investment can also be evaluated on the basis of a perpetual, or indefinite period of use by "capitalizing" renewal costs and expected benefits. Present value benefits of an investment in perpetuity are calculated by dividing the expected annual benefits by the discount rate. Present value costs in perpetuity are calculated by converting all costs other than first costs to an annual equivalent, dividing by the discount rate, and adding the amount to the first cost. Thus, if first costs were $500,000; operation costs $100,000 yearly; and renewal costs $400,000 every 5 years, the present value capitalized cost of the above turbine A in perpetual service would be equal to $500,000 + ($100,000 / i) + (($400,000 (USF, i, 5 years) / i).

The choice among these four approaches to measuring present value is often not critical to the outcome; the particular nature of the investment will generally determine which approach is used. For example, in the case of a short-lived investment with unrecoverable salvage, the third approach explained above would be preferred.

The use of the net present value method to compare alternative investments in a topping turbine is illustrated in Table D-5. It is assumed that the investment choice is between Plan X, the addition of a large topping turbine and Plan Y, the addition of a smaller turbine. It is further assumed that Plan Y has a longer expected economic life due to lower temperatures.

Plan Y is found to provide $340,900 more in net benefits than Plan X. The additional fuel savings from the larger turbine is, therefore, not sufficient to justify its additional costs.

**Table D-5 Illustration of net present value method**

| Investment Alternatives | First Cost | Expected Economic Life | Annual Savings | Net Present Value |
|---|---|---|---|---|
| Plan X | $2000000 | 6 Years | $700,000 | $12,284,700 |
| Plan Y | $1200000 | 10 Years | $500,000 | $ 1,625,600 |

In the case where alternative investments are expected to provide the same level of benefits, the net present value method becomes equivalent to a net present cost method, often referred to as life-cycle costing, or cost-effectiveness analysis. In this case, the most efficient alternative may be identified as the one with the least present value of costs alone. This approach is also often used when the benefit levels cannot be quantified.

A feature of the net present value method which may be a disadvantage in some applications is that, in focusing only on net benefits, it does not distinguish between a project involving relatively large benefits and costs, and one involving much smaller benefits and costs, as long as the two projects result in equal net benefits. Thus a project requiring $1,000,000 in present value costs, and resulting in $1,100,000 in present value benefits (i.e., $100,000 in net benefits) will yield the same net benefits as a project costing $100,000 and resulting in $200,000 in total benefits (again $100,000 in net benefits). A way to avoid this problem is to compute benefit-to-cost ratios for further evaluation of the projects. Choices will be most efficient if independent projects are chosen in the order of their benefit-cost ratios, starting with the highest and working down until the budget is exhausted.

Another possible disadvantage of the net present value method is that the results are quite sensitive to the discount rate, and failure to select the appropriate rate may alter or even reverse the efficiency ranking of the alternatives. For example, with too low a rate an alternative with benefits spread far into the future may unjustifiably appear more profitable than an alternative whose benefits are more quickly realized but of lower amount in undiscounted terms. Since changing the discount rate can change the outcome of the evaluation, the rate used should be considered carefully.

As was explained earlier, the discount rate which a firm should use to discount cash flows of an investment is the firm's opportunity cost of capital, expressed as an interest rate; i.e., the rate of return which will be foregone by using the funds (resources) for the investment under consideration instead of the next best investment opportunity available to it.

If the firm is uncertain as to the appropriate discount rate to use, it may wish to compute the net benefits of an investment based on several alternative discount rates to test for sensitivity of the outcome to the choice of rates.

The net present value method has the advantage of measuring the net effect of an investment over its life, taking into account the opportunity cost of capital. The method is particularly useful for determining the efficient scale or size of an investment project.

## NET ANNUAL VALUE (NET ANNUAL BENEFITS) METHOD

This method takes essentially the same form as the net present value method. The difference is that all costs and benefits of the net annual benefits method are converted to a uniform annual basis instead of to present value. The formula is as follows:

$$A = \sum_{j=1}^{n} \frac{(S_j + R_j) - (I_j - V_j + M_j))}{(l + i)^j} \times \frac{i\,(l + i)^n}{(l + i)^n - l}$$

where

A = Annual value of net benefits,

n = Number of time intervals over which the investment is analyzed,

$S_j$ = Energy cost savings in year j,

$R_j$ = Revenue received in year j,

$I_j$ = Investment costs in year j,

$V_j$ = Salvage value in year j,

$M_j$ = Maintenance and repair costs in year j,

$l / (l + i)^j$ = Single present value discount formula,

$(i\,(l + i)^n) / ((l + i)^n - l)$ = Uniform compound amount formula.

If alternative investments have different life expectancies, either of two approaches may be taken to compare the alternatives; (1) It may be assumed that whichever alternative is chosen will be needed for an indefinite period of time and, hence, will be renewed as needed. In this case, the annual cost of each system may be simply calculated for its expected economic life, regardless of the fact that the lives of the alternatives may be different. (2) If the use of the investment alternatives is required for only a limited time, it is necessary to calculate costs based on the planned investment period, estimating the salvage value of each alternative at the planned time of terminated use. (This is similar to third approach described under Net Present Value.)

With the cash flows of the preceding example (described in Table D-5), and the assumption that either plan will be used for an indefinite period of time, the net annual value of Plan X is $240,800 and of Plan Y is $304,700. As we would expect, this method shows Plan Y to be the most efficient choice; as did the present value example.

This method, like the net present value method, has the disadvantage of failing to distinguish between projects of unequal magnitudes which yield equal net benefits. However, analysis of benefit-cost ratios can be used to overcome this problem. A possible advantage of the net annual value method, as compared with the net present value method, is that the concept of an equivalent annual amount may be easier to understand than the concept of a present equivalent of all cash flows over the period of analysis.

## BENEFIT/COST RATIO METHOD

The benefit/cost ratio method expresses benefits as a proportion of costs, where benefits and costs are discounted to either a present value or an annual value equivalent. The formula (with benefits and costs discounted to either a present value or an annual value equivalent. The formula (with benefits and costs discounted present value) is as follows:

$$BC = \mathbf{E}_{j=l}^{n} \frac{(S_j + R_j)}{(l+i)^j} \Big/ \mathbf{E}_{j=l}^{n} \frac{(I_j - V_j + M_j))}{(l+i)^j}$$

where

B/C = Benefit-cost ratio

n = Number of time intervals over which the investment is analyzed

$S_j$ = Energy cost savings in year j

$R_j$ = Revenue received in year j

$I_j$ = Investment costs in year j

$V_j$ = Salvage value in year j

$M_j$ = Maintenance and repair costs in year j

While the net present value and net annual value methods require that discounted benefits minus discounted costs be positive in order for an investment to be worthwhile, the benefit-cost ratio method requires that the ratio of discounted benefits to costs be greater than 1.

A disadvantage of the benefit-cost ratio method is that the ratio is influenced by the decision as to whether an item is classified as a cost or as a benefit, i.e., whether it appears in the numerator or denominator of the ratio. For many cost or benefit items, this is simply an arbitrary decision, but one which can lead to confusion as to the real efficiency of a project.

Another problem with the benefit-cost ratio method is that it is subject to being misapplied in determining the efficient scale of a given project. It pays to expand a project up to the point that the ratio for the last increment of the investment is equal to 1.0, assuming no alternative investment is available with a higher ratio. Because the benefit-cost ratio for the overall investment declines as the investment is expanded towards the most efficient level, a smaller, less efficient size of a project may have a higher ratio for the total investment than a larger more efficiently sized project. This problem can be avoided by applying the benefit-cost ratio method to evaluate the efficiency of increments of an investment, rather than the total investment.

## INTERNAL RATE OF RETURN METHOD

This method (not to be confused with the ROI method evaluated earlier) calculates the rate of return an investment is expected to yield. This may be contrasted with the net present value, the net annual value, and the benefit-cost ratio methods which calculate the net dollar value of the investment based on a predetermined required rate of return. The internal rate of return method expresses each investment alternative in terms of a rate of return (a compound interest rate). The expected rate of return is the interest rate for which total discounted benefits become just equal to total discounted costs, i.e., net present benefits or net annual benefits are equal to zero, or for which the benefit-cost ratio equals one. The criterion for selection among alternatives is to choose the investment with the highest rate of return.

The rate of return is usually calculated by a process of trial and error, whereby the net cash flow is computed for various discount rates until its value is reduced to zero. As a simple example, let us calculate the internal rate of return for a turbine which will cost $1,000,000 to install, will last 10 years, and will result in electricity savings of $300,000 each year. We wish to find that i which will equate the following:

$$\$1,000,000 = \$300,000\ (\text{UPW}, i = ?, 10 \text{ years})$$

To do this, we calculate the net present value (NPV) for various i values, selected by visual inspection:

$$\begin{aligned} \text{NPV } 25\% &= (\$300,000)\ (3.571) - 1,000,000 \\ &= \$1,071,300 - \$1,000,000 \\ &= \$71,300 \\ \text{NPV } 30\% &= (\$300,000)\ (3.092) - \$1,000,000 \\ &= \$927,600 - \$1,000,000 \\ &= -\$72,400 \end{aligned}$$

For i = 25%, net present value is positive; for i = 30%, net present value is negative. Thus, for some discount rate between 25 and 30 percent, present value benefits are equated to present value costs. To find the rate more exactly, without the benefit of a complete set of discount tables, one may interpolate between the two rates as follows:

$$i = 0.25 + 0.05\ (\$71,300 / \$70,300 - (-\$72,400))$$

$$= 0.275, \text{ or } 27.5 \text{ percent.}$$

To decide whether or not to undertake this investment, it would be necessary for the firm to compare the expected rate of return of 27.5 percent with its minimum attractive rate of return.

Use of this method has the advantage of generally resulting in conclusions consistent with the three other comprehensive methods discussed in this section. However, there are several possible disadvantages which might arise. For one thing, under certain circumstances there may be either no rate of return solution or multiple solutions. (An example of a type of investment which gives rise to multiple solutions is one characterized by a net benefits stream which is first negative, then positive and finally negative again). Secondly, confusion may arise when this method is used to choose among mutually exclusive alternatives. For example, if the compared alternatives are different sizes of the same project (e.g., different capacity turbines), the rate of return on the larger scale project may be lower than on the smaller scale project, causing the larger to appear less efficient than the smaller. However, additional investment in the larger project may nonetheless yield a positive rate of return in excess of the minimum attractive rate of return. This problem which is comparable to the problem which was described for the benefit-cost method, can be avoided by analyzing incremental changes in an investment. A third problem is that the rate of return may be somewhat more cumbersome to calculate than the other methods.

## SPECIAL FACTORS TO CONSIDER IN INVESTMENT ANALYSIS

This section discusses the treatment of income taxes, inflation, and uncertainty in investment analysis.

### Treatment of Income Taxes

This section treats taxes as they appear in investment analysis. For a more detailed discussion of taxation of cogeneration equipment, see the Taxation sections of the Manual. Income taxes can have a profound impact on optimum investment decisions. By changing the effective values to the firm of the revenues and costs associated with an investment, taxes can reverse the relative profitability of alternative projects as evaluated apart from taxes, and they can alter the optimal size of investments. Taxes are, therefore, an important element in the economic evaluation of investment in cogeneration.

If, for example, an investment results in increased taxable revenue, the effective value of the additional revenue to the firm is reduced by taxation. Over the life of an investment, the effective present value of after-tax revenue (R) can be calculated as follows:

$$\bar{R} = \sum_{j=1}^{n} \frac{R_j (1 - t_j)}{(1+i)^j}$$

where

$n$ = the number of time intervals over which the investment is analyzed,

$R_j$ = the taxable revenue resulting from the investment in year j

$t_j$ = the firm's tax rate at the margin in year j

$i$ = the discount rate

Similarly, if an investment results in benefits in the form of cost saving, the after-tax value of the benefits will be lessened by the loss of any related income tax deduction; i.e., the cost savings are generally not tax free. The present value of after-tax cost-savings (C) from an investment can be calculated as follows;

$$C = \sum_{j=1}^{n} \frac{C_j (1 - a_j t_j)}{(1+i)^j}$$

where

$C_j$ = the cost savings from an investment in year j

$a_j$ = that percentage of the cost savings which were deductible from taxable revenue, and

$t_j$ = the firm's tax rate at the margin in year j.

On the other hand, investment costs are also reduced somewhat by the value of corresponding tax deductions.

An economic analysis of investment in cogeneration systems might involve each of these three tax effects. For instance, system equipment costs may give rise to tax deduction in terms of ACRS allowances and-or investment tax credits; savings in fuel costs will reduce deductions from taxable income of current operating costs; and sale of excess heat or electricity to other firms will result in additional taxes on the revenue received.

The following example compares the before-tax and after-tax annual net benefits from an investment in a gas turbine, given the assumptions below.

GIVEN

| | |
|---|---|
| Investing Firm's Tax Rate, at the Margin: | 40% |
| Opportunity Cost of Capital: | 10% after taxes |
| Source of Capital: | Equity Funds |
| Method of Equipment Depreciation: | Straight-line Method |
| First Cost of turbine: | $2,500,000 |
| Annual Maintenance Cost: | $50,000 |
| Expected Salvage Value in 10 years: | $0 |
| Expected Annual Savings: | $500,000 |

BENEFITS

| | |
|---|---|
| Annual Reductions in Electric Costs, Before Taxes | = $500,000 |
| (minus) | |
| Annual Value of Tax Deductions Lost | = $200,000 |
| = 0.40 ($500,000) | |
| (equals) | |
| Annual Reductions in Costs, After Taxes | = $300,000 |

COSTS:

| | |
|---|---|
| Annual Capital Cost of Equipment = $2500,000 (UCR: 10%; 10 years) = $25,000 (0.1628) | = $407,000 |
| (plus) | |
| Annual Maintenance Cost | = 50,000 |
| (equals) | |
| Total Annual Increase in Equipment and Maintenance Costs, Before Taxes | = $457,000 |
| (minus) | |
| Annual Value of Tax Deductions of Equipment and Maintenance Cost = | = $120,000 |
| (equals) | |
| Annual Increase in Equipment and Maintenance Costs, After Taxes | = $337,000 |

NET BENEFITS: Before and After Taxes:

| | |
|---|---|
| Net Benefits of Investment, Before Taxes = ($500,000 - $457,000) | = $43,000 |
| Net Benefits of Investment, After Taxes = ($300,000 - $337,000) | = -$37,000 |

In this case, before-tax evaluation of the investment in the turbine indicates a positive annual net cost savings of $43,000. But the after-tax evaluation indicates a negative cost savings of $37,000, which means the investment would not be profitable.

Even though depreciation is not itself a cash flow, it affects the firm's cash flow through the deductions from taxable income. By its choice of depreciation methods (e.g., the declining balance method as opposed to the straight-line method of depreciation), an investing firm can affect the timing of the tax deduction and thereby influence the effective amount of income tax payments. The choice of a depreciation method which yields higher depreciation allocations in the early years will increase the present value of depreciation deductions, thereby decreasing the present value of income tax payments and raising the profitability of the investment to the firm. (The choice of a depreciation method (i.e., the desired pattern of tax deduction) is also influenced by such factors as the presence or absence of carry-forward losses, the expected pattern of revenue, and the expected future tax rate. Hence, the most rapid depreciation may not always be most profitable).

### Inflation

For simplicity, changes in price were not considered in the preceding illustrations of methods of evaluation and treatment taxes. However, price changes may be very important in an investment analysis.

To take price changes into account, it is useful to distinguish between two kinds of price changes. One is a "nominal" price change which results from changes in the purchasing power of the dollar, i.e., inflation or deflation. The amount of this nominal price change is indicated by the change in the general level of prices as measured by price indices.

The other type of price change which can occur is a change in "real" terms. This means that the price of a given good or service rises proportionally more or less than the change in the general purchasing power of the dollar, such that its price changes over time relative to the prices of other goods and services. For example, given that the general price index rose over a 5-year period from 1.00 to 1.30, the price of a piece of equipment which also rose 30 percent over the same period, would in effect remain constant in terms of its price relative to prices in general; i.e., its price would change in nominal terms but would remain constant in real terms.

An approach often followed in investment analysis is to assume that all costs and revenues inflate at the same general rate, and that they therefore remain constant in real terms. With this assumption, renewal costs and other future expenses and benefits are evaluated at present prices. If there is reason to believe that certain items of costs or benefits will not inflate at the general rate, then their future values are adjusted only for the estimated real change, i.e., the effective change after taking into account the change in the general price level. If, for example, oil prices were expected to rise at a rate of 6 percent a year and the general price index to rise at a rate of 3 percent a year, the imputed real rise in oil prices would be 3 percent, i.e., half the rise would be considered due to general inflation, and half due to changing demand and supply conditions for oil relative to other commodities. Alternatively, if the future price of a good or service were, say fixed at current levels by lease arrangements (without a cost escalation clause) the price would decline in real terms, in the face of inflation. It would be necessary to adjust the future payments by a price index prior to discounting them to present value in order to express them in constant dollars. Employing this procedure, the appropriate discount rate is a real rate, that is, one from which the inflation factor has been removed.

Assuming constant future prices can greatly simplify an analysis and in many applications will result in reasonably accurate results. However, the success of the approach rests not only on the assumption that future receipts and expenditures will respond fully and evenly to inflation, but also on the assumptions that tax considerations and the source of investment funds do not importantly affect the outcome-two assumptions which may be quite unrealistic for private investments.

The real after-tax return to the firm may be substantially changed by inflation, even if pretax investment receipts and costs are assumed fully and equally responsive to inflation. Other things equal, inflation will tend to have a detrimental effect on an investment financed principally by equity funds, i.e., nonborrowed funds. Among the reasons are the following: (1) tax deductions for depreciation are unresponsive to inflation; (2) terminal value of equipment is responsive to inflation and will be reflected in the capital gains tax; (3) tax deductions for interest on the borrowed portion of capital are unresponsive to inflation, such that the present value of the deductions diminishes over time; (4) inflation in receipts tends to move a firm into higher tax brackets.

However, in the case of an investment funded by a large proportion of borrowed funds it is possible that the after-tax return to equity may be maintained, or even improved, other things equal, in the face of inflation due to the following factor: The real value of the debt charges (amortized principle and interest) to finance the investment may decline because of the fixed nature of these payments. This gain to the borrower may exceed his loss due to the fixed nature of depreciation and interest deductions.

From a practical standpoint, methods of simplifying the treatment of inflation are desirable whenever they can be implemented without significantly altering the results.

### Uncertainty

Evaluation results depend directly on both the data estimates and the assumptions employed in the analysis. Among the key factors affecting the outcome of evaluations of cogeneration systems are the following: (1) the cost estimates for planning, purchasing, installing, and operating and maintaining the cogeneration systems; (2) the additional costs imposed by the investment, such as labor downtime and production loss; (3) the future rate of real price escalation in energy/electricity sources; (4) the amount of electricity generated; (5) the economic lives of system components, the length of intended use of the system, the salvage values at termination; and (6) the discount rate used to convert future costs and benefits to a common time.

There will often be uncertainty as to the correct values to use in evaluating an investment. Uncertainty, which can be defined broadly as disparity between the predicted and the actual, encompasses two specific concepts: "risk" an event whose probability of occurrence can be predicted; and "uncertainty", an event whose expected chance of occurrence can not be predicted.

Following are brief descriptions of three general approaches to dealing with uncertainty (used in the broad sense): (1) probability analysis, (2) sensitivity analysis, and (3) break-even analysis.

Probability analysis is generally used for situations for which the probability of an expected occurrence can be estimated; i.e., for evaluating risk. By multiplying the probability that an event will occur by the resulting dollar value if it does occur, it is possible to express costs and benefits as "expected values", rather than simply as "point " estimates.

For an example, cost of production down time during installation may be roughly estimated at, say $200,000, but perhaps the following might be a more accurate statement of costs:

| Possible Situation | Proba-Bilities | Cost if | Expected Cost |
|---|---|---|---|
| No difficulties in installation | 0.20 | $150,000 | $ 30,000 |
| No serious difficulty | 0.70 | $ 30,000 | $210,000 |
| Serious difficulty | 0.10 | $750,000 | $ 75,000 |
| | | | $315,000 |

A more accurate measure of cost in this situation would be $315,000.

Probability or expected value analysis, therefore, provides a method of incorporating uncertainty into the investment evaluation in a quantitative way. It requires, however, determination of probabilities.

Sensitivity analysis can be used to assess the consequences of assuming alternative values for the significant variables in the analysis. By determining the effect on the outcome of potential variation in a factor, the analyst identifies the degree of importance of that estimate or assumption and can then seek more information about it if desired. For example, the profitability of a gas turbine might be tested for sensitivity to the expected utilization rate of the turbine.

Break-even analysis, a third technique for dealing with uncertainty, focuses on a single key variable which is regarded as uncertain, and calculates the minimum (or maximum) value of the variable which is required to achieve a given outcome. For example, one might solve for the rate of escalation in electricity prices required for an investment in a cogenerator to break even, other things given. To find the break-even escalation rate, an equation is developed which equates capital and maintenance costs with cost savings/income. The "uncertain variable", i.e., the electricity price escalation rate, is entered as an unknown and the equation is solved for the break-even rate.

## APPLICATION OF EVALUATION METHODS TO DIFFERENT KINDS OF DECISIONS

As Table D-6 shows, the manager of a plant may be confronted with different kinds of investment decisions. The nature of the decision will influence which method of evaluation is preferred.

One kind of decision is whether to accept or reject a particular investment project – where acceptance may mean implementation without further analysis or designation of the project as potentially profitable and worthy of further evaluation against other profitable projects. A simple accept-reject decision might be made, for example, if there were only one kind and size of cogeneration system feasible for a par-

**Table D-6**
**Kinds of investment decision problems**

| **Decisions** | **Examples** |
|---|---|
| To accept or reject a given project | Project A is accepted if its level of profitability meets the minimal acceptance level |
| To choose between mutually exclusive projects | If project A is accepted, Project B is rejected; if size 1 of Project A is accepted, size 2 of Project A is rejected. |
| To decide priority among independent | Project A is preferred to Project projects B, but both are undertaken if funding is adequate. |
| To determine the desirability of a project | Project A yields a low return, prerequisite but Projects B and C which yield high returns cannot be undertaken until A is done. |
| No decision needed – mandatory project | Project A is undertaken regardless of its expected return in order to meet mandated requirement. |

ticular plant application. Any of the comprehensive evaluation methods described here generally could be used to determine if the expected benefits (or cost savings) from the investment could exceed the expected costs.

Another kind of decision is choosing among mutually exclusive projects, where any of them would be expected to yield positive net benefits. This is the kind of choice which arises when doing one project precludes doing the others. For example, it might be determined that cogeneration would be desirable in a particular application via either of two competing topping turbine systems. Since only one system would be used, a choice must be made between them.

Determining the efficient scale or size of investment in a given project is a special case of mutual exclusions, in that choosing one size for a project usually means rejecting the other possible sizes. The choice of time for undertaking a single project, that is, whether to begin it now or to delay it, is another special case of mutual exclusion.

The appropriate approach to evaluating mutually exclusive projects is generally to determine the choice which maximizes net benefits by using either the present value or annual net benefit method of evaluation. Assuming that there is no budget constraint, the project which maximizes net benefits is the most efficient choice.

Yet another kind of decision that the manager may face is the choice among a set of project alternatives which are independent, that is, projects where the undertaking of one bears no influence on the acceptance of the others. In face of budgetary and other constraints, it is usually impossible for a firm to carry out all of its potentially profitable investment opportunities; a ranking mechanism is needed to identify those projects which are most profitable. For example, investment in cogeneration equipment may compete for funding with other cost-saving or revenue-generating investments.

When there is a budget constraint, the ranking of individual independent projects according to their net benefits cannot be relied upon to result in selection of that subset of projects from the group which will yield the overall maximum net benefits. This is because a project that yields higher total net benefits than another project may yield lower benefits per last dollar of investment spent, such that combined benefits for the two projects could be increased by transferring funds from the first project to the second. Choosing among independent projects according to either the benefit-cost ratio or the internal rate of return criterion until the available budget is exhausted will yield the maximum total net benefits for that budget. For example, Table D-7 shows a possible gain of $400,000 by using the benefit-cost ratio method rather than the net benefits criterion to rank Projects 1, 2, and 3, when

**Table D-7 Net benefits and benefit/cost ratio rankings for a set of independent projects**

| Project | Benefits (B) (thousands) | Costs (C) (thousands) | Net Benefits (thousands) | Project Ranking | B/C Ratio | Project Ranking |
|---|---|---|---|---|---|---|
| 1 | $3,000 | $2,000 | $1,000 | 1 | 1.50 | 3 |
| 2 | $2,000 | $1,200 | $ 800 | 2 | 1.67 | 2 |
| 3 | $1,400 | $ 800 | $ 600 | 3 | 1.75 | 1 |

the budget is fixed at $2,000,000 (Presumes choice of projects 2 and 3 instead of project 3 alone).

A fourth kind of decision is determining the desirability of a project when it is a prerequisite to the undertaking of other projects. This type of decision may be particularly relevant to the evaluation of cogeneration systems, in that investment in cogeneration may in some cases be necessary to preserve the overall operation of a plant. This situation might arise in the face of severe curtailment of electricity.

Without some means of supplementing available electric service, plants might have to reduce production levels and lay off workers, thereby causing substantial loss of revenue, without equal reductions in costs. Alternatively investment in cogeneration might be necessary in order to enable desired expansion of production in the face of unreliable electric service, or to avoid costly changeover to substitute sources of energy.

If the project in question is a prerequisite for other activities of the firm, the evaluation of the prerequisite project should take into account the overall return expected not only from it, but also from the other activities whose success depends upon it. Even if the project shows a very small or negative rate of return when evaluated in isolation, it may in fact produce a very high rate of return in terms of enabling other profitable activities to be undertaken. If the overall return to investment on the group of interdependent projects is attractive, funds should be allocated first to the prerequisite project.

In most cases, any of the comprehensive evaluation methods would be adequate to assess the importance of a project that is prerequisite to other functions of the firm. The essential factor is that all relevant costs and benefits attributable to the investment be included in the evaluation.

The final entry in Table D-6 does not involve a decision. It is the case in which there is no decision to be made because the investment project must be undertaken by mandate, regardless of its profitability to the firm. Legislated requirements for specified levels and-or types of pollution control investments are examples. An important point to remember in this case is that other investments which are being considered, perhaps in conjunction with the required investment, should not be saddled with its costs and-or benefits. For example, the decision to add cogeneration equipment while installing legally required pollution control equipment should be evaluated on its own merits; the investment evaluation should not lump the costs and benefits of the two systems together.

In summary, each of the investment evaluation methods has its particular advantages and disadvantages, and will be a useful decision criterion in certain cases. for most decision problems, the net present value or the net annual value method, supplemented by benefit-cost ratios or internal rates of return, will provide adequate measures for economically efficient investment decisions.

## CALCULATION OF SYSTEM ECONOMIC PERFORMANCE AND ENERGY SAVINGS

### Methodologies for Estimating Economic and Energy Savings Performance

The objective of the preliminary financial analysis is to determine whether the installation of a cogeneration system is justified economically for the potential cogenerator. Specifically, the analysis should identify those systems which meet or exceed the after-tax ROI required by an industrial firm. Based on the economic results, an optimum system can be selected.

The long-term nature of the investment and the interaction of fuel costs, electricity prices, capital costs, and other cost factors require a sound economic approach to identify and discount each set of annual cash flows. While the financial analysis can be performed using any methodology consistent with user practices, this section illustrates the use of a discounted cash flow (DCF)/internal rate of return (IRR) model. (MODEL AVAILABLE FROM ARGONNE CODE CENTER, ARGONNE NATIONAL LABORATORY, 9700 CASS AVENUE, ARGONNE, ILLINOIS 60439).

The DCF model identifies and discounts all relevant cash flows and computes the IRR of the proposed cogeneration system. Each element of the methodology is described in the following section. Although the discussion focuses on this particular DCF model, it is written as generally as possible, noting that this analysis can be performed by other models or even by hand calculations.

### Model Inputs

The economic and energy savings analysis first requires that the system under consideration be described by its technical characteristics, system costs, energy costs and financial parameters. These characteristics constitute the input data to the model, as shown in the sample model run in Table D-8. The calculations and sources of information for these inputs are described in more detail below. For purposes of comparing alternative cogeneration systems, the analysis uses 1985 as the initial year of operation for all systems, unless otherwise specified.

### Technical Characteristics

Technical characteristic inputs to the model are incremental fuel consumption, net electricity generation, and the utility heat rate.

*Incremental fuel consumption*, defined as the annual increase in industrial plant fuel consumption caused by operation of the new cogeneration system, is used to compute energy savings. It is also used with fuel price data to compute the annual increase in fuel cost to the industry for operating the system. This figure must be a composite for all fuel types previously used or planned for use by the plant. If the cogeneration system results in a fuel switch, all of the fuels replaced must be included in these calculations.

*Net electricity generation* is the annual kWh output of the cogeneration system, or the net increase in electric output for plants having existing levels of cogeneration. It is used to compute energy savings caused by displacement of utility electricity and is used with electricity price data to compute the industrial plant's annual savings on the electricity bill.

*Utility heat rate* is used to compute primary energy savings that result from displacement of central station power. The utility heat rates should include the T&D losses (estimated here at 7%) that the utility must accept in serving an industrial customer. For that portion of the cogenerated electricity that is sold back to the utility, no T&D adjustment is made if it is assumed that the electricity is consumed outside the load center containing the industrial plant.

### System Costs

The initial capital investment is the primary system cost. This is the capital cost above that for comparable alternatives. For example, if the potential cogenerator is currently using conventional steam boilers, the alternative is simply to continue operating the existing boilers, and the initial capital investment is the total cost of the retrofit cogeneration system. If the user is contemplating replacement of oil- or gas-fired conventional boilers with new boilers burning coal or alternative fuels (with no new cogeneration), then the latter may be considered the alternative, and only incremental cogeneration capital costs above and beyond this alternative cost need be considered. Where the installation of cogeneration is accompanied by a switch from oil or gas to coal or alternative fuels, the economics of fuel switching alone should be considered in addition to the fuel switching/cogeneration combination. The third possibility is in the case of a new facility, with no existing system, and the cogeneration option is again compared on an incremental basis with the cost of new steam boilers. Estimates of system capital investment are obtained from vendor quotes and should be further escalated to dollar figures for the year and period in which construction begins.

The construction cost distribution spreads the capital investment over the construction period to account for interest or burden during construction. Although construction times for industrial cogeneration systems are short compared with utility plants, the effect of the construction burden is by no means negligible. Applying this burden accounts for the fact that if a company were not making an investment in the cogeneration project, the funds would still be invested internally where they would be expected to earn a return at the IRR rate. This distribution consists of (1) the number of periods of construction per years, (2) the total number of construction periods, and (3) a construction fraction per period. The model will accept any distribution of construction costs as percentages per quarter for up to 20 quarters (5 years) although a uniform distribution of construction cash outlays has been assumed.

Other system cost inputs include the incremental O&M costs for the start-up year and O&M escalation rate. These are used by the model to generate O&M costs, are provided by the system designer and are based on the additional expenditures necessary for O&M of the cogeneration system. Typically, these are given in dollars per year and are escalated to a dollar figure for the first year of operation. The O&M escalation rate can be assumed equal to the inflation rate.

The system salvage value at the end of the analysis period can be estimated by the user.

### Energy Costs

The energy cost inputs are (1) the incremental fuel costs in the start-up year, (2) the fuel cost escalation rate, (3) the electricity cost savings, (4) the electricity escalation rate, and (5) the price of electricity sold to the utility. Note that the energy price forecasts used in the model are in current dollars. That is, they include the effects of inflation. The incremental fuel costs in the start-up year are obtained for each fuel type by multiplying the incremental fuel consumption (from cogeneration) by the price for that type of fuel. Calculations of incremental fuel consumption were shown previously. Projected prices should be initialized using local data, if available.

The fuel cost escalation rate is calculated using the incremental fuel cost in the start-up year together with the incremental fuel cost in 1995. Incremental fuel costs in 1995 are calculated like those in the startup year using 1995 fuel price projections. Where the fuel mix changes with the installation of cogeneration, the fuel cost is thus a composite number reflecting the cost of the fuel used for cogeneration less the cost of one or more other fuels that are replaced.

In such cases, the fuel cost escalation rate is also a composite. Normally, the fuel cost increase is characterized by a power curve, and the fuel cost escalation rate is calculated from this. In the case where composite fuel costs change sign over the analysis period; however, incremental fuel costs are calculated as discrete annual values. These discrete values must be tabulated for each year of the analysis period.

The model uses both incremental fuel costs and the fuel cost escalation rates to calculate incremental fuel costs for subsequent years of system operation.

The *electricity cost* in the start-up year is calculated as follows:

Electricity Cost = (Electricity generated for on-site use)
x (price of purchased electricity)
+ (Electricity sold to utility)
x (Price of electricity sold)
+ Capacity payments

The first term is the product of the electricity generated for use on-site and the electricity price. This term accounts for the savings the cogeneration system realizes by not purchasing this power from the utility. If local data are available, those data should be used to initialize the price projections.

The second term accounts for the revenue realized by the system through sale of power to the utility. A cogenerator is not restricted to the sale of excess power only. Under the provisions of the Public Utility Regulatory Policies Act (PURPA), all of the power produced by the new capacity may be sold to the utility at the full avoided cost (see Section A, REGULATIONS), and the cogenerator can purchase back electric power for internal needs at non-discriminatory retail rates. The prices of electricity sold under these provisions are available from the Public Utility Commission (PUC) of each state. PUC contacts are listed in the Appendix to Section A.

The third-term accounts for savings the system realizes through the utilities avoided capacity costs, which are the costs avoided by the utility as a result of not having to generate this power itself or purchase it from another source. Capacity payments by utilities are also subject to the provisions of PURPA (Section A).

The overall electricity cost is indicated as a negative number (i.e., negative costs) for input to the ICOP model.

The *electricity cost escalation rate* is calculated using electricity price projections for 1995 and 1985. The electricity cost escalation rate and the electricity cost in the startup year are used to calculate electricity cost savings from cogeneration in subsequent years. Both the price of electricity sold to the utility and the capacity payments are assumed to escalate at the same rate as the price of electricity purchased.

**Financial Parameters**

Financial parameters include the discount rate, (Table 2-7), entry 13L) or the IRR hurdle rate, defined as the rate of return which the project must earn to equal alternative investments. This rate of return includes some implied expectation of inflation.

Three financial parameters relate to the system capital cost if the investment is not entirely equity funded. These are the *down payment* as a percentage of total capital cost, the *loan life*, and the *interest rate* on borrowed capital.

The *investment tax credit* is ten percent. Coal and other alternate fuel fired facilities are granted an additional 10 percent credit. (See TAXATION section for more detail).

Other financial parameters include the *income tax rate*, and selection of a depreciation method (See TAXATION section).

*System life*, which is the period over which the firm desires to view the economic performance of the system, must be specified. The system life is not necessarily related to the actual physical life of the equipment. *Insurance, prop-*

*erty taxes*, and their *escalation rates* must be examined. The model can accommodate projected changes in the property tax rate, if these are likely.

### Analysis and Model Outputs

The economic and financial analysis may be further illustrated by referring to a sample printout of the ICOP model (Table D-9). The printout starts with an identification header and a summary of inputs. Next, undiscounted cash flows are calculated and tabulated.

The *unburdened capital investment* (principal) is spread over the construction period; *O&M and insurance costs* (increasing at the inflation rate) and *fuel costs* (increasing at a fuel escalation rate) are calculated for each year of system operation. Fuel cost is usually a positive number as in the sample, because extra fuel must be burned to cogenerate electricity not being generated previously. If, however, cogeneration is accompanied by a switch to lower cost fuels (e.g., coal or wood waste), fuel cost can be negative, even though more fuel is being consumed.

In most runs, the fuel cost increase is characterized by a power curve (i.e., using an initial value and an average annual escalation rate). The model has the capability, however, to accept discrete annual values. As noted, discrete annual values must be used in cases in which the composite fuel cost changes sign over the analysis period, because this cannot be accommodated by a power curve.

The *electricity cost* is always a negative number, as shown in the sample printout, since it represents the savings from cogeneration (reduced electricity bills plus any revenues from sales of electricity). In the sample run, these yearly savings are calculated using an escalation rate of 10.2% per year.

The printout also shows *depreciation* calculated according to the 175% DB/SYD schedule. The depreciation is used to determine the income tax. *Capital property tax* and *income tax* are also calculated. The income tax is small the first year as a result of the investment tax credit taken. (In many cases, the first year tax is negative and is thus an income tax credit that could be used to offset income taxes elsewhere in the firm).

*Net cost* is calculated by algebraically summing all the columns except for depreciation.

The *energy savings* computation is shown in the lower portion of the table. The utility heat rate, corrected for T&D losses, is multiplied by the net generation to compute primary energy savings in the utility system. Subtracting the incremental fuel consumption at the industrial plant gives the overall net energy savings.

While not an output of the ICOP computer model itself, the net savings of oil and gas can be easily determined through hand calculations based on the model inputs. Utility fuels displaced require estimates of the type(s) of primary energy saved at the utility (e.g., oil/gas or coal/nuclear). These inputs can be based on local utility data concerning current and projected fuel use.

The *Internal Rate of Return* is calculated as the discount rate that results in a net present value of zero (i.e., the discount rate that equates the present value of dollar savings to the present value of dollar costs). For cases in which the present value of savings is less than the present value of costs for all positive discount rates, the internal rate of return is taken to be zero.

Table D-10 presents the *discounted cash flows*. Note that in these calculations, the construction capital burden is included in the burdened capital cost in the principal and net cost columns. The burden on the construction cash flows is not subject to depreciation, because the burden does not represent an actual capital expenditure. Calculations are basically the same as those described previously except that discounted dollars are used. The final total at the bottom of the net cost column is the net present value. Where the present value of savings is less than the present value of costs for the internal rate of return hurdle used, the net present value will be a positive number. Conversely, if the present value of savings exceeds that of costs, the net present value will be negative. This is because costs in the model are represented by positive values and savings by negative values.

# Appendix D-1 Run Sheets

Run No. ____________

____________

____________

| Parameters | Variable Name | Value | Or Default |
|---|---|---|---|
| **Technical Characteristics:** | | | |
| Incremental Fuel Consumption, (106 Btu/hr) | FUELCN | ____ | 0.0 |
| Net Electricity Generation (106 kWh/hr) | NETGEN | ____ | 0.0 |
| Utility Heat Rate (Including T&D Losses) (Btu/kWh) | HTRATE | ____ | 0.0 |
| **System Costs:** | | | |
| Capital Investment ($ in 1985) | CAPINV | ____ | 0.0 |
| Construction Cost Distribution – | | | |
| Periods of Construction | NPER | ____ | 0 |
| Periods Per year | IPER | ____ | 0 |
| Construction Fraction Per Period | FCONST | ____ | 0.0 |
| Operations & Maintenance Cost, Initial ($ / yr in 1985) | CANDM | ____ | 0.0 |
| O&M Cost Escalation Rate (Fraction) | OMRATE | ____ | 0.0 |
| Salvage Value ($) | SALVAG | ____ | 0.0 |
| **Energy Costs** | | | |
| Incremental Fuel Cost, Initial ($ / yr in 1985) | FUEL | ____ | 0.0 |
| Incremental Fuel Cost Escalation rate (Fraction) | FRATE | ____ | 0.0 |
| Electricity Cost, Initial ($ / yr in 1985) | ELECT | ____ | 0.0 |
| Electricity Cost Escalation Rate (Fraction) | ERATE | ____ | 0.0 |

Table D-8

## Run Sheet (Continued)

| Parameters | Variable Name | Value | Or Default |
|---|---|---|---|
| Financial Parameters: | | | |
| Discount Rate (Fraction) | DISRTE | ____________ | 0.2 |
| Downpayment (Fraction) | DWNPMT | ____________ | 1.0 |
| Loan Life (Up To 40 Years) | LIFLON | ____________ | 12 |
| System Life (Up to 40 Years) | LIFSYS | ____________ | 12 |
| Interest Rate (Fraction) | IRATE | ____________ | 0.2 |
| Income Tax Rate (Fraction) | TXRATE | ____________ | 0.48 |
| Tax Credit Rate (Fraction) | TAXCRD | ____________ | 0.20 |
| Insurance Rate (Fraction) | INS | ____________ | 0.005 |
| Insurance Escalation Rate (Fraction) | INSRAT | ____________ | 0.0 |
| Property Tax Rate (Fraction) | PCTAX | ____________ | 0.025 |
| Property Tax Escalation Rate (Fraction) | PCRATE | ____________ | 0.0 |
| Depreciation Life (Up to 40 Years) | DEPYRS | ____________ | 5 |
| Depreciation Mode<br>= 1 – Sum-of-Years-Digits<br>= 2 – Straight Line<br>= 3 – Double Declining Balance<br>= 4 – 150% Declining Balance Changing to Straight Line<br>= 5 – 175% Declining Balance Changing to Sum-of-Years Digits<br>= 6 – 200% Declining Balance Changing to Sum-of-Years Digits | MODDEP | ____________ | 1 |

**Table D-8**

Table D-9

**ICOP Model – Version 3.5 (08/17/81)**

**ICOP Run No. u- 1- 2 (08/17/81)**
**Union Carbide, Institute, WV**
**7900 KW Steam Turbine System**
**Depcreciation on 1985 Schedule**

## SYSTEM COST ANALYSIS

| | | | | | | | |
|---|---|---|---|---|---|---|---|
| Capital Investment | = $ 4142286 | Equity | = $ 4142286 | Analysis Period | = 12 Years |
| Initial O and M | = $ 54438 | Initial Fuel Cost | = $ 945536 | Initial Elect. Cost | = $ -4247040 |
| O and M Esc. Rate | = 7.10 Percent | Fuel Cost Esc. Rate | = 10.98 Percent | Elect. Esc. Rate | = 10.20 Percent |
| Initial Insurance | = $ 0.00/$1000. | Property Tax Rate | = 2.50 Percent | Loan Interest Rate | = 000.00 Percent |
| Insurance Esc. Rate | = 7.10 Percent | Prop. Tax Esc. Rate | = 7.10 Percent | Loan Period | = 0 Years |
| Discount Rate | = 20.00 Percent | Income Tax Rate | = 48.00 Percent | Inv. Tax Credit Rate | = 20.00 Percent |
| Declining Balance Depreciation | | Depreciation Life | = 5 Years | Salvage Value | = $ 0 |
| | | Annual Loan Payment | = $ 0 | | |

| Year | Principal | Interest on Capital | O and M | Insurance | Replacement Cost | Fuel Cost | Electricity Cost | Deprec-iation | Tax on Capital | Delta Income Tax | Net Cost * |
|---|---|---|---|---|---|---|---|---|---|---|---|
| -1 | 4142286 | | | | | | | | | | 4142286 |
| 1 | 0 | 0 | 54438 | 0 | 0 | 945536 | -4247040 | 1449800 | 103557 | -15477 | -3158986 |
| 2 | 0 | 0 | 58303 | 0 | 0 | 1049356 | -4680238 | 1076944 | 110910 | 1144644 | -2317025 |
| 3 | 0 | 0 | 62443 | 0 | 0 | 1164575 | -5157622 | 807746 | 118784 | 1441956 | -2369865 |
| 4 | 0 | 0 | 66876 | 0 | 0 | 1292445 | -5683700 | 538497 | 127218 | 1756150 | -2441002 |
| 5 | 0 | 0 | 71624 | 0 | 0 | 1434356 | -6263437 | 269249 | 136250 | 2088940 | -2532267 |
| 6 | 0 | 0 | 76710 | 0 | 0 | 1591848 | -6902308 | 0 | 145924 | 2442156 | -2645669 |
| 7 | 0 | 0 | 82156 | 0 | 0 | 176633 | -7606343 | 0 | 156285 | 2688609 | -2912660 |
| 8 | 0 | 0 | 87989 | 0 | 0 | 1960610 | -8382190 | 0 | 167381 | 2959781 | -3206430 |
| 9 | 0 | 0 | 94236 | 0 | 0 | 2175884 | -9237174 | 0 | 179265 | 3258138 | -3529650 |
| 10 | 0 | 0 | 100927 | 0 | 0 | 2414797 | -10179865 | 0 | 191993 | 3586391 | -3805257 |
| 11 | 0 | 0 | 108093 | 0 | 0 | 2679941 | -11217661 | 0 | 205624 | 3947521 | 4276481 |
| 12 | 0 | 0 | 115767 | 0 | 0 | 2974199 | -12361862 | 0 | 220224 | 4344803 | -4706868 |
| TOT | 4142286 | 0 | 979562 | 0 | 0 | 21450181 | -91918940 | 4142286 | 1863416 | 29643621 | -33839875 |

* First Operational Year Net Cost Includes Tax Credit.

| Utility Heat Rate X | Net Generation – | Incr. Fuel Cons = | Energy Saved |
|---|---|---|---|
| Btu/KwH | Mil. Kwh / Yr. | Mil. Btu/Yr. | Mil. Btu/ Yr |
| 10818 | 63.20 | 384800 | 348338 |

Internal Rate of Return is 50.42 Percent.

## SYSTEM COST ANALYSIS

## PRESENT VALUE COSTS

| Year | Principal | Interest on Capital | O and M | Insurance | Replacement Cost | Fuel Cost | Electricity Cost | Deprec-iation | Tax on Capital | Delta Income Tax | Net Cost * |
|---|---|---|---|---|---|---|---|---|---|---|---|
| -1 | 4648278 | | | | | | | | | | 4648278 |
| 1 | 0 | 0 | 45365 | 0 | 0 | 787947 | -3539200 | 1208167 | 86298 | -12898 | -2632488 |
| 2 | 0 | 0 | 40488 | 0 | 0 | 728719 | -3250165 | 747913 | 77021 | 794892 | -160945 |
| 3 | 0 | 0 | 36136 | 0 | 0 | 673944 | -2984735 | 467445 | 68741 | 834465 | -1371449 |
| 4 | 0 | 0 | 32251 | 0 | 0 | 623286 | -2740982 | 259692 | 61351 | 846913 | -1177181 |
| 5 | 0 | 0 | 28784 | 0 | 0 | 576436 | -2517135 | 108205 | 54756 | 839498 | -1017661 |
| 6 | 0 | 0 | 25690 | 0 | 0 | 533107 | -2311569 | 0 | 48870 | 817873 | -886029 |
| 7 | 0 | 0 | 22928 | 0 | 0 | 493035 | -2122791 | 0 | 43616 | 750341 | -812870 |
| 8 | 0 | 0 | 20463 | 0 | 0 | 455975 | -1949430 | 0 | 38927 | 688350 | -745713 |
| 9 | 0 | 0 | 18264 | 0 | 0 | 421701 | -1790226 | 0 | 34743 | 631449 | -684070 |
| 10 | 0 | 0 | 16300 | 0 | 0 | 390003 | -1644024 | 0 | 31008 | 579222 | -627491 |
| 11 | 0 | 0 | 14548 | 0 | 0 | 360688 | -1509762 | 0 | 27675 | 531289 | -575563 |
| 12 | 0 | 0 | 12984 | 0 | 0 | 333576 | -1386465 | 0 | 24700 | 487299 | -527907 |
| TOT | 4648278 | | | | | | | | | | |

* First Operational Year Net Cost Includes Tax Credit.

Net Present Value is $ -8019189 at Discount Rate of 20.00 Percent

Table D-10

# Appendix D-2 Interest Tables

10% Compound interest factors

| | Single payment | | Uniform series | | | | |
|---|---|---|---|---|---|---|---|
| *n* | Compound amount factor *F/P* | Present worth factor *P/F* | Sinking fund factor *A/F* | Capital recovery factor *A/P* | Compound amount factor *F/A* | Present worth factor *P/A* | |
| 1 | 1.1000 | 0.9091 | 1.000 00 | 1.100 00 | 1.000 | 0.909 | 1 |
| 2 | 1.2100 | .8264 | 0.476 19 | 0.576 19 | 2.100 | 1.736 | 2 |
| 3 | 1.3310 | .7513 | .302 11 | .402 11 | 3.310 | 2.487 | 3 |
| 4 | 1.4641 | .6830 | .215 47 | .315 47 | 4.641 | 3.170 | 4 |
| 5 | 1.6105 | .6209 | .163 80 | .263 80 | 6.105 | 3.791 | 5 |
| 6 | 1.7716 | .5645 | .129 61 | .229 61 | 7.716 | 4.355 | 6 |
| 7 | 1.9487 | .5132 | .105 41 | .205 41 | 9.487 | 4.868 | 7 |
| 8 | 2.1436 | .4665 | .087 44 | .187 44 | 11.436 | 5.335 | 8 |
| 9 | 2.3579 | .4241 | .073 64 | .173 64 | 13.579 | 5.759 | 9 |
| 10 | 2.5937 | .3855 | .062 75 | .162 75 | 15.937 | 6.144 | 10 |
| 11 | 2.8531 | .3505 | .053 96 | .153 96 | 18.531 | 6.495 | 11 |
| 12 | 3.1384 | .3186 | .046 76 | .146 76 | 21.384 | 6.814 | 12 |
| 13 | 3.4523 | .2897 | .040 78 | .140 78 | 24.523 | 7.103 | 13 |
| 14 | 3.7975 | .2633 | .035 75 | .135 75 | 27.975 | 7.367 | 14 |
| 15 | 4.1772 | .2394 | .031 47 | .131 47 | 31.772 | 7.606 | 15 |
| 16 | 4.5950 | .2176 | .027 82 | .127 82 | 35.950 | 7.824 | 16 |
| 17 | 5.0545 | .1978 | .024 66 | .124 66 | 40.545 | 8.022 | 17 |
| 18 | 5.5599 | .1799 | .021 93 | .121 93 | 45.599 | 8.201 | 18 |
| 19 | 6.1159 | .1635 | .019 55 | .119 55 | 51.159 | 8.365 | 19 |
| 20 | 6.7275 | .1486 | .017 46 | .117 46 | 57.275 | 8.514 | 20 |
| 21 | 7.4002 | .1351 | .015 62 | .115 62 | 64.002 | 8.649 | 21 |
| 22 | 8.1403 | .1228 | .014 01 | .114 01 | 71.403 | 8.772 | 22 |
| 23 | 8.9543 | .1117 | .012 57 | .112 57 | 79.543 | 8.883 | 23 |
| 24 | 9.8497 | .1015 | .011 30 | .111 30 | 88.497 | 8.985 | 24 |
| 25 | 10.8347 | .0923 | .010 17 | .110 17 | 98.347 | 9.077 | 25 |
| 26 | 11.9182 | .0839 | .009 16 | .109 16 | 109.182 | 9.161 | 26 |
| 27 | 13.1100 | .0763 | .008 26 | .108 26 | 121.100 | 9.237 | 27 |
| 28 | 14.4210 | .0693 | .007 45 | .107 45 | 134.210 | 9.307 | 28 |
| 29 | 15.8631 | .0630 | .006 73 | .106 73 | 148.631 | 9.370 | 29 |
| 30 | 17.4494 | .0573 | .006 08 | .106 08 | 164.494 | 9.427 | 30 |
| 31 | 19.1943 | .0521 | .005 50 | .105 50 | 181.943 | 9.479 | 31 |
| 32 | 21.1138 | .0474 | .004 97 | .104 97 | 201.138 | 9.526 | 32 |
| 33 | 23.2252 | .0431 | .004 50 | .104 50 | 222.252 | 9.569 | 33 |
| 34 | 25.5477 | .0391 | .004 07 | .104 07 | 245.477 | 9.609 | 34 |
| 35 | 28.1024 | .0356 | .033 69 | .103 69 | 271.024 | 9.644 | 35 |
| 40 | 45.2593 | .0221 | .002 26 | .102 26 | 442.593 | 9.779 | 40 |
| 45 | 72.8905 | .0137 | .001 39 | .101 39 | 718.905 | 9.863 | 45 |
| 50 | 117.3909 | .0085 | .000 86 | .100 86 | 1 163.909 | 9.915 | 50 |
| 55 | 189.0591 | .0053 | .000 53 | .100 53 | 1 880.591 | 9.947 | 55 |
| 60 | 304.4816 | .0033 | .000 33 | .100 33 | 3 034.816 | 9.967 | 60 |
| 65 | 490.3707 | .0020 | .000 20 | .100 20 | 4 893,707 | 9.980 | 65 |
| 70 | 789.7470 | .0013 | .000 13 | .100 13 | 7 887.470 | 9.987 | 70 |
| 75 | 1 271.8952 | .0008 | .000 08 | .100 08 | 12 708.954 | 9.992 | 75 |
| 80 | 2 048.4002 | .0005 | .000 05 | .100 05 | 20 474.002 | 9.995 | 80 |
| 85 | 3 298.9690 | .0003 | .000 03 | .100 03 | 32 979.690 | 9.997 | 85 |
| 90 | 5 313.0226 | .0002 | .000 02 | .100 02 | 53 120.226 | 9.998 | 90 |
| 95 | 8 556.6760 | .0001 | .000 01 | .100 01 | 85 556.760 | 9.999 | 95 |
| 100 | 13 780.6123 | .0001 | .000 01 | .100 01 | 137 796.123 | 9.999 | 100 |

12% Compound interest factors

| | Single payment | | Uniform series | | | | |
|---|---|---|---|---|---|---|---|
| n | Compound amount factor F/P | Present worth factor P/F | Sinking fund factor A/F | Capital recovery factor A/P | Compound amount factor F/A | Present worth factor P/A | n |
| 1 | 1.1200 | 0.8929 | 1.000 00 | 1.120 00 | 1.000 | 0.893 | 1 |
| 2 | 1.2544 | .7972 | 0.471 70 | 0.591 70 | 2.120 | 1.690 | 2 |
| 3 | 1.4049 | .7118 | .296 35 | .416 35 | 3.374 | 2.402 | 3 |
| 4 | 1.5735 | .6355 | .209 23 | .329 23 | 4.779 | 3.037 | 4 |
| 5 | 1.7623 | .5674 | .157 41 | .277 41 | 6.353 | 3.605 | 5 |
| 6 | 1.9738 | .5066 | .123 23 | .243 23 | 8.115 | 4.111 | 6 |
| 7 | 2.2107 | .4523 | .099 12 | .219 12 | 10.089 | 4.564 | 7 |
| 8 | 2.4760 | .4039 | .081 30 | .201 30 | 12.300 | 4.968 | 8 |
| 9 | 2.7731 | .3606 | .067 68 | .187 68 | 14.776 | 5.328 | 9 |
| 10 | 3.1058 | .3220 | .056 98 | .176 98 | 17.549 | 5.650 | 10 |
| 11 | 3.4785 | .2875 | .048 42 | .168 42 | 20.655 | 5.938 | 11 |
| 12 | 3.8960 | .2567 | .041 44 | .161 44 | 24.133 | 6.194 | 12 |
| 13 | 4.3635 | .2292 | .035 68 | .155 68 | 28.029 | 6.424 | 13 |
| 14 | 4.8871 | .2046 | .030 87 | .150 87 | 32.393 | 6.628 | 14 |
| 15 | 5.4736 | .1827 | .026 82 | .146 82 | 37.280 | 6.811 | 15 |
| 16 | 6.1304 | .1631 | .023 39 | .143 39 | 42.753 | 6.974 | 16 |
| 17 | 6.8660 | .1456 | .020 46 | .140 46 | 48.884 | 7.120 | 17 |
| 18 | 7.6900 | .1300 | .017 94 | .137 94 | 55.750 | 7.250 | 18 |
| 19 | 8.6128 | .1161 | .015 76 | .135 76 | 63.440 | 7.366 | 19 |
| 20 | 9.6463 | .1037 | .013 88 | .133 88 | 72.052 | 7.469 | 20 |
| 21 | 10.8038 | .0926 | .012 24 | .132 24 | 81.699 | 7.562 | 21 |
| 22 | 12.1003 | .0826 | .010 81 | .130 81 | 92.503 | 7.645 | 22 |
| 23 | 13.5523 | .0738 | .009 56 | .129 56 | 104.603 | 7.718 | 23 |
| 24 | 15.1786 | .0659 | .008 46 | .128 46 | 118.155 | 7.784 | 24 |
| 25 | 17.0001 | .0588 | .007 50 | .127 50 | 133.334 | 7.843 | 25 |
| 26 | 19.0401 | .0525 | .006 65 | .126 65 | 150.334 | 7.896 | 26 |
| 27 | 21.3249 | .0469 | .005 90 | .125 90 | 169.374 | 7.943 | 27 |
| 28 | 23.8839 | .0419 | .005 24 | .125 24 | 190.699 | 7.984 | 28 |
| 29 | 26.7499 | .0374 | .004 66 | .124 66 | 214.583 | 8.022 | 29 |
| 30 | 29.9599 | .0334 | .004 14 | .124 14 | 241.333 | 8.055 | 30 |
| 31 | 33.5551 | .0298 | .003 69 | .123 69 | 271.292 | 8.085 | 31 |
| 32 | 37.5817 | .0266 | .003 28 | .123 28 | 304.847 | 8.112 | 32 |
| 33 | 42.0915 | .0238 | .002 92 | .122 92 | 342.429 | 8.135 | 33 |
| 34 | 47.1425 | .0212 | .002 60 | .122 60 | 384.520 | 8.157 | 34 |
| 35 | 52.7996 | .0189 | .002 32 | .122 32 | 431.663 | 8.176 | 35 |
| 40 | 93.0510 | .0107 | .001 30 | .121 30 | 767.091 | 8.244 | 40 |
| 45 | 163.9876 | .0061 | .000 74 | .120 74 | 1 358.230 | 8.283 | 45 |
| 50 | 289.0022 | .0035 | .000 42 | .120 42 | 2 400.018 | 8.305 | 50 |
| ∞ | | | | .120 00 | | 8.333 | ∞ |

15% Compound interest factors

| | Single payment | | Uniform series | | | | |
|---|---|---|---|---|---|---|---|
| *n* | Compound amount factor *F/P* | Present worth factor *P/F* | Sinking fund factor *A/F* | Capital recovery factor *A/P* | Compound amount factor *F/A* | Present worth factor *P/A* | *n* |
| 1 | 1.1500 | 0.8696 | 1.000 00 | 1.150 00 | 1.000 | 0.870 | 1 |
| 2 | 1.3225 | .7561 | .465 12 | .615 12 | 2.150 | 1.626 | 2 |
| 3 | 1.5209 | .6575 | .287 98 | .437 98 | 3.472 | 2.283 | 3 |
| 4 | 1.7490 | .5718 | .200 26 | .350 27 | 4.993 | 2.855 | 4 |
| 5 | 2.0114 | .4972 | .148 32 | .298 32 | 6.742 | 3.352 | 5 |
| 6 | 2.3131 | .4323 | .114 24 | .264 24 | 8.754 | 3.784 | 6 |
| 7 | 2.6600 | .3759 | .090 36 | .240 36 | 11.067 | 4.160 | 7 |
| 8 | 3.0590 | .3269 | .072 85 | .222 85 | 13.727 | 4.487 | 8 |
| 9 | 3.5179 | .2843 | .059 57 | .290 57 | 16.786 | 4.772 | 9 |
| 10 | 4.0456 | .2472 | .049 25 | .199 25 | 20.304 | 5.019 | 10 |
| 11 | 4.6524 | .2149 | .041 07 | .191 07 | 24.349 | 5.234 | 11 |
| 12 | 5.3503 | .1869 | .034 48 | .184 48 | 29.002 | 5.421 | 12 |
| 13 | 6.1528 | .1625 | .029 11 | .179 11 | 34.352 | 5.583 | 13 |
| 14 | 7.0757 | .1413 | .024 69 | .174 69 | 40.505 | 5.724 | 14 |
| 15 | 8.1371 | .1229 | .021 02 | .171 02 | 47.580 | 5.847 | 15 |
| 16 | 9.3576 | .1069 | .017 95 | .167 95 | 55.717 | 5.954 | 16 |
| 17 | 10.7613 | .0929 | .015 37 | .165 37 | 65.075 | 6.047 | 17 |
| 18 | 12.3755 | .0808 | .013 19 | .163 19 | 75.836 | 6.128 | 18 |
| 19 | 14.2318 | .0703 | .011 34 | .161 34 | 88.212 | 6.198 | 19 |
| 20 | 16.3665 | .0611 | .009 76 | .159 76 | 102.444 | 6.259 | 20 |
| 21 | 18.8215 | .0531 | .008 42 | .158 42 | 118.810 | 6.312 | 21 |
| 22 | 21.6447 | .0462 | .007 27 | .157 27 | 137.632 | 6.359 | 22 |
| 23 | 24.8915 | .0402 | .006 28 | .156 28 | 159.276 | 6.399 | 23 |
| 24 | 28.6252 | .0349 | .005 43 | .155 43 | 184.168 | 6.434 | 24 |
| 25 | 32.9190 | .0304 | .004 70 | .154 70 | 212.793 | 6.464 | 25 |
| 26 | 37.8568 | .0264 | .004 07 | .154 07 | 245.712 | 6.491 | 26 |
| 27 | 45.5353 | .0230 | .003 53 | .153 53 | 283.569 | 6.514 | 27 |
| 28 | 50.0656 | .0200 | .003 06 | .153 06 | 327.104 | 6.534 | 28 |
| 29 | 57.5755 | .0174 | .002 65 | .152 65 | 377.170 | 6.551 | 29 |
| 30 | 66.2118 | .0151 | .002 30 | .152 30 | 434.745 | 6.566 | 30 |
| 31 | 76.1435 | .0131 | .002 00 | .152 00 | 500.957 | 6.579 | 31 |
| 32 | 87.5651 | .0114 | .001 73 | .151 73 | 577.100 | 6.591 | 32 |
| 33 | 100.6998 | .0099 | .001 50 | .151 50 | 664.666 | 6.600 | 33 |
| 34 | 115.8048 | .0086 | .001 31 | .151 31 | 765.365 | 6.609 | 34 |
| 35 | 133.1755 | .0075 | .001 13 | .151 13 | 881.170 | 6.617 | 35 |
| 40 | 267.8635 | .0037 | .00 56 | .150 56 | 1 779.090 | 6.642 | 40 |
| 45 | 538.7693 | .0019 | .000 28 | .150 28 | 3 585.128 | 6.654 | 45 |
| 50 | 1 083.6574 | .0009 | .000 14 | .150 14 | 7 217.716 | 6.661 | 50 |
| ∞ | | | | .150 00 | | 6.667 | ∞ |

20% Compound interest factors

| | Single payment | | Uniform series | | | | |
|---|---|---|---|---|---|---|---|
| n | Compound amount factor F/P | Present worth factor P/F | Sinking fund factor A/F | Capital recovery factor A/P | Compound amount factor F/A | Present worth factor P/A | n |
| 1 | 1.2000 | 0.8333 | 1.000 00 | 1.200 00 | 1.000 | 0.833 | 1 |
| 2 | 1.4400 | .6944 | 0.454 55 | 0.654 55 | 2.200 | 1.528 | 2 |
| 3 | 1.7280 | .5787 | .274 73 | .474 73 | 3.640 | 2.106 | 3 |
| 4 | 2.0736 | .4823 | .186 29 | .386 29 | 5.368 | 2.589 | 4 |
| 5 | 2.4883 | .4019 | .134 38 | .334 38 | 7.442 | 2.991 | 5 |
| 6 | 2.9860 | .3349 | .100 71 | .300 71 | 9.930 | 3.326 | 6 |
| 7 | 3.5832 | .2791 | .077 42 | .277 42 | 12.916 | 3.605 | 7 |
| 8 | 4.2998 | .2326 | .060 61 | .260 61 | 16.499 | 3.837 | 8 |
| 9 | 5.1598 | .1938 | .048 08 | .248 08 | 20.799 | 4.031 | 9 |
| 10 | 6.1917 | .1615 | .038 52 | .238 52 | 25.959 | 4.192 | 10 |
| 11 | 7.4301 | .1346 | .031 10 | .231 10 | 32.150 | 4.327 | 11 |
| 12 | 8.9161 | .1122 | .025 26 | .225 26 | 39.581 | 4.439 | 12 |
| 13 | 10.6993 | .0935 | .020 62 | .220 62 | 48.497 | 4.533 | 13 |
| 14 | 12.8392 | .0779 | .016 89 | .216 89 | 59.196 | 4.611 | 14 |
| 15 | 15.4070 | .0649 | .013 88 | .213 88 | 72.035 | 4.675 | 15 |
| 16 | 18.4884 | .0541 | .011 44 | .211 44 | 87.442 | 4.730 | 16 |
| 17 | 22.1861 | .0451 | .009 44 | .209 44 | 105.931 | 4.775 | 17 |
| 18 | 26.6233 | .0376 | .007 81 | .207 81 | 128.117 | 4.812 | 18 |
| 19 | 31.9480 | .0313 | .006 46 | .206 46 | 154.740 | 4.844 | 19 |
| 20 | 38.3376 | .0261 | .005 36 | .205 36 | 186.688 | 4.870 | 20 |
| 21 | 46.0051 | .0217 | .004 44 | .204 44 | 225.026 | 4.891 | 21 |
| 22 | 55.2061 | .0181 | .003 69 | .203 69 | 271.031 | 4.909 | 22 |
| 23 | 66.2474 | .0151 | .003 07 | .203 07 | 326.237 | 4.925 | 23 |
| 24 | 79.4968 | .0126 | .002 55 | .202 55 | 392.484 | 4.937 | 24 |
| 25 | 95.3962 | .0105 | .002 12 | .202 12 | 471.981 | 4.948 | 25 |
| 26 | 114.4755 | .0087 | .001 76 | .201 76 | 567.377 | 4.956 | 26 |
| 27 | 137.3706 | .0073 | .001 47 | .201 47 | 681.853 | 4.964 | 27 |
| 28 | 164.8447 | .0061 | .001 22 | .201 22 | 819.223 | 4.970 | 28 |
| 29 | 197.8136 | .0051 | .001 02 | .201 02 | 984.068 | 4.975 | 29 |
| 30 | 237.3763 | .0042 | .000 85 | .200 85 | 1 181.882 | 4.979 | 30 |
| 31 | 284.8516 | .0035 | .000 70 | .200 70 | 1 419.258 | 4.982 | 31 |
| 32 | 341.8219 | .0029 | .000 59 | .200 59 | 1 704.109 | 4.985 | 32 |
| 33 | 410.1863 | .0024 | .000 49 | .200 49 | 2 045.931 | 4.988 | 33 |
| 34 | 492.2235 | .0020 | .000 41 | .200 41 | 2 456.118 | 4.990 | 34 |
| 35 | 590.6682 | .0017 | .000 34 | .200 34 | 2 948.341 | 4.992 | 35 |
| 40 | 1 469.7716 | .0007 | .000 14 | .200 14 | 7 343.858 | 4.997 | 40 |
| 45 | 3 657.2620 | .0003 | .000 05 | .200 05 | 18 281.310 | 4.999 | 45 |
| 50 | 9 100.4382 | .0001 | .000 02 | .200 02 | 45 497.191 | 4.999 | 50 |
| ∞ | | | | .200 00 | | 5.000 | ∞ |

25% Compound interest factors

| | Single payment | | Uniform series | | | | |
|---|---|---|---|---|---|---|---|
| n | Compound amount factor F/P | Present worth factor P/F | Sinking fund factor A/F | Capital recovery factor A/P | Compound amount factor F/A | Present worth factor P/A | n |
| 1 | 1.2500 | 0.8000 | 1.000 00 | 1.250 00 | 1.000 | 0.800 | 1 |
| 2 | 1.5625 | .6400 | .444 44 | .694 44 | 2.250 | 1.440 | 2 |
| 3 | 1.9531 | .5120 | .262 30 | .512 30 | 3.813 | 1.952 | 3 |
| 4 | 2.4414 | .4096 | .173 44 | .423 44 | 5.766 | 2.362 | 4 |
| 5 | 3.0518 | .3277 | .121 85 | .371 85 | 8.207 | 2.689 | 5 |
| 6 | 3.8147 | .2621 | .088 82 | .338 82 | 11.259 | 2.951 | 6 |
| 7 | 4.7684 | .2097 | .066 34 | .316 34 | 15.073 | 3.161 | 7 |
| 8 | 5.9605 | .1678 | .050 40 | .300 40 | 19.842 | 3.329 | 8 |
| 9 | 7.4506 | .1342 | .038 76 | .288 76 | 25.802 | 2.463 | 9 |
| 10 | 9.3132 | .1074 | .030 07 | .280 07 | 33.253 | 3.571 | 10 |
| 11 | 11.6415 | .0859 | .023 49 | .273 49 | 42.566 | 3.656 | 11 |
| 12 | 14.5519 | .0687 | .018 45 | .268 45 | 54.208 | 3.725 | 12 |
| 13 | 18.1899 | .0550 | .014 54 | .264 54 | 68.760 | 3.780 | 13 |
| 14 | 22.7374 | .0440 | .011 50 | .261 50 | 86.949 | 3.824 | 14 |
| 15 | 28.4217 | .0352 | .009 12 | .259 12 | 109.687 | 3.859 | 15 |
| 16 | 35.5271 | .0281 | .007 24 | .257 24 | 138.109 | 3.887 | 16 |
| 17 | 44.4089 | .0225 | .005 76 | .255 76 | 173,636 | 3,910 | 17 |
| 18 | 55.5112 | .0180 | .004 59 | .254 59 | 218.045 | 3.928 | 18 |
| 19 | 69.3889 | .0144 | .003 66 | .253 66 | 273.556 | 3.942 | 19 |
| 20 | 86.7362 | .0115 | .002 92 | .252 92 | 342.945 | 3.954 | 20 |
| 21 | 108.4202 | .0092 | .002 33 | .252 33 | 429.681 | 3.963 | 21 |
| 22 | 135.5253 | .0074 | .001 86 | .251 86 | 538.101 | 3.970 | 22 |
| 23 | 169.4066 | .0059 | .001 48 | .251 48 | 673.626 | 3.976 | 23 |
| 24 | 211.7582 | .0047 | .001 19 | .251 19 | 843.033 | 3.981 | 24 |
| 25 | 264.6978 | .0038 | .000 95 | .250 95 | 1 054.791 | 3.985 | 25 |
| 26 | 330.8722 | .0030 | .000 76 | .250 76 | 1 319.489 | 3.988 | 26 |
| 27 | 413.5903 | .0024 | .000 61 | .250 61 | 1 650.361 | 3.990 | 27 |
| 28 | 516.9879 | .0019 | .000 48 | .250 48 | 2 063.952 | 3.992 | 28 |
| 29 | 646.2349 | .0015 | .000 39 | .250 39 | 2 580.939 | 3.994 | 29 |
| 30 | 807.7936 | .0012 | .000 31 | .250 31 | 3 227.174 | 3.995 | 30 |
| 31 | 1 009.7420 | .0010 | .000 25 | .250 25 | 4 034.968 | 3.996 | 31 |
| 32 | 1 262.1774 | .0008 | .000 20 | .250 20 | 5 044.710 | 3.997 | 32 |
| 33 | 1 577.7218 | .0006 | .000 16 | .250 16 | 6 306.887 | 3.997 | 33 |
| 34 | 1 972.1523 | .0005 | .000 13 | .250 13 | 7 884.609 | 3.998 | 34 |
| 35 | 2 465.1903 | .0004 | .000 10 | .250 10 | 9 856.761 | 3.998 | 35 |
| 40 | 7 523.1638 | .0001 | .000 03 | .250 03 | 30 088.655 | 3.999 | 40 |
| 45 | 22 958.8740 | .0001 | .000 01 | .250 01 | 91 831.496 | 4.000 | 45 |
| 50 | 70 064.9232 | .0000 | .000 00 | .250 00 | 280 255.693 | 4.000 | 50 |
| ∞ | | | | 250 00 | | 4.000 | ∞ |

# Appendix D-3 Third-Party Financing

## INCREASING INVESTMENT IN ENERGY EFFICIENT INDUSTRIAL PROJECTS

---

A project of the Alliance to Save Energy, a private, nonprofit coalition of business, government, public interest and labor representatives dedicated to increasing the efficiency of energy use. Alliance to Save Energy. 1925 K Street, N.W., Washington, D.D. 20006.

---

### CONTENTS:



### CONSIDERATIONS IN STRUCTURING A THIRD-PARTY FINANCING APPROACH

In considering third-party financing for a specific project, an industrial firm will be strongly influenced by its own financial position and objectives, its willingness and ability to manage the risks associated with the project in a way that capital can be attracted, and specific limitations imposed by the project's technical and operating characteristics. In deciding whether third-party financing represents a viable alternative that might enable the firm to undertake an energy-efficiency project for which funds might not otherwise be available, the industrial decisionmaker must consider two basic questions:

1. Will third-party financing provide valuable financing advantages?
2. Can risks be managed in a way that third-party capital can be attracted at an acceptable cost?

In evaluating these questions, the industrial decision-maker will be influenced by two additional considerations:

1. What are the implications of federal tax laws for the viability of the financing structure and the distribution of the project's benefits?
2. Can operating arrangements that will meet both the firm's and the investor's requirements be formulated?

The interrelationships among these factors are complex. The feasibility and effectiveness of a particular third-party financing approach will depend largely on the answers to these questions and how well the approach resolves any related conflicts.

## FINANCING ADVANTAGES

The principal advantage of third-party financing for an industrial firm is expanded access to capital. In the ideal case, third-party financing will enable the firm to raise capital without having an offsetting effect on either the amount or cost of capital the firm can raise through its conventional debt and equity sources. In all cases, the industrial firm should focus on determining whether third-party financing expands its access to capital and helps reduce overall capital cost. Whether third-party financing accomplishes these objectives in a particular case will be largely a function of the fundamental soundness of the project being financed and the effectiveness of the financing structure in allocating specific types of risk.

A key determinant of the firm's access to and cost of capital is the credit rating assigned to its securities by Moody's Investor Services and Standard & Poor's. Accordingly, a firm will be particularly concerned about the credit rating institution's evaluation of the impacts of third-party financing. In rating a company's security, these institutions focus on the firm's overall credit-worthiness; that is, its fundamental economic viability and, in particular, its downside protection.

A widely used indicator in the industrial and financial sectors of the nature of a third-party financing and its potential impact on a firm's creditworthiness is whether the financing is "off balance sheet". Accounting rules promulgated by the Financial Accounting Standards Board (FASB) govern the reporting requirements for third-party financing. These rules are designed to reflect the fundamental character of the the financing and to make available to interest parties the information needed to assess the firm's financial performance and position. If a third-party financing is off balance sheet, the inference is that it does not affect the firm's creditworthiness.

With off balance sheet financing, the firm is not required, according to generally accepted accounting principles, to report its obligations associated with third-party financing on its balance sheet. Since reporting is not required, the presumption is that the financing is not of a nature or materiality that might affect the firm's financial position.

The financial community often uses the term "off credit" to refer to financing that does not affect the firm's creditworthiness and, hence, its access to other capital. With off credit financing, the firm's obligations associated with a financing approach are not debt-equivalent (i.e., not fixed and not determinable), and the firm's overall financial risk and flexibility are not affected.

In evaluating the creditworthiness of a firm's security, the rating institutions in effect put off balance sheet financing back on the balance sheet. In practice, the effect of off balance sheet financing on rating decisions depends on the following factors:

- *Materiality.* Unless a third-party financing arrangement is significant in relation to the firm's overall operations, it will not normally be considered an important factor in determining creditworthiness.
- *Contractual obligations.* If obligations are contingent rather than fixed and determinable, third-party financing is less likely to affect creditworthiness, particularly if the contingency is unlikely to occur.

- *Project soundness.* If the project is expected to produce measurable efficiencies, reduce the firm's costs, and improve its competitiveness, then some off balance sheet financing will not affect the firm's creditworthiness.

Third-party financing of industrial process energy projects is, in many respects, analogous to fuel supply contracts (e.g., for natural gas), which may involve contractural commitments by the firm to make minimum purchases or payments. The credit rating institutions do not normally consider such fuel supply contracts, including those with "hell-or-high-water" or "take-or-pay" clauses, to be debt-equivalent. (A hell-or-high-water clause requires the firm to make minimum payments for specified services regardless of whether these services are actually tendered or delivered. A take-or-pay contract, on the other hand, may be more accurately referred to as a "take-if-tendered" contract. It requires the firm to pay for a minimum specified quantity of service, whether the firm needs the service or not, if it has been tendered by the seller.) Of course, if the supply contract is a poor one (e.g., involves quantities in excess of the firm's reasonable require ments, includes unfavorable price terms) – and if the potential financial impact is material – it can become a factor in the credit decision.

Similarly, to the extent that an industrial process energy-efficiency project saves energy at an attractive cost and the firm has a continuing need to operate the related facilities – that is, the project is economically sound – third-party financing will not adversely affect the firm's creditworthiness.

Rating institutions may scrutinize third-party off balance sheet financing more closely when the firm's capital structure is materially changing as a result of external financing activities or chronic adverse operating performance. Even in this case, however, third-party financing will not necessarily affect creditworthiness adversely. Again, to the extent that projects financed by third parties are good business arrangements that help reduce the company's cost of operations, they may improve creditworthiness.

In short, if a third-party financing approach requires no reporting on the industrial user's balance sheet and is otherwise evaluated as being off credit, it will not have any potential adverse effects on its overall capacity to raise capital. Even if full reporting is required, the approach may still meet the industrial firm's requirements. In either case, the key issue is the transaction's real impact on the firm's operations and financial stability.

The accounting profession, through FASB, is in the process of revising and expanding accounting and reporting standards for off balance sheet financing. These revised standards may affect the firm's and the financial community's perception and evaluation of the impacts of off balance sheet financing on the firm's creditworthiness. Even if future standards require additional reporting of transactions on the balance sheet, however, soundly financed projects should not adversely affect the firm's creditworthiness.

In some cases, third-party financing provides another advantage by reducing the cost of capital to the industrial firm. Specifically, if the third-party investor is in a higher marginal tax bracket and/or can take fuller advantage of the investment tax credits than the firm can, part of the investor's tax savings can be reflected in the effective financing rate to the industrial firm. In addition, the investor may have higher overall creditworthiness and be able to attract capital at a lower before-tax cost.

In summary, in evaluating the attractiveness of third-party financing, the industrial firm should focus on determining (1) whether its overall access to capital is expanded and (2) whether third-party financing can help reduce its overall capital cost.

## RISK MANAGEMENT

Third-party financing of an industrial process modification is tied to a firm's specific product lines as well as to specific equipment; therefore, different dimensions of risk are involved than in conventional financing. Moreover, the industrial firm and the investor will probably have different perceptions of the same risk. For example, the industrial firm may perceive little risk in the technical performance of a project or in the market for its output, while an investor who is unfamiliar with the technology or the market may perceive more risk. In negotiating project financing, the participants should seek to place a specific risk with the party who is most experienced in dealing with that type of risk. In this way, the overall cost of financing can be minimized.

In developing a financing arrangement, the firm and the investor must resolve risk-sharing the following areas:

- *Technical performances.* In most cases, the technical performance of a process modification involves some degree of uncertainty. To the extent that third-party financing is structured before technical performance is demonstrated on a commercial scale, the firm and the investor must resolve how the associated financial risk will be shared.
- *Construction cost.* For most projects construction costs and schedule will also be subject to some degree of uncertainty. Again, if third-party financing is structured prior to project completion, it will be necessary to formulate a risk-sharing arrangement.
- *Operating and maintenance costs.* Like construction cost, operating and maintenance costs are subject estimation error and uncertainty.
- *Utilization.* Generally, the financial returns that are based on energy and other savings resulting from a project are influenced directly by the rate of utilization of the related projection facilities. Because the rate of utilization is a function of many unpredictable or, at least, variable factors, significant risk can result.
- *Measurement and valuation of returns.* Determining the financial returns from a project requires both measuring and pricing energy savings. To the extent that the energy returns from the project are tied to market prices, significant risk can be involved.
- *Contingencies.* Events such as fires or strikes can jeopardize the financial returns from the project.

If the financial benefits of third-party financing are to be realized, these risks must be managed and allocated. The more that risks are shifted to the industrial user, for example, the more likely that the financing will not qualify as off balance sheet or off credit. Similarly, as risks are shifted to the investor, the cost of financing increases, particularly if any of the risks are unfamiliar to the investor and are therefore perceived as significant.

Again, the goal of third-party financing is to obtain capital for the firm at a reasonable cost. Negotiating a financing arrangement may require making a trade-off between off balance sheet or off credit financing and risks to the investor.

At one extreme, the industrial firm could forgo third-party financing, use its own funds, and retain all the investment returns – but also retain all the risks and receive none of the third-party financing advantages. At the other extreme, the firm could shift substantially all the risks to the investor and achieve off balance sheet and off credit financing – but relinquish a substantially larger share of the investment returns.

## TAX IMPLICATIONS

The most fundamental effect of federal tax laws on third-party financing arrangements relates to the allocation of tax benefits. For a project that qualifies for both the investment tax credit (ITC) and the energy tax credit (ETC) as well as accelerated 5-year cost recovery, the present value of the expected tax benefits can represent as much as 60 cents of each dollar invested in the project.

To the extent that the tax benefits are allocated to the party that can use them most effectively, the overall cost of financing the project should be reduced. Prospective investors in industrial efficiency projects are primarily institutions, including insurance companies and banks; specialized industrial finance companies, including leasing companies; and corporations. These investors, who usually have marginal tax rates comparable to the user firm, will normally be able to use tax credits on a current basis, rather than defer them, as many industrial forms are forced to do. Thus, allocation of the tax benefits to the investor is usually an advantageous way to reduce financing cost. Moreover, the availability of tax benefits to third-party investors is particularly important because such benefits may reduce their perceived risk by ensuring the return of a substantial portion of the capital invested in the project.

The fundamental problem that arises in allocating tax benefits under third-party financing arrangements is the Internal Revenue Service (IRS) requirements that must be met to establish ownership of the project facilities for tax purposes. Ownership largely determines to whom the ITC, ETC, and accelerated cost recovery depreciation associated with the project are available.

Complying with federal requirements to establish ownership can be particularly difficult with regard to industrial process investments, which tend to involve specialized property that has limited value for anyone other than the industrial user. This limited-use characteristic presents difficulties under certain of the third-party approaches and effects the participants' flexibility in structuring the details of the investment.

Not only may tax laws preclude an approach for a specific case, but they also have important implications for structuring financing arrangements. For example, to avoid having a financing arrangement be construed as a lease or installment sale, participants in a third-party financing arrangement must account for tax laws in structuring a contract and sharing risks.

In some cases, the parties to a third-party financing arrangement may seek to make special allocations of the benefits from the project. Again, tax laws will largely prescribe the flexibility and conditions for making these special allocations.

In short, considering the potential importance of tax benefits, especially to the investor, the parties should make certain that the transaction is structured in such a way that IRS will to challenge claims of ownership and associated rights to tax credits or other allocations of project benefits.

The availability of the ITC, ETC, and accelerated cost recovery for a specific project will have a significant effect on overall project economics and the ability to attract third-party capital. For this reason, the firm should determine the applicability of these tax benefits to the project at an early stage.

The availability of the ITC, ETC, and accelerated cost recovery is governed by the provisions of the Energy Recovery Tax Act (ERTA) of 1981, the Tax Equity and Fiscal Responsibility Act (TEFRA) of 1982, and a relatively complex body of law consisting of IRS rules, revenue rulings, and court decisions. (For more information, see Tax Section, Industrial Cogeneration Manual).

Some of the principal issues to be resolved include:

1. Which parts of the project qualify for the ITC? Generally, the ITC is not available for buildings, structural components of buildings, or inherently permanent structures. However, if property other than buildings and their structural components is used as an integral part of manufacturing, production, or extraction, or as an integral part of furnishing transportation, communications, electrical energy, gas, water, or sewage disposal services by a person engaged in a trade or business of furnishing any such service, the ITC may apply. Buildings and their structural components are excluded under this broader definition. However, single-purpose structures used to house machinery or equipment that are specifically designed to provide for the purpose and satisfy the requirements of such equipment and that could not be economically used for other purposes qualify for the ITC.

It is unclear whether a third-party investor who is engaged in providing energy conservation services fits within these service provisions.

2. Which parts of the project qualify for the ETC? The Energy Tax Act of 1978, supplemented by the Crude Oil Windfall Profit Tax Act of 1980, lists 12 items eligible for the ETC and the criteria for determining their availability (see Tax Section, Industrial Cogeneration Manual).

3. Is accelerated cost recovery available? In general, most industrial equipment, except for structural components of buildings, will qualify for the accelerated 5-year cost recovery provided by ERTA. However, TEFRA lengthens the Accelerated Cost Recovery Schedule (ACRS) period for depreciation for any property that has been safe-harbor leased. A special-purpose structure to house a production operation will also qualify.

Editor's note: As regards cogenerators of electricity, failure to qualify under PURPA may result in application of longer recovery periods. See Tax Section, Industrial Cogeneration Manual.

## OPERATING ARRANGEMENTS

Third-party financing or process-related projects will, in most cases, affect the firm's production operations. Specifically, the equipment being financed is likely to be an important component of the firm's production, and it may be a key component on which many or all of the other production activities depend. For this reason, the industrial firm will seek to maintain as much operational control over the equipment as possible.

Critical areas of control to the industrial firm include the skills of the operators; operating procedures; the timing, nature, and expense of maintenance; and additional capital improvements. Likewise, the investor has a vested interest in the efficient operation and maintenance of the equipment, given that it is a source of economic return for the investment.

In the broadest sense, then, the investor and the firm have a common objective: to operate and maintain the equipment as efficiently as possible. However, a difference of opinion could arise on how best to accomplish this objective. Moreover, protection of confidential technology or operating practices, or unfamiliarity with or insufficient confidence in the operation of production equipment by a third party, may present difficulties in structuring an operating role for the third party.

In a third-party financing arrangement where the investor prefers to employ an on-site operator or manager, problems may arise with the union at the industrial firm's plant. To avoid labor problems, this individual might have to come from the union and a separate contract for his services might have to be negotiated with the union.

Another operating factor that must be considered in establishing the financing arrangement is the measurement and valuation of energy and any other economic benefits resulting from the project. In some cases, measurement may be straightforward. For example, if the project involves recovering and upgrading waste gases, the upgraded gas can simply be metered. The gas can then be valued in a number of ways (e.g., based on a market reference price, the firm's other fuel costs, or an agreed-upon base plus escalation). In other cases, the measurement task may be relatively complex. For example, the project may require that a regression model be constructed to account for the significant variables (e.g., product mix, capacity utilization affecting energy usage. A baseline energy consumption level would be established to forecast energy use in the absence of the improvement, and, on an ongoing basis, savings from the baseline would be estimated using the regression model. Alternatively, the investor and the firm could simply agree on assumed energy savings based on the engineering studies performed to determine project feasibility.

In addition to energy savings, the project may yield other important economic benefits, such as reduced labor costs, reduced material costs, improved product quality, and increased production capacity. Three general approaches are available for accounting for such benefits in the returns due the investor. In the most simple case, they can be ignored. If the energy savings are substantial, the associated investment return may be all that is required to attract third-party investors. In this case, the other added benefits will accrue to the industrial company. Alternatively, the value of these nonenergy savings can be established beforehand on the basis of engineering estimates and management judgment. A portion of their value can then be allocated to the investor on an ongoing basis; for example, per unit to production output. Under the third general approach, a method is developed to estimate actual savings on an ongoing basis. For example, as in the case of energy savings, a regression model can be constructed to account for the various production inputs and outputs. On a current basis, the model wold estimate the various economic savings relative to a baseline level before the process modification was implemented. As with energy, these savings would then be valued; for example, by using the firm's current cost of these inputs.

Depending on which approaches are adopted to measure and value energy and nonenergy savings, various provisions for the parties to audit and verify and measurement and valuation of savings might be required.

## CHAPTER 2

## THIRD-PARTY FINANCING APPROACHES

Four basic third-party financing approaches are available to an industrial firm and third-party investors:

1. Leasing
2. Joint venture
3. Shared-savings plan
4. Energy services contract

Each approach differs in terms of financing advantages, risk management, tax implications, and operating arrangements. Depending on the interests and priorities of the industrial firm and the investors, each approach can be structured to provide important benefits to all participants.

## LEASING

Leasing is a widely practiced means of third-party financing of industrial equipment. In lease financing, the investor (i.e., the lessor) finances and owns the assets involved in the efficiency improvement. In return for the privilege of using the assets, the industrial firm (i.e., the lessee) agrees to make payments at some specified interval for the duration of the lease. The lease payments, which are negotiated by the investor and the industrial firm, are set to recover the capital cost of the asset, net of salvage value at the end of the lease, and to provide a reasonable return to the investor. Traditionally, the payments do not vary except in accordance with escalation provisions. At the end of the lease period, the industrial firm can purchase the assets, renew the lease, or withdraw from further involvement in the transactions.

Prior to 1981, the tax law restricted leasing to financing equipment that can be readily removed from the premises at which it is being leased and subsequently installed and used at another premise. For this reason, only a limited range of energy-efficiency investments could be financed under a "traditional" lease financing approach. In fact, most investments in which the efficiency improvement becomes an integral component of the production process are not suitable for traditional leasing.

In 1981, however, ERTA broadened the applicability of leasing to industrial energy projects. Under the act's "safe harbor" leasing provisions, leases can be used to finance energy-efficiency improvements that are integral to a production process. In particular, a lease may be used to finance equipment that is permanently fixed to a user's premises and that cannot be removed without causing substantial damage and economic loss. The safe harbor leasing provisions also permit firms to take advantage of unused tax benefits through a tax lease.

Under the traditional lease, the investor can be a corporation, a partnership of corporations or individuals, or an individual. Safe harbor lease rules, however, require the lessor to be a corporation or partnership of corporations.

In 1982, because of alleged "abuses" of the safe harbor leasing provisions by some corporations which dramatically reduced their taxes, Congress enacted TEFRA. TEFRA significantly changed the statutory provisions applicable to safe harbor leasing. Among its most significant effects on safe harbor leasing are: (1) its elimination of safe harbor leasing at the end of 1983; (2) reduction in the value of the benefits of the ITC and ACRS that are available when using a safe harbor lease; (3) a limitation on the amount that a corporation may reduce its taxes by employing safe harbor leases; and (4) a limitation on the amount of property that a corporation may safe harbor lease in any one year.

TEFRA, however, created a new category of lease that will be important to the financing of energy-efficiency projects. It established a "finance" lease, which is a hybrid of traditional and safe harbor leases. The finance lease provisions become effective in 1984, after the expiration of safe harbor leasing. Finance leases can be used to lease limited-use property. for an arrangement to qualify as a finance lease, it must qualify as a lease under the non-safe harbor rules, with two important exceptions: a fixed price purchase option may be specified, and leasing of limited-use property is allowable. In addition, several of the restrictions applicable to safe harbor leases were continued, such as the requirement that the lessor be a corporation or partnership of corporations.

## FINANCIAL ADVANTAGES

The financing advantages of "traditional" "safe harbor" and "finance leases" have different implications.

"Traditional" leasing does not qualify as off balance sheet or off credit financing because it involves fixed and determinable payment obligations on the part of the firm to the investor. Under Financial Accounting Standards Nos. 13 and 47, leases are essentially treated as equivalent to debt. Accordingly, most traditional leases are reported in the firm's financial statements; in addition, the stream of lease payments is capitalized at the firm's incremental borrowing rate and reported as a liability on the firm's balance sheet. Thus, lease payments are recognized as firm liabilities by sophisticated investors, bond rating institutions, and lenders. The transaction is only off balance sheet if the fixed obligation to make lease payments is avoided; in this case, it does not qualify as a true lease for tax purposes.

Although not off balance sheet, traditional lease financing can provide valuable financial advantages to an industrial firm. In particular, it is advantageous to the firm that it is unable to commit the cash to acquire an improvement. Unlike most traditional forms of direct financing of capital asset, a traditional lease is for 100 percent of the cost of the equipment. Moreover, the net cost of capital may be lower than with traditional debt and equity capital, particularly if the industrial firm has a lower marginal tax rate than the investor or cannot use immediately the tax benefits that flow from the investment. In addition, where the investor is a well-established, diversified financial intermediary (e.g., leasing corporation, bank subsidiary), the investor may have higher overall creditworthiness than the industrial firm and be able to access debt and equity at a lower before-tax cost of capital.

The "safe harbor" leasing provisions enable less profitable firms to take advantage of unused depreciation and tax credits. Through a safe harbor "tax" lease, a firm that could not otherwise benefit from the depreciation and tax credits on a fixed asset can sell these benefits. Typically, the lease involves the firm "selling" the asset to another corporation and leasing it back. The only real dollar transfer is the payment, by the purchasing company, of an amount equal to the discounted value of the tax benefits. The buyer issues a note for the balance of the purchase price. In return, the selling company enters into a lease agreement, the payments under which equal the payments on the note. A standard provision of these tax leases is that both the buyer and seller have the right of offset with respect to the amounts due each other under the agreements. An offset agreement usually provides that the lessee's obligation to make lease payments offsets the lessor's obligation to make loan installment payments. In addition, the seller has the option of purchasing the equipment for a nominal amount upon expiration of the lease. Under proposed accounting standards, the seller and buyer would not report in their balance sheets any receivables or payables arising from the transaction if the right to offset exists.

The safe harbor tax lease can be illustrated by a simple example:

- An industrial firm invests $1 million in new production equipment.
- The industrial firm sells the equipment to another corporation (the lessor) for $1 million and leases it back for 15 years at $137,000 per year.
- The lessor pays the purchase price by giving industrial firm (the lease) $200,000 in cash and $800,000 in a nonrecourse note with 15 annual installments of $137,000. The lessor and lessee enter into an offset agree-

ment; that is, the lessee's obligation to make lease payments offsets the lessor's obligation to make loan installment payments.

- At the end of 15 years, the industrial firm purchases the equipment for a nominal amount (e.g., one dollar).

The only cash involved between the parties in the transaction is the $200,000. The lessor earns a return on this $200,000 by using the tax credits and depreciation to offset tax liabilities with a present value greater than $200,000.

Although the safe harbor tax lease can assist a company in realizing cash from unused tax benefits, it does not meet the objective of financing all or a substantial portion of the project. Specifically, under the tax lease, the amount of funds that the selling firm receives for the tax benefits is typically 15 to 30 percent of the cost of the asset, based on a review of recent transactions. Therefore, the company must raise permanent financing for the remaining 70 to 85 percent.

Leases can also be implemented under the safe harbor provisions to provide 100-percent financing of the equipment (as opposed to only 15 to 30 percent with the safe harbor tax lease). Such safe harbor leases have the potential to expand the benefits of traditional leasing to projects for which traditional leasing is not possible because of tax restrictions. In this respect, safe harbor leases provide the basis for an effective third-party financing approach.

Indeed, the safe harbor lease can provide the basic structure for implementing a shared-savings financing arrangement. Under such an approach, the lease payments are variable and depend on the actual energy and possibly the other cost savings resulting from the project. The shared-savings formula determines the amount of the lease payment. This safe harbor lease/shared savings arrangement can be an important third-party financing approach in that it solves the limited-use property problem: safe harbor leases are not subject to limited-use property restrictions. In this respect, it allows the parties wide flexibility to structure the financing without the risk of IRS interpreting it as an installment sale. In particular, it allows the parties to vest operating control with the industrial firm if this is a critical requirement; this may be difficult to accomplish otherwise without incurring significant tax risks. Moreover, the parties can specify a fixed or nominal purchase option price for the industrial firm. Under other approaches (i.e., energy services contract or shared-savings plan), this is not generally practicable or, at least, is surrounded by uncertainty regarding its tax status.

Finance leases can be used to lease energy-efficiency property commencing in 1984. Generally, this type of lease must meet the requirements for lease treatment under non-safe harbor rules. With finance leases, however, fixed price purchase options will be allowed so long as the purchase price exceeds 10 percent of the cost of the property. Moreover, limited-use property restrictions do not apply. A financing transaction must have economic substance independent of its tax benefits and not merely be cast in the form of a lease for purposes of utilizing the lessor's tax base. The transaction cannot be in substance a financing arrangement or conditional sale in which the lessee has an investment in the property. Finance lease arrangements should, however, be adaptable to be a shared-savings approach to the leasing of limited-use property.

Finance leases will be able to provide 100 percent "upfront" financing for the industrial firm. However, because there is a requirement that any option to purchase be at least 10 percent of the cost of the asset, finance leases probably will not be able to provide 100 percent financing over the entire life of the asset for limited-use property. The industrial firm would not incur this cost until the end of the lease term, so the present value of the cost is less than 10 percent of the cost of the asset. If the property were limited-use property and the investor group simply abandoned the property without having the industrial firm exercise the option (or the lease did not provide any option to purchase) the IRS may treat the transaction as a conditional sale.

## RISK MANAGEMENT

Of the four general approaches for structuring third-party financing, lease financing shifts the greatest amount of risk to the industrial firm. In particular, because lease payments usually have not been structured to depend on the energy savings produced by the improvement, the industrial firm bears the risks associated with uncertainty about the technical performance of the improvement, the cost of operating and maintaining the improvement, the rate of utilization of the improvement, and the value of the energy-savings benefits. However, an arrangement could be structured to make the lease payments vary with actual energy savings. Also, the industrial user can shift some of the risk associated with uncertain technical performance and uncertain operating and maintenance costs to the vendor or installer of the improvement through warranty agreements.

Industrial process efficiency projects, even if they involve some off-the-shelf equipment, will be subject to final total cost and schedule uncertainty. One means of dealing with these risks is for the industrial firm to complete the project on a turnkey basis, including commercial-scale verification of operating performance. The investor then would pay a final, installed price to purchase the improvement and the fixed lease payments would be set on the basis of that price.

Alternatively, the firm could seek to shift some of the construction cost and schedule risk externally. For example, a fixed-price contract for the improvement would shift cost-escalation risk to the vendor. In the case where the investor is responsible for installation and construction, the user could negotiate for the investor to bear a relatively higher share of risks by adopting a formula to set lease payments that would only partially account for differences in final delivered price from expected price. In return for the higher risk entailed in such an arrangement, an investor would expect a commensurately higher return in the lease payment. At the same time, the investor could diversify risks by investing in several projects.

Leases are normally structured to provide a return to the investor during the fixed lease term. In this respect, the industrial user bears the risk of equipment obsolescence. (The leasing of fungible commercial items such as aircraft or railways cars is an exception, but industrial process improvements will almost never be fungible commercial items.) Thus, investors in industrial processes are not likely to look to residual value as a source of either return or security for the lease. The lack of security represented by customized equipment is a factor that may prevent the application of leasing to industrial process projects, unless alternative forms of security can be developed.

Given the higher degree of risk borne by the industrial firm, an investor should be willing to accept a lower return on investment than required under alternative third-party financing arrangements. In short, compared to the other financing structures, a lease should permit the industrial firm to retain a greater share of the dollar value of energy-saving benefits.

Risk-sharing under the safe harbor or finance lease/shared-savings approach can be made more flexible and subject to negotiation by the parties. In particular, lease payment obligations can be made variable to reflect the actual performance of the process. The parties then share certain performance risks. (The risk management characteristics of the shared-savings approach are discussed more fully in the section on that approach.)

## TAX IMPLICATIONS

Federal tax laws place several restrictions on traditional or true leases. Although safe harbor leases are free from these restrictions, they are available only to corporate lessors. "Finance leases" carry over some of the restrictions applicable to safe harbor leases as well as many of those applicable to traditional leases. However, finance leases can be used to lease limited-use property with a fixed purchase price option.

To qualify as a "true" lease, a transaction must meet the criteria specified in Revenue Rulings 55-540 and 55-541. Revenue Ruling 75-21, which delineates the procedures for obtaining advance rulings from IRS, specifies the most stringent criteria that have been used to determine what is a true lease. This ruling has been suspended, and IRS no longer will issue advance rulings because of safe harbor leasing. Nevertheless, IRS is still likely to consider the principles underlying that ruling in determining whether a transaction is a true lease.

From the standpoint of third-party financing of industrial energy-efficiency projects, the most critical condition is the requirement that the lessor maintain a minimum 20-percent at-risk investment in the leased property through-out the term of the lease. At-risk investment is defined as cash or debt for which the lessor is personally liable. Nonrecourse loans for which the lessor is not personally liable do not qualify as funds at risk. Under IRS requirements, the lessor must demonstrate that an amount equal to at least 20 percent of the original cost of the property is a reasonable projection of the fair market value of the property at the end of the lease term. Fair market value is calculated after subtracting any cost for removal and delivery of the leased property to the lessor. In addition, the lessor must demonstrate that a reasonable estimate of the remaining useful life of the property at the end of the lease term is longer than 1 year or 20 percent of the originally estimated useful life of the property.

This minimum at-risk requirement is a significant constraint because the cost of installing and removing many kinds of conservation equipment is a relatively high percentage of total project cost. In addition, many kinds of conservation equipment must be custom-designed and engineered; as a result, the equipment has little or no use to anyone other than the lessee at the end of the lease term.

The installation of heat exchangers and associated piping illustrates both of these problems. In many cases, the cost of installing this equipment may be as much as 50 to 80 percent of the cost of the equipment itself. For example, the heat exchanger itself may cost $100,000, and the labor to install it may run another $50,000 to $80,000. If the cost of removal is another $25,000, the length of the lease that a lessor can offer may be so short that the transaction is not economically viable. If the equipment has a useful life of 10 years and total initial cost for equipment and installation is $150,000, the value of the equipment at the end of the term must be 20 percent of $150,000 (or $30,000), after subtracting the $25,000 removable cost.

In effect, this means that the equipment alone must be worth $55,000 (or over one-half of the initial capital cost) at the end of the lease term to comply with IRS rules. This requirement may effectively preclude the leasing of such equipment because the length of the term would be only 2 to 3 years.

A related problem is that the piping required to move heated water or gases may have salvage value to no one other than the lessee at the end of the lease term.

IRS classifies such property as limited-use property and will not issue advance rulings for transactions involving purported leases of such equipment. IRS defines such property as that which is expected not to be useful or usable by the lessor at the end of the lease term except for purposes of continued leasing or transfer to any member of the lessee group. An attempt by a taxpayer to lease such equipment will almost certainly be treated as a conditional sale. This can be a particularly difficult problem to overcome for industrial process projects, which tend to involve limited-use property.

IRS recognizes, however, that it may be possible to lease electric generating or manufacturing plants as long as the output can be sold to someone other than the lessee at the end of the lease term and owner has access to the necessary expertise and raw materials to operate the facility on its own. Unfortunately, energy that is saved or reclaimed from projects involving limited-use property is not likely to be salable to anyone other than the lessee because of the prohibitive cost and the infeasibility of transmitting such energy to other users.

Editor's note: This, of course, is not true as to those portions of a cogeneration facility which is used in the production of electricity, which is easily transmitted, nor is it true of the steam or heat generating components of such a project where district heating or other sales of steam or heat are envisioned.

This prohibition on leasing limited-use property effectively restricts the leasing of energy conservation equipment to easily removable devices with low installation and removal costs relative to capital cost. Items such as energy management systems, particularly those that use radio signals rather than relying on hard wiring, are leased routinely. The life of these systems is relatively long (15-30 years), and the installation and removal cost is modest compared to their capital cost. Other items that can probably be leased successfully are "plug in" lighting controls, adjustable frequency drives for electric motors, and power factor controllers. These relatively fungible items can be installed and removed fairly easily. On the other hand, most types of insulation, solar window film, caulking, and similar items cannot be leased, at least under the traditional rules. The same is true for heat recovery and combustible gas recovery projects where a significant fraction of the cost is for fixed piping.

However, conservation equipment that has a relatively short useful life and is leasable under the IRS criteria may not be attractive to noncorporate lessors. Section 46 (e) (3) of the Internal Revenue Code allows a noncorporate lessor to take the ITC and ETC only if the property is leased for less than 50 percent of its useful life. The only exception is property manufactured by the lessor. In addition, the lessor's expenses (those deductible under Section 162) during the first 12 months of the lease must exceed 15 percent of the rental income produced by the property.

Under ERTA, "useful life" is defined as the class life or asset depreciation range (ADR) midpoint as of January 1, 1981. In many cases, this figure will be shorter than the true economic life of an asset. In addition, the class life may depend on the type of facility in which the equipment is installed. It is often difficult to determine the appropriate class life for conservation equipment because the ADR system generally does not list energy conservation equipment as a separate category; rather it classifies equipment by specific industries. (See Tax Section, Industrial Cogeneration Manual). The effect of this provision is to reduce even further the allowable lease term and the potential economic viability of such a transaction.

In addition to the minimum at-risk provision, true leases are subject to several other IRS restrictions:

- *Calculation of lease term.* In calculating the length of the lease term, all renewal or extension periods must be included except renewals at the option of the lessee. In this case, the lease rate is based on the fair market rate at the time of renewal.
- *Purchase and sale rights.* The lessor cannot require the lessee to purchase the leased property at any time during the term, nor does the lessor have the right to abandon the property. If the lessee is given a purchase option, it must be based on fair market value at the time the option is exercised.
- *No investment by lessee.* No part of the purchase price of the leased property may be paid by the lessee. In addition, the lessee cannot pay the cost of any improvements or additions to the property, unless such improvements can be removed readily without significant damage. However, ordinary maintenance performed by the lessee is generally not considered an improvement.
- *No lessee loans or guarantees.* The lessee may not lend to the lessor any part of the funds necessary to purchase the property nor provide any loan guarantees.
- *Profit requirement.* The lessor must be able to demonstrate that it expects to make a profit on the transaction, apart from any tax benefits arising from the transaction. The test is met if "the aggregate amount required to be paid by the lessee...over the lease term plus the value of the residual investment... exceeds an amount equal to the sum of the aggregate disbursements required to be paid by or for the lessor in connection with the ownership of the property and the lessor's equity investment in the property, including any direct costs to finance the equity investment..."

If the at-risk provision or any of these other tests is not satisfied, IRS will treat the transaction as a conditional sale, thereby triggering the loss of some or all of the associated tax benefits claimed by the purported lessor.

The tax consequences to both the purported lessor and lessee of a transaction that takes the form of a lease and is subsequently determined to be a sale are outlined in Revenue Ruling 72-408. First, the lessor can except to lose some or all of the ITC and ETC. Because the transaction is treated as a sale rather than a lease, the lessor faces loss of these credits under the recapture provisions governing early disposition of Section 38 property. In general, for 5-year recovery property, the lessor will lose all of the ITC and ETC if the equipment is sold within the first year after its purchase. On the other hand, the lessor will be able to retain 20 percent of the ITC and 20 percent of the ETC for each full year the property is held prior to the conditional sale.

Similarly, the purported lessor will lose any depreciation allowances claimed during the lease period, since he no longer owns the equipment. Finally, any gain realized on the disposition of the property by the purported lessor will be treated as ordinary income rather than capital gain to the extent of depreciation already taken (that is, prior to entering the purported lease). In short, the tax consequences for the purported lessor are adverse.

On the other hand, the purported lessee in a transaction later found to a conditional sale may be eligible for tax benefits that were not anticipated. First, the purported lessee may be eligible to take the ITC on that portion of the payments that is construed to be principal. The amount of the ITC, however, will be limited to that available for used property, which is currently $125,000. Similarly, the purported lessee will be entitled to depreciation deductions to the extent of the principal payments. However, the payments made by the purported lessee will no longer be deductible as ordinary and necessary business expenses under Section 162, except for the portion of such payments that IRS construes as imputed interest.

In conclusion, the characterization of a purported lease as a conditional sale can create real problems for both lessor and lessee. To reduce the risk to the parties, the agreement should probably contain provisions providing for a recalculation of lease payments if the transaction is treated as a conditional sale.

Virtually all of the obstacles can be overcome by using the safe harbor leasing provision of ERTA. Under these rules, there is no restriction on the leasing of limited-use property. In addition, the leased asset may be sold to the lessee for a nominal purchase price at the end of the lease. Moreover, the lessee can assist in the financing of the leased property. The maximum term of the lease (including extension) was changed in TEFRA to not exceed the applicable recovery period for the leased property (5 years for 3-year property, 8 years for 5-year property, and 15 years for 10-year property), or if greater, 120 percent of the property's present (ADR) class life. The minimum at-risk investment that must be maintained by the lessor is reduced by ERTA to 10 percent (from 20 percent under the true lease rules).

However, these provisions are available only to corporate lessors and cannot be used by individuals, partnerships (of individuals), Subchapter S corporations, or corporations that are personal holding companies. Other requirements include:

- All parties to the lease must characterize it as a lease and elect to treat the lessor as the owner of the property.
- Lessor and lessee must file an information return indicating use of the safe harbor provisions.
- Leased property must be new Section 38 property (i.e., personal rather than real property).

This last requirement is particularly important. If property is classified as a structural component of a building, it cannot be leased under even the safe harbor provisions. Similarly, if it is classified as an inherently permanent structure and does not qualify as other tangible property, it cannot be leased under the safe harbor leased (See Tax Section, Industrial Cogeneration Manual).

One problem that IRS has not addressed is how to handle purported safe harbor leases of conservation equipment, where a percentage of the equipment is construed to be a building component and the balance is classified as tangible personal property. It is not clear whether the whole transaction would be disqualified because some of the property involved is non-Section 38 property.

The safe harbor rules do alleviate the limited-use property problem, thereby removing a particularly troublesome obstacle to lease financing for many types of equipment. The rules also eliminate, for qualified leases, restrictions on transfer of ownership. For example, the lease agreement can specify a fixed transfer price; even a nominal sum such as $1.00 is sufficient. Furthermore, these rules provide the parties with greater flexibility in pricing the lease. For example, the lessor need not necessarily show a positive return from the lease payments alone.

In short, traditional leasing of most types of energy conservation equipment is relatively difficult from a tax standpoint. Under safe harbor rules, however, virtually any type of equipment can be leased. Safe harbor leases are not available, however, to noncorporate lessors or for non-Section 38 property.

TEFRA significantly changed the value of the tax benefits that are realized by employing safe harbor leases. The value of ACRS and ITC to the lessor are reduced. The recovery period for the property is changed to 5 years for 3-year property, 8 years for 5-year property and 15 years for 10-year property. The method of cost recovery is changed to 150-percent declining-balance method, changing to straight-line where appropriate and using a half-year convention in the first year. Instead of being able to claim the full ITC or ETC in the first year, the amount of the ITC or ETC in the first year, the amount of the ITC or ETC allowed is prorated over 5 taxable years, and only 20 percent can be claimed in each year. Finally, the tax basis for the property for cost recovery purposes must be reduced by 50 percent of the total ITC and ETC allowable on the property. An election to reduce the amount of the ITC as an alternative to the basis reduction is available; however, this election does not apply to the ETC (See Tax Section, Industrial Cogeneration Manual).

The new category of finance leases created by TEFRA to commence in 1984 will be significant for leasing industrial process improvements. This category allows the leasing of limited-use property with a fixed purchase price option, but the purchase price must exceed 10 percent of the cost of the equipment to the lessor. TEFRA requires that a finance lease be characterized by the parties as a lease and have economic substance independent of the transfer of tax benefits. A finance lease must generally meet the requirements of a lease under the non-safe harbor rules, disregarding the fact that there is an option to purchase the property at a fixed price or it is limited use property. Additional requirements and limitations include: (1) lessors must be corporations or partnerships of corporations; (2) a lessor cannot reduce its tax liability by more than 50 percent for property placed in service prior to September 30, 1985; and (3) the ITC and ETC must be spread over 5 years for property placed in service prior to September 30, 1985. As with all allowable ITCs and ETCs under any financing arrangement, a reduction of the tax basis of the property by 50 percent of the amount of the ITC and ETC allowable, or an election to reduce the allowable ITC must be made. Lessors will be allowed regular ACRS periods and depreciation methods.

## OPERATING ARRANGEMENTS

Of the four third-party financing arrangements, leasing presents the lease difficulty in making operating arrangements. Because the industrial firm bears essentially all of the risk, the investor has little reason to intervene in the operations.

Generally, the lessee decides on the equipment needed, selects the manufacturer, negotiates the price, arranges for delivery, and supervises and approves installation. Operation and maintenance of the equipment are normally performed by the user or, where applicable, the equipment manufacturer. The user also pays taxes and insurance. In essence, the lessor provides capital but is not responsible for operating or maintenance expertise or service.

The investor only needs to ensure that the equipment is maintained sufficiently so that it has a reasonable residual value at the end of the lease (this is less critical for safe harbor leases). However, this requirement should not prove burdensome because it can usually be met by a lease clause defining the required equipment condition at the end of the lease. Moreover, the industrial firm will have a substantial incentive to maintain the improvement at a high level of operating efficiency in order to realize the maximum energy savings and economic value.

## JOINT VENTURE

Under the joint venture approach, the industrial user undertakes the project with one or more partners. The partners may be investor groups whose primary role is to furnish the required capital, or other independent firms who furnish either capital or other services (e.g., operation and maintenance). A joint venture can be either a corporation or a partnership. A partnership, which offers broad flexibility in structuring the roles, risks, and returns, appears to be the most advantageous joint venture arrangement for third-party financing.

In establishing a partnership, the industrial firm and the investor negotiate specific sharing arrangements regarding the upfront cash requirements of the project; the ongoing costs of operating and maintaining the project; the partnership's debt and other monetary obligations; and the revenues or profits expected to accrue as a result of the project. The industrial firm establishes its equity interest in the project by providing the opportunity and site for the energy-efficiency investment and may have little or no cash interest in the capital cost of the project. The industrial firm and the investor also negotiate a basis for defining the revenue flow associated with the project, based on either the measured through-put or utilization of the facility, or other revenue sources (e.g., labor or material savings). Typically, these revenues are first allocated to operating and maintenance expenses, debt payments, and any other monetary obligations of the partnership. The residual is then divided between the industrial firm and the investor according to the negotiated sharing formula.

A joint venture can be open-ended or can terminate after a specified number of years. At that time, the energy user can either renew the joint venture under renegotiated terms or purchase the improvement at residual value from the joint venture partnership.

In contrast to leasing, a joint venture partnership can be structured to provide off balance sheet, off credit financing. Moreover, because the partnership owns the equipment, and the industrial firm is one of the partners, the joint venture partnership avoids many of the tax drawbacks of leasing, in particular, the restriction that prevents the leasing of limited-use property. However, the joint venture arrangement may be most appropriate for larger projects, especially if the transaction costs of setting up and managing the venture become substantial.

### Financing Advantages

Joint ventures can provide off balance sheet, off credit financing. Moreover, the industrial firm may be able to achieve higher than normal leverage; that is, if the venture is structured properly, the industrial firm may not need to commit any permanent capital.

The impact of a joint venture financing on the firm's balance sheet, creditworthiness, and access to other capital is a function of the principles of consolidation the firm follows for reporting purposes, the nature and extent of its contractual commitments, and the associated reporting requirements and practices.

The preferred and most common method of reporting joint ventures is the one-line equity method. Under this method, the firm includes in its balance sheet a single amount representing its percentage share of the net equity of the venture; in each reporting period, the firm records it share of the net income of the venture. The debt and other liabilities of the joint venture do not appear in the firm's balance sheet.

The one-line method can be used where the firm owns 50 percent or less of the joint venture. A majority owner normally consolidates the joint venture, showing a minority interest, and includes the joint venture liabilities in its balance sheet. For this reason, the firm may seek to structure a joint venture financing arrangement so that it retains a 50-percent or less interest.

The cost method of consolidation can also be used. Under this method, the firm carries the joint venture on its balance sheet at its original investment.

However, this method is consistent with generally accepted accounting principles only if the investment is immaterial to the firm's financial statements or if the firm's share is so small as to preclude any influence over the joint venture's operations. A share of less than 20 percent normally implies that the firm does not have the ability to exercise significant influence.

In structuring a third-party joint venture financing approach, the firm will probably find it necessary to enter into contractual arrangements with the joint venture. To the extent that the firm's obligations under these contracts are conditional and do not involve the joint venture's liabilities, no reporting is warranted. Moreover, such an arrangement is off credit.

More likely, however, the firm will find it necessary to enter into a more stringent contract to enhance the creditworthiness of the venture and meet the risk protection requirement of the third-party investor. These contracts fall into two categories:

1. *Unconditional purchase obligations.* The two most common obligations are "take-or-pay" and throughput (deficiency) contracts. There are two varieties of take-or-pay contracts. A "take-if-tendered" contract requires the firm to make payments for a specified amount of product that is offered even if it does not take delivery of it. A "hell-or-high-water" contract requires the firm to make minimum payments for specified amounts of product, regardless of whether the product is offered for sale. The specified product may be recovered waste heat, recovered gas, or any other product produced by the joint venture. The minimum payments are designed to cover the joint venture's debt service requirements. Throughput contracts are similar in nature to "hell-or-high-water" clauses but usually relate to minimum purchases of a specified amount of service (e.g., transportation, refining) as opposed to a purchase of a product.
2. *Guarantees of the joint venture's debt.* A guarantee of the joint venture's debt can be direct or indirect. In an indirect guarantee, the firm agrees to keep the joint venture whole by making payments to the venture under specified conditions; for example, if the joint venture's working capital, income, or coverage of fixed charges falls below a specified minimum. Such guarantees are referred to as "keep well" or "equity maintenance and comfort agreements".

The primary issue raised by these purchase obligations is whether and under what situations they result in the firm, in effect, acquiring an ownership interest in assets and incurring liabilities that should be recognized on its balance sheet. FASB has adopted interim measures that require certain reporting of unconditional purchase obligations. (Currently, FASB is evaluating and may issue revised standards on the consolidation of joint ventures and the equity method, and on when assets and liabilities should be recognized on the balance sheet. However, to the extent that the third-party financing enables the firm to undertake a worthwhile energy-saving project, the firm's creditworthiness should not be adversely affected by revised reporting requirements.) In particular, FASB Statement No. 47, Disclosure of Long-Term Obligations, specifies that unconditional purchase obligations must be reported if they meet the following conditions:

- Noncancelable, or cancelable only upon the occurrence of some remote event
- Negotiated as part of arranging financing for the facilities that will provide the production service
- A term greater than 1 year.

For obligations meeting these requirements, the firm must report:

- Nature and term of the obligation
- Fixed and determinable portion of the obligation in the aggregate and, if determinable, for each of the next five years
- Description of any of the variable components of the obligation
- Amounts purchased under the obligation for each year for which an income statement is presented.

Key factors that are considered when deciding on reporting requirements are whether the obligations are determinable, the probability of their occurrence, and their materiality.

In summary, the reporting requirements for joint ventures are minimal, particularly if the firm is a 50-percent or less owner and provides only indirect guarantees to support the creditworthiness of the joint venture. Moreover, a joint venture financing arrangement should qualify as off credit. To the extent that the improvement is a sound investment, any minimum payment obligations should not materially affect the firm's creditworthiness. In this case, credit rating institutions are not likely to believe that the obligation has an adverse impact on the firm's creditworthiness, especially if the obligation is not large relative to the firm's total capitalization.

In addition to the benefit of off balance sheet financing, a joint venture can be structured in a way that the industrial firm has no initial, unrecoverable cash investment as equity in the project. Its contribution of the site and opportunity for undertaking the energy-efficiency improvement establishes its equity interest in the venture. Even then the industrial firm finances the improvement to the point of operation, it still has no ongoing commitment of cash because the investor is essentially responsible for contributing the cash on behalf of the joint venture once the project is operational.

**Risk Management**

Of the four third-party financing arrangements, the joint venture partnership arrangement provides the greatest opportunity and flexibility for risk-sharing between the industrial firm and the investor. In addition, under this arrangement, the industrial firm and the investor should be able to share the incentives and interests in project management to the greatest extent.

The terms of the joint venture agreement regarding each party's share of individual cost items, liabilities, and revenues will define the specific degree of risk assumed by each party in each risk area. To the extent that the specific terms of a joint venture shift a relatively greater degree of risk to the investor than other financing arrangements would, the investor will require a higher expected return for participation. Conversely, the industrial firm will expect to retain a lower share of the expected dollar value of the energy savings associated with the improvement. Because of its strong risk-sharing capability and flexibility, a joint venture is potentially adaptable to a wide range of projects with diverse operating and risk characteristics.

In structuring risk-sharing arrangements in a joint venture, the uncertainties associated with the following factors should be considered:

- *Project cost and construction schedule.* Project cost and schedule uncertainties are important considerations in all energy-efficiency projects other than those involving readily installed off-the-shelf equipment.

  Schedule uncertainty creates risk because a delay can affect the project cost where the cost is defined to include the accrued opportunity cost of money tied up in the investment before the project begins operations. Delays in project completion increase the opportunity cost or return that might have been earned had the money been invested in some other income-producing activity. If the tied-up money is borrowed, the cost becomes a real out-of-pocket expense as additional interest accrues prior to project completion.

  A range of alternatives is available for managing this risk factor. At one extreme, if the industrial firm has little or no cash available for the capital investment, the investor can assume full responsibility for the project cost and completion risk. This will be the case if the investor is responsible for the capital cost of the improvement and if all other cost- and revenue-sharing terms were negotiated and fixed prior to full knowledge of the project's cost and completion date. At the other extreme, the industrial firm can assume full responsibility for project cost risk if it agrees to deliver the improvement to the joint venture at a fixed price. Under both arrangements, it may be possible to shift risk externally by negotiating fixed-price or limited escalation agreements with the vendor or installer of the efficiency improvement.

  Another alternative is for the industrial firm and the investor to share jointly in the project cost by having a fixed-share responsibility for its final cost; however, under this arrangement, the industrial firm has to make an upfront, nonrecoverable cash outlay. Yet another possibility that permits risk-sharing and eliminates the industrial firm's cash obligation is to specify the investor's percentage share of revenues on the basis of the final cost of the project (i.e., the higher the cost, the higher the investor's percentage share). This approach essentially involves assigning a dollar value to the industrial firm's contribution of the site and opportunity for the improvement. The investor's dollar contribution for purchasing the capital improvement can then be pooled with the industrial firm's implied investment to establish equity participation percentages for sharing the project's revenues and subsequent costs.

- *Technical performance.* In general, the risk associated with performance uncertainty is shared according to the arrangements specified for sharing project operating costs and revenues. Again, this risk can be shifted to one or the other party by requiring an adjustment to the revenue- or profit-sharing formula based on the technical performance of the improvement measured at the time that operation begins. For example, if the industrial firm is responsible for delivering a completed improvement to the joint venture, then its share of project revenues or profits can be adjusted on the basis of the technical performance achieved at delivery. Presumably, higher than target performance would earn a higher share of revenues or profits for the user and vice versa.

  In addition, some of the risk associated with uncertainty about project technical performance may be shifted externally by negotiating performance warranties with the vendor or cost discounts with the vendor/installer if the improvement does not meet specified operating efficiency.

- *Operating and maintenance costs.* Unless special terms are included in the agreement, the risk of uncertain operating and maintenance costs is shared according to the negotiated cost-sharing formula. However, this risk can be shifted between participating parties by having one party enter a fixed-price agreement to cover operating and maintenance costs. Alternatively, the reimbursement to either party for operating and maintenance costs can be set according to a specified formula, where the reimbursement in any time period increases with the operating efficiency and availability realized during that period.
- *Rate of utilization.* To manage this risk, it may be prudent to enter into agreements in which the efficiency improvement is installed at a facility that has historically had, and may be expected to continue to have, a high utilization rate. For example, in negotiating an agreement with an industrial firm that has several operating facilities, an investor can enter into an agreement for installing an improvement at the "flagship" facility. In this way, the investor can reduce the risk that economic weakness in the economy, the industry, or the firm will lead to a reduced operating rate for the efficiency improvement and, consequently, reduce the generation of revenues. The joint venture can also negotiate a "run-first" agreement with the industrial firm; that is, if several facilities are available to meet production requirements at any time, the user commits to assign production priority to the facility on which the improvement has been installed. Alternatively, the industrial firm can guarantee minimum revenues to the joint venture independent of the actual utilization rate. These revenues can be set at a level to meet debt payments and other fixed project expenses.
- *Measured energy savings or other economic benefits.* In some instances it may be easy to measure the physical savings in energy requirements or other factor inputs that result from an efficiency improvement (e.g., a manufacturing plant with well-known, unvarying energy requirements prior to installation of the efficiency improvement). More often, however, there will be uncertainty regarding the physical quantity of energy or other factor input savings. In this case, the industrial firm and the investor must negotiate some measurement formula on the basis of engineering analyses and projections of energy and factor input savings. For example, it may be possible to agree that a specified reduction in energy requirements will occur as a result of implementing the improvement. The supposed physical quantity of energy savings can then be directly inferred from the actual consumption that occurs following installation of the efficiency improvement. Alternatively, it may be possible to agree on the physical savings of energy or other factor inputs per unit of plant output.
- *Value of energy-efficiency benefits.* The price of some form of energy or some energy price index is generally used to specify the price that must be paid by the industrial firm to the joint venture. Depending on the energy user's and the investor's energy price expectations, either party may prefer to use a different formula than straight percentage sharing of project revenues to allocate the risk associated with energy price uncertainty. For instance, the industrial firm can agree to a minimum, maximum, or fixed rate of increase in the energy price on which project revenues are based. Another price risk concerns the possibility that project revenues may be based on a specific

energy price that does not match, in the future, the so-called average energy price. To hedge against this possibility resulting in large gains or losses to one or the other party, the energy-saving revenues can be determined on the basis of movements in an index of composite energy prices.

- *Major unforeseen events.* A joint venture might deal with the risk of unforeseen events (e.g., strikes, environmental damage) preventing operation of the efficiency improvement by obtaining conventional business insurance to cover lost revenues. In this way, the risk is shifted externally.

**Tax Implications**

For tax purposes, the term "partnership" includes "a syndicate, group, pool, joint venture or other unincorporated organization through or by means of which any business, financial operation, or venture is carried out, and which is not within the meaning of this title a corporation or a trust or estate" (Section 761(a) of the Internal Revenue Code). As such, a partnership is not subject to federal taxation; rather, the partners themselves are liable for income tax, but only in their individual capacities. In effect, all items of income or loss are effectively "flowed through" the partnership to the individual partners, whether they are individuals or corporations.

Unlike a lease, the term of the partnership can cover the expected economic life of the energy conservation equipment. If the equipment has residual value at the end of the partnership agreement, the partnership can sell the property to the energy user partner and distribute the proceeds a part of its liquidation. (Under IRS rules, a partnership must terminate when 50% or more of the total interest in partnership capital and profits is sold or exchanged within a period of 12 consecutive months. Such a sale or exchange includes the sale or exchange to another member of the partnership). Even more importantly, the joint venture can own limited-use property. Thus, the associated tax benefits are available to the joint venture and can be allocated to its partners.

A key issue, however, is whether the allocation of income, losses, and tax benefits must match the respective interest of each partner in a joint venture arrangement. For example, if an energy user forms a joint venture partnership with an investor who provides all or most of the capital needed to finance energy conservation equipment, the investor may require a relatively large share of the energy savings, at least in the early years of the partnership. Once the investor has recovered most or all of the initial capital, the parties could then shift to a different distribution of savings. Alternatively, the industrial firm may seek a proportionately large share of the savings, if such savings are allocated on a "net" basis (i.e., after payment of joint venture debt services).

A change in the allocation of profits and losses can create what is known as a "special allocation" under the Internal Revenue Code. Section 704(b) of the code provides that a special allocation in the partnership agreement will not be recognized for federal income tax purposes if it does not have a "substantial economic effect". Specifically, "if the principal purpose of any provision in the partnership agreement determining a partner's distributive share of a particular item is to avoid or evade the federal income tax, the provision shall be disregarded and the partner's distributive shares of that item shall be determined in accordance with the ratio in which the partners divide the general profits or losses of the partnership (as described in Section 702(a)(9)." (IRS Rules Section 1.704-1(b)(2).

Among the factors that IRS examines in determining whether a particular allocation of profits or losses is designed to avoid or evade the federal income tax are:

- Whether the partnership or an individual partner has a business purpose for the allocation
- Whether the allocation has "substantial economic effect" (i.e., whether the allocation may affect the dollar amount of the partner's shares of the total dollar partnership income or loss independently of tax consequences)
- Whether related items of income gain, loss, deduction, or credit are from the same source or are subject to the same allocation.
- Whether the allocation was made without recognition of normal business factors and only after the amount of the specially allocated item could reasonably be estimated.
- The duration of the allocation.
- The overall tax consequences of the allocation.

Essentially, Section 704(b) is designed to prevent the partners from allocating income or losses in a way that will maximize the resulting tax benefits. The IRS rules accompanying Section 704(b) provide an example that is reasonably close to the type of shifting allocation that may be preferred by participants in joint venture partnerships involving energy conservation equipment.

For example, consider two parties entering into a partnership to develop and market electronic devices. One party (H) contributes $2,500 in cash and agrees to devote full-time services to the partnership. The other party (G) contributes $100,000 in cash and agrees to obtain a loan for the partnership for any additional capital required. The partnership agreement provides that the full amount of any research and experimental expenditures and any interest on partnership loans are to be allocated to G. In addition, G's distributive share is to be 90 percent of partnership income or loss until all loans have been repaid and the amount of research and experimental expenditures is recovered. During this "capital recovery period", H's distributive share is to be 10 percent. Thereafter, however, G and H are to share profits and losses equally. In the absence of other circumstances demonstrating that the arrangement's primary purpose is tax avoidance, IRS would probably recognize such a special allocation as having substantial economic effect.

This example, while not conclusive, suggests that bottom-line allocations of income and losses can be shifted under appropriate circumstances. However, depreciation, tax credits, and other deductions cannot be allocated to certain members of the partnership simply because they can make the greatest use of such deductions. In general, the safest approach is to allocate all items of gain, loss, deduction, or credit in the same proportions as general profits or losses are divided under the partnership agreement.

### Operating Arrangements

Two basic factors must be addressed in establishing operating arrangements in a joint venture:

1. Management of the operation and maintenance of the energy-efficiency improvement.
2. Measurement and validation of the savings in energy or other factor inputs on which the joint venture revenues are based.

With regard to the operation and maintenance of the improvement, the primary management responsibility should be assigned to the industrial firm, which will be responsible for day-to-day operations and maintenance and will be expected to operate the improvement in a manner consistent with its overall requirements for efficient conduct of its production activities. To protect its investment, the investor may require that the joint venture retain an independent project manager to oversee the activities of the industrial firm in operating and maintaining the improvement. However, for the most part, the incentives and management decisions of the industrial firm in operating and maintaining the efficiency improvement should match the objectives of the investor.

With regard to measuring and validating the savings energy or other factor inputs, the joint venture agreement must specify the basis for establishing the physical quantity of savings. In negotiating the joint venture, the industrial firm and investor must agree on some observable measure as a basis for establishing the physical quantity of savings. For example, physical energy savings could be established as a percentage of actual energy consumption after installation of the improvement or as a specific quantity per unit of plant production. Once a measure is agreed upon, the joint venture should have access to the energy consumption or plant production records that are needed to establish the physical quantity of energy or other factor input savings. This responsibility may be fulfilled by an independent project manager retained by the joint venture.

## SHARED-SAVINGS PLANS

Under the shared-savings approach, the investor finances and owns an energy-efficiency improvement undertaken on a facility of the industrial firm. The investor is responsible for the capital and all maintenance and operating costs of the efficiency improvement. The investor earns a return by receiving a negotiated share of the gross dollar value (i.e., before deduction of project costs) of the energy savings realized from the project. Generally, the payments by the industrial firm to the investor depend on the economic benefits resulting from the operation of the efficiency improvement.

The selection of a sharing formula should be based on the economic merits of the specific investment opportunity and the risk/return preferences and objectives of the industrial firm and the investor. Most shared-savings plans formulated to date have been based on a 50-percent sharing formula (i.e., 50 percent of the dollar value of the energy savings is retained by the industrial firm and 50 percent goes to the investor). However, there is no *a priori* basis for expecting that 50 percent is an optimum or preferred sharing formula. Indeed, rigid reliance on a 50-percent formula will result in an arbitrary elimination of projects that do not offer sufficiency economic return at that sharing rate to attract external capital. On the other hand, reliance on a 50-percent formula for extremely lucrative projects might transfer an unusually large return to the investors. In the long run, as the market for these investments becomes more competitive, the sharing formula for a specific project will be driven to a level that provides a return to investors that is competitive with the return provided by capital markets for projects of comparable risk.

In a shared-savings plan, the industrial firm probably must accept some involvement in equipment servicing and maintenance by the investor. Because the investor's return is based entirely on results, there is a continuing basis and incentive for ongoing involvement in the energy project.

Shared-savings plans are offered by numerous firms, typically for various types of retrofit applications (e.g., economizers) and energy management systems. Most of these applications focus on the commercial and industrial sector. As practices by these firms, a full-service shared-savings plan entails the following:

- The investor, through an independent engineer or its own staff, performs a preliminary audit of the user's operations, usually at no cost to the user.
- If the results indicate significant and attractive investment opportunities, the potential investor reviews the audit with the user. At this point, the user is under no financial obligation. However, if the user decides to proceed to a detailed audit, he agrees either to allow the investor to implement identified investments or to reimburse the cost of the audit.
- The investor performs a detailed audit, working closely with the user's engineering staff. At the user's option, the investor will use an engineer selected by the user to perform the audit. On the basis of the audit, the engineer presents technical and financial evaluations of various energy-efficiency investment projects.
- The investor selects those projects that he is willing to fund and reviews them with the user. If the user elects to proceed, he signs an agreement with the investor covering the projects involved and specifying the obligations of each.
- The investor retains an independent firm to develop the methods to be used to calculate the monthly savings. Selection of methods will depend on the potential projects and involve the establishment of baseline consumption (i.e., consumption before equipment installation). This baseline may account for variations in production and weather patterns.
- The independent engineer or the investor's staff engineers prepare plans and specifications, take bids and award the contracts, and supervise installation. The investor is also responsible for servicing and maintaining the entire project for the duration of the agreement.
- On a monthly basis, energy consumption is compared with the appropriate baseline to calculate energy savings. Using current utility and energy prices, cost savings are estimated, and the investor bills the user for his negotiated share.
- At the expiration of the agreement, often 7 years under current practices, the user may purchase the equipment at fair market value.

The investor can fund the project with his own funds or sell the project to a syndicator for placement with other investors, perhaps through limited partnerships. If the project is syndicated, the original investor may retain responsibility for administering the calculation of monthly savings and servicing and maintaining of equipment. Alternatively, the investor might arrange for an independent engineer to perform maintenance and servicing over the life of the agreement and be compensated from the actual energy savings. In any case, the investor should remain fully committed to the success of the project, given that his return is based entirely on results.

In the case of industrial energy-efficiency projects tied closely to or integral with the firm's production operations, many of these activities are performed by the industrial firm. In particular, an industrial firm interest in third-party financing will probably have already identified and evaluated the potential project. Moreover, because of its expertise and experience with appropriate technical contractors, the firm may have strong and value reasons for performing the detailed engineering audit and managing construction. In this case, the third-party investor needs to validate the economic viability of the project as a basis for negotiating a specific ownership, operating, risk-sharing, and investment return arrangement.

In any case, compared to leasing and typical joint ventures, the investor is much more involved in detailed project evaluations and subsequently in operations, particularly if the shared-savings plan does not contain guarantees by the industrial firm that reduce the investor's risk.

**Financial Advantages**

Generally, a shared-savings plan should not result in fixed financial obligations for the industrial firm; as a result, it qualifies as off balance sheet, off credit financing. Because the user incurs no fixed obligations to make payments, he has not incurred any debt-equivalent liabilities. Consequently, no reporting should be required in financial reports.

From the standpoint of investors, bond rating institutions, and lenders, the payments under a shared-savings plan are operating expenses. Although the user might experience even lower operating expenses if he invested with his own capital, he has reduced his cost and essentially tapped a new source of capital through this approach.

A shared-savings plan does not require the industrial firm to commit cash to the investment. Generally, the investor provides 100 percent of the capital required for the energy-efficiency improvement.

As with a joint venture, if the specific shared-savings plan contains provisions in which the energy user firm guarantees payments to the investor, then the arrangement might not be strictly off balance sheet or off credit. The principal risk incurred by the investor is that the industrial firm may utilize the project at a rate lower than projected. For example, if a waste heat recovery project is financed through a shared-savings plan, the investor will earn a reduced return or even incur a loss if the industrial firm fails either to operate the source of the waste heat or to utilize the recovered waste heat at the projected level. To guard against this contingency, the investor may seek related guarantees from the industrial firm. If these obligations are noncancelable and unconditional, the firm will probably be required to report them in its financial statements. In this respect, their effect on the off balance sheet, off credit status of the financing arrangement is analogous to that of the unconditional purchase obligations that are often associated with a joint venture. Depending on the likelihood that these obligations will apply to the firm and their materiality, such unconditional obligations may affect the creditworthiness of the firm as assessed by astute investors, credit rating agencies, and prospective creditors. However, if the project is a sound cost reduction project in one of the firm's long-term, stable production facilities, the financing should not adversely affect the firm's access to other sources of capital.

A feasible way of structuring a shared-savings plan is to execute a safe harbor lease that provides 100 percent financing but provides that the level of lease payments is variable according to actual energy savings. A shared-savings plan could also be implemented using a finance lease, which would provide 100 percent "up-front" financing. As discussed, the requirement that any purchase price option be at least 10 percent of the cost of the property means users may have to incur a cost at the end of the lease term if the asset is limited-use property. Under these arrangements, the financing advantages would be the same as with the shared-savings plan. However, this structure would avoid virtually all the potential problems of the shared-savings approach under federal tax laws. In particular, the issues of whether the project involves limited-use property and whether the shared-savings plan is, in fact, a service contract and not a lease would be avoided (see Tax Implications).

### Risk Management

The shared-savings plan is similar to a joint venture with respect to risk management. However, there is one major difference: in a joint venture, the dollar distributions to the investor and energy user are made *after* deducting project costs and debt payments from the gross revenues assigned to the project; in a shared-savings plan, the dollar distributions are made *before* deducting project expenses and debt payments. The investor pays these expenses from his share of the gross dollar value of energy savings realized from the project and bears all the risk associated with uncertainty about project capital cost, construction schedule, and project operating and maintenance costs. Other risk areas that relate to uncertainty about the gross value of the energy savings are shared by the energy user and the investor according to the negotiated shared-savings formula. These risk areas include uncertainty about technical performance, project utilization rate, energy-saving measurement and value, and unforeseen contingencies.

In view of this risk profile, the investor typically bears more risk in a shared-savings plan than in leasing and joint ventures. Accordingly, for a given investment opportunity, the investor will probably require a greater expected return. In turn, the industrial firm will probably retain a smaller share of the expected dollar value of the energy savings.

As in the case of joint venture financing of energy-efficiency improvements, many possibilities exist for varying the basic structure of a shared-savings plan to achieve a different risk distribution. Generally, the risk management mechanisms applicable to the joint venture arrangement are also applicable to a shared-savings plan. In addition, a number of mechanisms are appropriate specifically for a shared-savings plan.

One risk management mechanism that bears particular scrutiny is the possibility of varying the sharing formula over time to increase the probability that the investor breaks even on his cash investment within a relatively brief period of time (e.g., 2 years). Specifically, the share of the dollar value of energy savings that the investor receives is set at a relatively high level in the early years of an agreement and then decreases after, say, the second or third year. This approach can significantly reduce the investor's risk.

Other pertinent risk management mechanisms include:

- Sliding-scale formula that increases the investor's percentage of savings as the rate of project utilization decreases from the planned rate. This procedure shifts to the industrial firm a greater share of the risk associated with an uncertain rate of facility operation.
- Minimum payment obligation by the industrial firm to reduce the risk of uncertain revenues to the investor.
- Sliding-scale formula that varies the investor's percentage of savings as the growth rate of energy prices exceeds or falls below a pre-negotiated target rate of increase. As the growth rate in energy prices exceeds the target growth rate, the investor's savings share is reduced and vice versa. This arrangement reduces the risk associated with uncertain energy prices.

### Tax Implications

Under ideal conditions, the shared-savings plan involves no unusual tax restrictions or special treatment. The third-party investor is considered (1) to be the owner of the equipment for tax purposes and, as such, is allowed the associated tax benefits, including ITC, cost recovery, and interest deductions, and (2) to have entered into a valid contract to furnish specified services to the industrial user. The user is considered to have received a reduction in operating expenses equal to the share of the savings received from the third-party investor. Unlike leases, shared-savings plans do not require the energy user to incur an obligation to make fixed payments for the use of energy-saving equipment, regardless of operating results. The user is obligated to pay only if the savings are achieved.

In practice, however, tax laws can potentially impose limitations on both the investor and the user. From a legal standpoint, these arrangements must qualify as service contracts if they are to avoid treatment as leases and hence be subject to all of the restrictions that apply to leases.

The key criteria that distinguish a service contract from a lease have been set out by the United State Court of Claims in a recent case, *Xerox Corporation v. United States*, 656 F.2d 659 (1981). In this case, Xerox Corporation claimed that copying machines furnished to the U.S. government were provided as part of a service contract and not leased. The dispute arose because IRS had disallowed an ITC on the machines claimed by Xerox on the grounds that equipment leased or used by the federal government is not eligible for an ITC under Section 48(a)(5) of the Internal Revenue Code.

In holding that the machines were furnished as part of a service contract and were not leased, the court identified the following factors as critical to its decision:

- The taxpayer's retention of property ownership.
- The taxpayer's retention of possession and control of the property.
- The taxpayer's retention of risk of loss.
- The taxpayer's reservation of the right to remove the property and replace it with comparable property.

In addition to these key "service arrangement" factors, the court relied on a number of other findings in its determination. These included:

- The amount paid by the government was largely determined by the number of copies made; thus, the customer paid for an end result rather than for the use of a machine for some duration.
- Repairs were made and replacement equipment provided at no additional charge to the customer.
- Government personnel were not allowed to alter or move the machine without Xerox's permission.
- Xerox incurred additional economic risk because the agreement could be cancelled on 15 days' notice.
- "Retrofit" improvements were installed on existing machines even if they were operating satisfactorily.
- Like-for-like exchanges were made when a machine could not be repaired.
- Although the government could deny access to the plaintiff's equipment, there was no evidence that such access was denied.

Although these service factors were important in the trial court's holding, the court of claims panel that reviewed the case on appeal emphasized one factor above all others:

> ...this is *not* a case in which the cost or value of the machine itself overwhelmingly dominates the price of the total arrangement. It appears (using the instances of two of the most important machines) that the cost of a separate full maintenance agreement would be from 26 percent to 36 percent of the cost of the full copy service, but this maintenance charge does not include training, risk of loss, retrofits (of which there were a goodly number), like-

for-like exchanges and the other aspects of full service. If these other aspects of full service (additional to maintenance) were included, there would be a substantial additional percentage of the total cost of the arrangement which is not attributable to the cost or value of the machine itself.

Obviously, any determination of whether a shared-savings plan satisfies the foregoing criteria depends on the specific plan. On the other hand, some general observations can be made.

A shared-savings plan has only one key element in common with the agreement in the *Xerox* case: the user is obligated to pay only for end results. Unless otherwise negotiated, no payment is due for use of the equipment if no savings are produced or if savings fall below projections.

Although there is a strong indicator of a service contract, rather than a lease, it probably is not sufficient, on its own, to ensure service contract status. If the shared-savings contract calls for the installation of a large amount of expensive capital equipment and the cost of servicing that equipment is a relatively small percentage of the total contract cost, or if the operation and maintenance responsibilities are turned over to the energy user, problems could develop. Undoubtedly, IRS would argue that, in such a case, the cost or value of the equipment "overwhelmingly dominates" the price of the total arrangement and that the service element is immaterial. If operation and maintenance were the user's responsibility, IRS would undoubtedly argue that both possession and use had been turned over to the user. These two key elements indicate the existence of a lease rather than a service contract.

To increase the probability that the shared-savings plan will be treated as a service contract, as many of the following elements as possible should be included:

- Unrestricted right of access to the site and equipment.
- Right to replace and/or add new energy-saving equipment to maximize total savings.
- Operation and maintenance of the equipment by persons employed by the investor group.
- All risk of loss assumed by the investor, except for gross negligence or willful damage by user.
- No charges to user of any kind during equipment down time.
- All taxes and insurance costs paid by owner/investor.
- No "bargain" purchase option by user.

Even if all these provisions are included in the shared-savings agreement, there can be no assurance that IRS or the courts will treat the transaction as a service contract rather than a lease.

Of the various IRS rulings reviewed by the court in the *Xerox* case, probably the most favorable for shared-savings plans is letter ruling 7913003 (November 28, 1978), in which IRS ruled that a pipeline constructed and operated by a taxpayer to transport natural gas from a tax-exempt organization's gas field to its electric-generating plant could qualify as a service contract. The key standard identified by IRS in that ruling was "whether the owner of the property utilizes such property to provide a service to another party who uses such property to provide service to itself". This letter ruling is particularly significant because the property involved appears to have been limited-use property; that is, it would have little value at the end of the service contract other than as salvage since the pipeline was specifically constructed to carry gas from one gathering field to one generating plant. If the generating plant shut down or the gas fields ran dry, there would be little or no alternative use for the pipeline.

It is difficult to conclude on the basis of one letter ruling that shared-savings arrangements involving limited-use property will, in fact, be treated as service contracts rather than leases. There is no indication in the letter ruling or in the court's discussion whether operation and maintenance costs were a relatively significant part of the cost of the total arrangement. It is also possible that the tax-exempt organization had no personnel who would be capable of operating the pipeline in the absence of personnel supplied by the taxpayer.

Although neither IRS nor the courts appear to have ruled on shared-savings plans, somewhat similar arrangements have been used by the gas industry for over 30 years without IRS challenge. For example, companies that specialize in furnishing industrial gases to firms have routinely constructed air separation units at industrial sites to provide oxygen and other industrial gases to industrial users. A large proportion of the equipment installed in such industrial gas units is nonrecoverable; as a result, IRS would construe it as limited-use property. Moreover, it is generally not feasible to transport the output of these plants (i.e., oxygen or nitrogen) to other industrial users if the original industrial user no longer needs the output. In addition, the labor component of operation and maintenance costs is a fairly small percentage of total costs because the highly automated plants are designed to run for relatively long period without maintenance.

On the other hand, the industrial gas companies generally perform the necessary operation and maintenance and assume full responsibility for providing a reliable supply of product. In this respect, the arrangements differ from the typical shared-savings plan. Normally, the industrial gas supplier guarantees delivery of the amount of gas contracted for even if the air separation plant is shut down for repairs. To fulfill this commitment, the industrial gas supplier generally incurs substantial additional costs for trucking in liquid oxygen or nitrogen to meet the industrial user's needs during these periods.

In this respect, the commitment taken on by the industrial gas supplier to provide backup liquified gas in the event that its plant out of operation is similar to Xerox's commitment to provide backup copying machines in the event that the machines provided to the government could not be repaired within a short time. On the other hand, a shared-savings plan normally does not provide for "guaranteed" energy savings; the investor has no obligation to pay the industrial firm if the improvement fails to provide the projected level of savings. In short, the fact that IRS has not challenged such gas contracts is perhaps some indication, but no assurance, that shared-savings plans will not be challenged as leases.

A final problem is the manner in which the investment is valued for tax reporting purposes. Specifically, if the value of the equipment is simply set at its installed cost, no problem with IRS is likely. However, some syndicators have followed a valuation approach that yields a much higher value and, therefore, a larger depreciation base for the investor. Briefly, this approach entails basing the value of the investment on projected escalated energy savings and on the projected cost of money over the life of the agreement. It is not yet known whether such an approach will stand up under IRS scrutiny. Moreover, the user has an important stake in the valuation method and the source of financing: avoiding potential tax liens.

### Operating Arrangement

Under the shared-savings plan, the investor receives a return only if the equipment functions efficiently and achieves the projected savings. As a result, he is strongly motivated to ensure continued operations and proper maintenance.

Ideally, the investor can provide a range of services to the user. These services can include any combination of the following: audits of the plant to identify prospective projects; economic and engineering evaluations; architectural and engineering design; equipment procurement, construction, and project management; ongoing servicing and maintenance; and financing. In practice, the investor can pursue a number of approaches to ensure proper functioning of the equipment. depending on the type of equipment and the location, the investor may provide the necessary service directly. Alternatively, the investor can contract with a local firm to handle service and maintenance.

Some industrial firms may prefer or even require that their personnel maintain and service the equipment. In these cases, the user and the investor may negotiate service contracts, under which the investor compensates the user for the maintenance services provided. However, maintenance services by the industrial firm may jeopardize the investor's ability to have the agreement characterized by IRS as a service contract and not a lease. User of a safe harbor or finance lease as the basic structure for the shared-savings arrangement would eliminate this concern over IRS interpretation.

Under any of these operating arrangements, however, it is important to recognize that the investor will be more actively involved in ensuring efficient equipment performance than would be the case under a normal leasing arrangement. The investor will also be actively involved in the user's operations in calculating the energy savings. This calculation involves a number of steps. First, a baseline consumption reflecting energy consumption prior to equipment installation must be established. Although the methods employed to estimate this baseline will be affected by the specific facts surrounding a project, multiple regression techniques have proven to be effective and acceptable to both parties. These techniques can be used to establish a baseline for a range of production levels and weather factors; in most cases, they can account for 97 to 98 percent of consumption.

On a monthly basis, the user must provide the investor, or a selected independent party, with utility bills or other measures of production output or fuel utilization. Using the baseline for the month and comparing it with actual consumption, the firm can calculate energy savings (e.g., therms of gas, kWh). The savings are then multiplied by current energy prices, and the user is billed for the investor's percent of the cost savings.

Maintaining an updated, accurate measurement of energy savings requires effective communication between the user and the investor. The user must inform the investor of any changes (e.g., shift changes, process modifications) that materially affect energy consumption. A new baseline must then be calculated to reflect such changes.

To date, most shared-savings plans in industry have involved equipment tangential to the main production process, where an equipment failure does not have an immediate impact on production. When the shared-savings plan is applied to the mainstream production process, the savings realized may extend beyond energy. For example, raw material consumption may be reduced or manpower productivity improved. Estimating these savings, which may be substantial in relation to energy savings, may nonetheless be more problematical; moreover, the third-party's involvement in the user's operations will increase.

## ENERGY SERVICES CONTRACT

In many respects, an energy services contract is similar to a shared-savings plan. However, it differs in two important ways:

1. The investor contracts to provide specified services at an agreed-upon price, either in aggregate or per unit of service provided, rather than tying the price to actual energy savings.
2. The investor is solely responsible for project operation and maintenance.

The energy services contract is generally used as a vehicle for providing all energy requirements for a commercial building. In these cases, the third-party investor (generally a corporation organized for the express purpose of energy service contracting) agrees to purchase, install, operate, and maintain the heating, ventilation, and air conditioning (HVAC) equipment, lighting equipment, and other components of the building's energy systems. Normally, the energy services company also assumes responsibility for paying all utility bills. In return for these services, the building owner or energy user generally pays a flat annual fee (e.g., 90 percent of the energy costs it incurred in the preceding year). The user is thus effectively guaranteed some level of energy cost savings, regardless of equipment performance. To earn its return, the investor must produce the energy services at a cost less than the fee charged the user.

An energy services contract typically has a duration of 7 to 10 years. At the end of this period, the energy user can renew the contract or, if the contract permits, purchase the energy-efficiency improvement.

This approach follows the same basic pattern for an industrial energy-efficiency improvement project, except that the contract is usually limited to a specific project rather than a firm's total energy requirements. As in the shared-savings plan, the industrial firm may perform many of the actual design, engineering, and construction activities and turn the project over the third-party investor upon completion. Under the energy services contract, a price per unit of service provided, rather than the overall fixed fee, is usually established, to reflect the more variable energy requirements of an industrial operating facility compared to a commercial building. Upon completion of construction, the third-party investor operates and maintains the facilities to provide the specified services.

### Financing Advantages

Energy services contracts can enable the industrial firm to implement energy and other cost-saving projects without adversely affecting its access to other capital. Moreover, such a contract can provide 100-percent financing.

If the energy services contract requires the industrial firm to purchase services on an as-needed basis only, the financing is clearly off balance sheet, off credit. The third-party is more likely, however, to specify a minimum level of service that the firm must purchase, whether actually taken or not. In this case, the firm incurs a contingent liability that may have to be reported, particularly if the potential effects of the contingency are material. In this respect, the energy services contract is analogous to a commercial fuel supply contract. A natural gas supply contract, for example, typically includes minimum quantities of gas that the industrial firm must either take or pay for. Liabilities of this sort, which are integrally tied to the ongoing operations of the production facility, are normally not reported in the firm's financial statements unless some unusual problem that threatens to affect the firm adversely has arisen.

Under normal circumstances, it appears to be appropriate to treat the energy services contract like any other energy supply contract for reporting purposes. To the extent that the contract provides an economic supply of energy or other important production services to the firm, it will serve, if anything, to enhance the firm's creditworthiness.

In short, properly executed energy services contracts can provide off balance sheet, off credit financing, thereby expanding the firm's total access to capital. Moreover, because the energy services contract arrangement typically involves 100 percent financing by the third-party investor, it allows the industrial firm to conserve its cash for other uses.

### Risk Management

In its purest form, the energy services contract shifts most of the project risk to the investor. In turn, the investor requires a commensurate return. Unless risk-sharing mechanisms are included in the contract, the industrial firm's share of the project savings is smaller than it is under the other three financing approaches.

The investor in industrial projects tied closely to the firm's production operations will seek risk-sharing in three areas: project cost and construction schedule; level of equipment utilization (translated into the level of service purchases); and technical performance of the equipment. If the industrial firm, which is more familiar with these risks, assumes them, or at least shares them with the investor, a more favorable arrangement for both parties can result.

In the case where the industrial firm prefers to play an integral role in project design, engineering, and construction, an acceptable arrangement for managing construction cost and schedule risks might be to have the firm assume the obligation to deliver the project to the investor at a fixed price. Alternatively, the agreed-upon price for services can contain an adjustment to reflect construction cost increases over those originally projected. If the investor is responsible for construction, it can still adopt this latter risk-sharing technique, perhaps within prescribed limits.

The technical performance risk can be shared in a number of ways. If the industrial firm is responsible for construction, the parties can agree that the project must perform according to a specified set of criteria before the investor assumes control. Alternatively, the agreed-upon price can be subject to adjustment, within specified ranges, to reflect actual technical efficiency. However, after the initial phase of operation is complete and efficiency expectations have been determined, the investor normally bears the risk of subsequent technical performance of the equipment.

The third area where risk-sharing may be critical is in protection against lower-than-forecast utilization of the services by the energy user. As under the joint venture and shared-savings approaches, the most straightforward mechanism is for the firm to commit to an unconditional purchase obligation at a specified level (i.e., take-or-pay contract). The pricing formula variants discussed under the shared-savings approach may be an effective means of sharing this risk.

Future energy prices may also be an area for risk-sharing, depending on the specific nature of the project. If the the project depends on the industrial firm for its base source of energy supply (e.g., waste heat, combustible gases), such risk-sharing is not needed. However, if the project involves purchasing energy on the open market, it would be appropriate to share the risk of future energy price movements.

### Tax Implications

The legal and tax issues associated with energy services contracts are similar to those associated with shared-savings plans. However, the results, particularly on the lease vs. service contract question, may be different because the transactions are structured somewhat differently.

A full energy services contract covers all the energy needs of the industrial firm. However, such a contract is inappropriate for financing a specific process modification to improve energy efficiency. Instead, a hybrid contract covering only the specific project must be developed.

To understand the tax implications of hybrid agreements, it is necessary to understand the tax issues associated with the full energy services contract.

Three distinctive features of a full energy services contract make it likely that the transaction will be treated as a true service contract rather than a lease:

1. If it covers a facility's total energy needs, the operation and maintenance component of the transaction probably represents a significant portion of the cost of the total arrangement. Unlike a shared-savings plan, where the investor's responsibility is limited to specific pieces of energy conservation equipment, the responsibility of the energy services company under a full agreement extends to all energy-using equipment in the user's facility.
2. The fact that the energy services company assumes complete responsibility for paying energy bills, regardless of whether they exceed past levels, demonstrates that it assumes the risk of providing specific energy services regardless of its own "costs of production". On the other hand, under the shared-savings plan, if no savings are realized, the investor has no further responsibility to the energy user. For example, if energy costs after installation of equipment are higher than previous costs, the energy user must bear the increase.
3. Most full energy services contracts are open-ended with regard to the type of energy-saving equipment to be installed. Unlike a standard shared-savings plan, which specifies clearly what equipment will be installed, the energy services contract simply gives the energy services company the right to install various types of energy-saving equipment as well as to institute operating and management practices designed to save energy. In other words, the contract does not focus on equipment per se but allows the energy services company to use a wide range of methods to reduce energy consumption. Under these circumstances, it would be difficult for IRS to argue that the energy services company is simply providing energy conservation equipment to enable the user to provide its own services.

   In contrast, under a typical shared-savings agreement, the savings are generated from specific items of equipment installed by the investors and there is little or no opportunity to generate additional savings by altering operation and maintenance practices elsewhere in the facility.

A hybrid energy services/shared-savings approach can be adapted to projects where the investor recovers waste energy, meters it, and sells it back to the firm. Under this arrangement, the investor and the firm first estimate the savings that will result from the project and agree on the share for the investor. The price at which the investor sells the recovered energy back to the firm is set to reflect this agreed-upon share of savings. This arrangement is feasible where, for example, heat or a combustible gas is being recovered. Although such an arrangement involves the sale of some form of energy to the user, it stops short of a full energy services contract in that the investor typically does not assume responsibility for furnishing all required forms of energy to the user.

From a tax and legal standpoint, hybrid energy services contracts probably fall somewhere between a shared-savings plan and a full energy services contract. Unlike the shared-savings plan, a physical commodity is actually being metered and sold to the energy user, and the user pays only for the energy delivered. Under a typical shared-savings plan, the savings are measured by comparing the amount of energy actually used. No product is actually sold to the energy user; the savings are only an estimate developed by comparing a baseline with current energy usage.

On the other hand, hybrid agreements do not require the third-party investor to provide a backup source of energy if equipment is down. Under a full energy services contract, the outside company is required to maintain a given level of service to satisfy the customer's needs regardless of the ability of its own equipment either to provide the service or to provide the service at a cost that ensures a profit. Clearly, the outside company assumes a much greater risk in this case.

For these reasons, hybrid energy services contracts are more likely than shared-savings plans but less likely than full energy services contracts to be accorded service contract status by IRS. Once again, other factors, such as the ratio of operating and maintenance to capital costs, will weigh heavily in any IRS or court determination.

**Operating Arrangements**

A key feature of the energy services contract approach is that the third party is directly responsible for operating and maintaining the equipment. Under the other approaches, the industrial firm is more likely to perform operation and maintenance.

Because equipment operation directly affects production operations, procedures to ensure close coordination and compatible operation are required. In particular, agreement must be reached on such critical practices as production scheduling, preventive maintenance policy and scheduled maintenance, repairs, production quality, and contingencies. Moreover, labor relations and the firm's policies regarding its workers' union, where applicable, can become important factors in negotiating acceptable operating arrangements. Protection of proprietary technology and operating practices and data may also be of concern to the firm.

The specific details of an operating arrangement will, of course, depend on the characteristics of the project, the firm's policies regarding outside involvement in its operations, and the firm's confidence in the day-to-day operating dependability of the third party.

# Appendix D-4 Tax and Cogeneration

## MAJOR TAX CHANGES FOR COGENERATION AND SMALL POWER PRODUCTION FACILITIES

by Lee Goodwin
Reid & Priest

On September 28, 1986, the Senate passed the compromise version of the Tax Reform Act of 1986. The Tax Reform Act makes major changes in the treatment of cogeneration and alternative energy projects, and many of the significant tax advantages currently enjoyed by the cogeneration and alternative energy industries are lost under the Act. This article summarizes the key provisions of the Tax Reform Act for cogeneration and alternative energy projects.

### TAX RATES

The Tax Reform Act significantly reduces tax rates. Under current law, the top individual tax rate is 50%, while the top corporate tax rate is 46%. The Act custs the top individual tax rate to 38.5% in 1987 and 28% beginning in 1988. The top corporate rate is reduced to 34% for tax years beginning on or after July 1, 1987, with a blended rate for tax years beginning after July 1, 1986 and before July 1, 1987. The revenue loss caused by these reductions in rates is made up by significant reductions in tax benefits for both individual and corporate taxpayers.

### DEPRECIATION AND TAX CREDITS

Under current law, most cogeneration and alternative energy projects which are Qualifying Facilities under the Public Utility Regulatory Policies Act of 1978 (PURPA) are included in the 5 year ACRS category. Utility owned generating projects are included in the 10 and 15-year categories, while cogeneration and alternative energy projects which are classified as "building components" are included in the 19- year real property category. Rates for property in the 5-, 10-, and 15- year categories are based on the 150% declining balance depreciation method.

Under the Tax Reform Act, depreciation for most cogeneration and alternative energy projects is substantially less favorable than under current law. Many cogeneration projects, including coal-fired cogeneration projects and gas-turbine projects operated in a combined-cycle configuration, are included in the 20-year depreciation category, with depreciation rates based on the 150% declining balance method. Hydroelectric projects and utility-owned biomass and waste-to-energy projects are also included in the 20- year category. Diesel cogeneration projects and single cycle gas turbine cogeneration projects are included in the 15- year category, with depreciation also based on the 150% declining balance method. Large scale (greater than 500 kW) industrial cogeneration and alternative energy projects whose output is for the owner's internal consumption, and not primarily for sale to third parties, are also included in the 15- year category. For cogeneration and alternative energy projects which are classified as building components, depreciation is based on the straight line method, over a 27.5 year life in the case of residential rental property, and a 31.5 year life in the case of other real property. These new depreciation rules generally apply equally to utility and non-utility projects.

Special rules are provided for non-utility solar, wind, geothermal and ocean thermal projects, and for biomass projects which are qualifying small power production facilities under PURPA. These projects are included n the 5-year category, with depreciation rates based on the 200% declining balance method. For these types of assets, the Tax Reform Act actually provides more favorable depreciation than is available under current law.

## INVESTMENT TAX CREDIT

Under current law, a regular investment tax credit of 10% is allowed for investments in many items of business property, including most cogeneration and alternative energy projects. The Tax Reform Act repeals the regular investment tax credit for most property placed in service after December 31, 1985, although transition rules are provided, and carry-forwards from prior years may still be used after 1985. However, for tax years beginning after July 30, 1986, the Act reduces the regular investment credit for both unused credits which are carried forward from prior years and credits allowed for transition period property. The regular investment tax credit allowable for a taxable year beginning on or after July 1, 1987 is reduced by 35%, and a prorated reduction applies to credits allowable for a taxable year beginning between July 1, 1986 and July 1, 1987. The amount by which the credit is reduced cannot be used as a credit for any other taxable year. In addition, for property placed in service after December 31, 1985, the property's depreciable basis must be reduced by the full amount of the investment credit allowed for the property (after application of the credit reduction described above).

The Act extends the biomass energy tax credit for two years and the solar, geothermal and ocean thermal energy tax credits for three years, all at declining rates. Under prior law, all of these credits expired at the end of 1985. Although other energy tax credits were not extended, the existing affirmative commitment rules for certain other types of energy property, including hydroelectric projects, were retained. However, energy credits claimed pursuant to these affirmative commitment rules are subject to the the percentage reduction and basis adjustment rules described above with respect to the regular investment credit.

## EFFECTIVE DATES

The Depreciation provisions of the Tax Reform Act generally apply to property placed in service after December 31, 1986, and the repeal of the investment tax credit is generally effective for all property placed in service after December 31, 1985. However, the Act provides transition relief for a variety of items, including:

1. Property acquired, constructed, or reconstructed pursuant to a written contract that was binding as of March 1, 1986 (December 31, 1985 for purposes of the investment credit);

2. Property constructed or reconstructed by the taxpayer is the lesser of $1,000,000 or 5% of the cost of such property has been incurred or committed by March 1, 1986 (December 31, 1985 for purposes of the investment tax credit), and if construction or reconstruction of the property has begun by that date;

3. Property which is part of a project certified by FERC as a qualifying facility on or before March 1, 1986. The legislative history expressly states that a project which has sought self-certification – i.e., a project which "a developer has simply put FERC on notice is a qualifying facility: – is "not certified as a qualifying facility" for purposes of this rule.

4. Property which is part of a project which was granted a hydroelectric license by FERC on or before March 1, 1986;
5. Property which is part of a hydroelectric project of less than 80 MW, for which an application for a preliminary permit, a license or an examption from licensing was filed with FERC on or before March 1, 186; or
6. Property necessary to satisfy a service contract or lease which was binding on March 1, 1986 (December 31, 1985 for purposes of the investment tax credit). While electric power sales contracts qualify as service contracts, it is not clear whether this exception will apply to typical PURPA-based power sales contracts, since such contracts generally do not impose a penalty on the developer if no power is actually delivered.

The transition rules for depreciation do not apply to any property unless the property has an ADR (asset depreciation range) midpoint life of seven years or more and is placed in service before January 1, 1989, in the case of property with an ADR midpoint life of less than twenty years, and January 1, 1991, in the case of property with an ADR midpoint life of twenty years or more. Most cogeneration and alternative energy projects which had any ADR life had a life of 20 years or more. For purposes of the investment tax credit, the applicable placed in-service dates are:

1. For property with an ADR midpoint life of less than five years, July 1, 1986;
2. For property with an ADR midpoint life of at least five years, but less than seven years, January 1, 1987;
3. For property with an ADR midpoint life of at least seven but less than twenty years, January 1, 1989;
4. For property with an ADR midpoint life of twenty years or more, residential real property, and nonresidential real property, January 1, 1991.

For transition period projects, current law ACRS depreciation deductions and the reduced investment tax credit would remain available. Property which qualifies for transition relief may generally be transferred to a third party before it is placed in service, or may be sold and leased back within 90 days of being placed in service, without losing the benefit of transition period status.

## MISCELLANEOUS

The Tax Reform Act makes other changes to the tax law, such as changes to the minimum tax and tax exempt finance provisions, which could also have a significant impact on cogeneration and alternative energy project development. Of particular interest is the limitation on the deduction of passive losses by individual taxpayers. Under this provision, passive losses, including losses of partners in limited partnerships which invest in cogeneration and alternative energy projects, can only be deducted against income from passive investments, and cannot be deducted against income from other sources. Similarly, tax credits (including the investment and energy tax credits) can only be claimed against the tax on passive income. Losses and credits which cannot be used in the year in which they are incurred must be carried forward, and can only be used to offset income from, or tax on, the investment itself or other passive investments. This provision applies to losses and credits from any source, including losses and credits from transition period projects and other activities commenced before the consideration of the current round of tax reform legislation. However, for investments made before the date of enactment, the provision is phased in, so that it only applies to 35% of losses in excess of passive income for tax years beginning in 1987, 60% in 1988, 80% in 1989, and

90% in 1990. Beginning in 1991, all losses are covered. This proposal could significantly alter the pattern of capital formation for all types of capital investments, including energy projects, and could diminish the benefit of many of the transition rules.

## IMPACT OF TAX REFORM ON COGENERATION AND ALTERNATIVE ENERGY PROJECT DEVELOPMENT

At this point, it is obvious that the Tax Reform Act deprives the cogeneration and alternative energy industries of many important tax advantages that they enjoy under current law, it is too early to tell the ultimate impact of tax reform on cogeneration and alternative energy project development. Nevertheless, there are several general observations which can be made:

### The impact of tax reform varies depending upon the type of technology involved.

The Tax Reform Act can be expected to have the greatest impact on capital-intensive cogeneration and alternative energy projects, such as hydroelectric and coal-fired projects, for which capital costs are a major portion of overall project costs. By contrast, for projects for the capital costs are low relative to operating costs, such as diesel or gas turbine cogeneration projects, the impact can be expected to be much less significant. In fact, depending on the relationship between effective income tax rates and capital cost recovery allowances, certain cogeneration and alternative energy projects could actually be more attractive after tax reform than they are now. The impact of tax reform on particular technologies will also be affected by the extension of the energy tax credit and other incentives for certain alternative energy technologies, such as biomass and renewable energy facilities.

Cogeneration and alternative energy projects are generally worse off under tax reform than comparable capital investments.

One of the novel features of ACRS, when compared with prior depreciation systems such as ADR, is that it provides a uniform 5- year depreciation period for all capital investments, regardless of their actual economic useful life. Thus, under ACRS as it was originally enacted, most types of capital investments do not enjoy a tax advantage relative to other competing types of capital investments. However, the Tax Reform Act returns to the concept of useful lives as a measure of allowable depreciation. Accordingly, cogeneration and alternative energy projects, which have traditionally been assigned longer useful lives than comparable investments, have been assigned longer depreciation periods than other unregulated capital investments. As a result, the Tax Reform Act will make these cogeneration and alternative energy projects relatively less attractive, compared with other capital investments.

### Lower rates and a stiff minimum tax diminish the overall significance of all tax benefits

Much concern has been expressed regarding the impact of extended depreciation periods on cogeneration and alternative energy projects. However, this concern should be placed in context with the impact of reduced tax rates. Under current law, each dollar of tax deduction is worth $.46 of tax savings to a corporate taxpayer, and $.50 of tax savings to an individual taxpayer. The Tax Reform Act reduces this to $.34 and $.28, respectively. A stiff minimum tax could further reduce this savings. Because the reduced tax rates will reduce the benefit from all tax incentives, all projects, including those which have retained more favorable depreciation rates, will have to rely more heavily on actual economic return, and less on tax incentives to attract investors.

**The ultimate impact of tax reform depends on many factors not apparent from the face of the final legislation.**

The relatively generous tax benefits available for capital investments under ACRS have produced an investment climate in which high after-tax returns have become the business norm. It is not possible to predict the ultimate impact of tax reform on capital formation until it can be seen whether the capital markets will adjust their expectations to accept lower after-tax returns. If such an adjustment occurs, the loss of tax benefits may not be a significant impediment to the development of cogeneration and alternative energy projects.

## THE END OF COGENERATION

by Scott Spiewak

Two developers approach a utility which is in need of new generating capacity, and which seeks power from outside sources. One developer plans a cogeneration facility, and is offering to deliver electricity at an average of five cents per kWh. The other developer plans a facility which is not a cogenerator, but offers to sell electricity at an average of four cents per kWh. Which wins?

Prior to the advent of the new tax law, this scenario didn't arise. It couldn't because the "nonqualifying facility" was under an impossible handicap – it was treated as "public utility property" under the tax code, which meant that it received no tax credits, and was depreciated over ten years (gas turbines), or even fifteen years (steam turbines), while the "qualifying cogeneration facility" received a 10% tax credit and was depreciated over five years. Return on investment for a qualifying facility could run twice as high as it would for a nonqualifying facility. As an economic matter, a nonqualifying facility never had a chance.

The tax bill changes all that, creating the "level playing field" which utilities had asked for since PURPA was passed. While this is unlikely to aid electric utilities to regain their position as the builders of new powerplants, it *is* likely to very quickly end the dominance of cogenerators in new powerplant construction.

### The economic value of cogeneration

When PURPA was passed, cogeneration was looked upon as an efficient technology which could reduce the use of fossil fuels. As oil prices fell, the reality was that the efficiency of cogeneration was almost irrelevant. It is not uncommon for the cogeneration developer today to offer extremely low rates to steam purchasers in order to fulfill the "bona fide useful thermal output" requirement for qualifying status. In many cases, steam was literally given away for this purpose. Obviously the economic motive was the opportunity to sell electricity profitably.

As I wrote in the March 1986 issue, the need for new generating capacity is driving the development of new cogeneration projects today – not the efficiency of the technology. With $15 per barrel oil, efficiency is no longer terribly important.

Cogeneration will tend to go where new capacity is needed. However, in areas with high growth in electrical demand, there may not be commensurate opportunities to engage in steam sales. Without the steam, no other thermal application, there can be no "cogeneration". But there will still be a need for new powerplants.

### An opportunity for the electric utilities?

This would seem to create an opportunity for electric utilities to get back into the powerplant construction business. Yet that is unlikely. In the 1970's the utilities found themselves burdened with excess capacity due to the oil shocks and the recessions. State utility commissions, faced for the first time with rapidly rising electric rates, were forced, as a political matter, to find ways of keeping rate increases down. They couldn't do anything about fuel prices in most cases, but they could, and did start "cracking down" on utility "excesses" in construction. Unneeded generating capacity was simply "disallowed", in whole or in part, resulting in losses to the shareholders. This broke what had been a compact between the regulators and the utilities, under which utilities would, in the absence of gross behavior, be entitled to a reasonable return on their investment, but would not be permitted to ever make unusually high profits.

Under the rules as they appear to utilities today, powerplant construction, particularly for capital intensive projects, is a heads-I-win-tails-you-lose proposition. If the plant is constructed, and it is needed when it is ready to go into operation, a "reasonable" return on investment is allowed. If the plant isn't needed, the ratepayer will be protected from the "incompetents" who guessed incorrectly that additional capacity would be required some time down the road, and recovery will be disallowed.

These are exceptionally bad odds, and have led many utilities to seek to maximize shareholder value by placing their capital in cable TV, insurance companies and the like. We are starting to see what might become a wave of takeovers of electric utilities by nonutility firms anxious to use their fat cash flow for more lucrative purposes than acting as regulated power generators.

### The deregulation of wholesale power generation

Given that the capacity-short utility is not going to be anxious to build new plants itself, what will it do to obtain it? It will act like any other industrial firm in search of product – it will issue an RFP (request for proposals).

In the July 1986 issue I explained how utilities are moving away from the avoided cost concept toward one in which cogenerators engage in competitive bidding in order to obtain capacity sales contracts. In the August issue, Massachusett's PURPA implementation plan, which expressly relies upon such bidding, was described.

This brings us back to the question with which this article opened – two developers bid to provide generating capacity. One has a qualifying facility, one doesn't. The non-QF has a lower bid. What is there to stop the utility from accepting it and closing out the QF?

The major hurdle is that wholesale sales of electricity by non-QFs are subject to rate of return type regulation by the FERC. Theoretically, as an ordinary "utility" under the Federal Power Act, the bidder would have to submit its costs of generation to the FERC, and obtain an approved tariff for power sales based upon a reasonable return on investment. This is the same game that utilities engaging in retail sales are required to play at the state level.

However, FERC has shown a willingness to bend the rules requiring that it regulate electric rates, particularly when it results in increased competition.

For example, technically, small power producers with a capacity in excess of 30 MW are subject to rate regulation under the Federal Power Act. Yet in its 1983 Statement of Policy on PURPA Enforcement, FERC said:

"The Commission has determined that State-established rates which are consistent with the Commission's regulations will generally be accepted as the 'just and reasonable' rate for purchases by electric utilities from Federal Power Act jurisdictional qualifying facilities under Section 205 of the Federal Power Act." See Monthly, June 1983.

In other words, the Commission has since 1983 deferred t o the states to set the appropriate rate for utility purchases from qualifying small power producers over 30 MW, even though this is a matter of exclusive FERC jurisdiction under the Federal Power Act.

Is it likely that FERC would muck around with a State Commission-approved contract with a nonqualifying facility which beat the bids of all qualifying facilities?

### The End of Cogeneration

The "level playing field" utilities have lobbied for is a reality under the tax bill. But the results are likely to be different from those expected.

PURPA has been an interesting experiment in deregulation of power generation, and a very successful one. Yet it has led to identifiable economic inefficiencies, as developers of power projects reach to fit within the FERC qualifying status regulations. With the tax barriers down, these inefficiencies are going to become more obvious, and less tolerable.

Success in the cogeneration business has always required speed, in order to take advantage of short-lived windows of opportunity. The removal of the tax barrier to nonqualifying facilities is going to make the game faster and more competitive. Without the artificial benefit of tax relief, cogeneration projects of dubious economic efficiency will be unlikely to succeed against well-conceived nonqualifying powerplants.

It is unlikely that bureaucratic, overregulated electric utilities, will be able to compete against the nonregulated, competitive power generation companies which have been nurtured by PURPA. However, it is also unlikely that developers will continue to feel constrained to build only "qualifying facilities" now that the tax barriers are down.

Certainly there are a few applications in which serious revenue have been obtained from steam sales. But these have always been the exception, not the rule. The tax bill heralds the end of cogeneration, but the beginning of a more competitive, and ultimately more rewarding nonregulated power generation market.

## TAX REFORM AND AIR CONDITIONERS

by Scott Spiewak

The big sleeper in the tax reform bill is not related to its impact on capital investment, but rather its impact on consumption of electricity.

If electric consumption increases, then new generating capacity will be needed. The marginal increase in the cost per kWh caused by a change in the tax laws is not likely to change that fact.

Current utility demand forecasts generally project small increases in electric demand through the end of the century. These forecasts are based on what are called "econometric models", which attempt to explain the demand for electricity in terms of various economic indicators such as the projected level of electric prices, income, population and other factors. These models are typically used for more long-term forecasting of electric power demand. In the past decade, they have proven spectacularly unsuccessful in projecting the need for new generating capacity.

Thus, largely due to the oil price shocks and recessions of the '70s and early '80s, large increases in electric power demand predicted by econometric forecasting models failed to appear. Utilities which built new capacity, particularly those which built expensive nuclear powerplants, are generally having a devil of a time getting their respective state public utility commissions to allow them to boost rates to recover the cost of the in-retrospect, unnecessary construction.

Once burned, twice shy, electric utilities today tend not to want to build new plants. Because econometric forecasting is such a fuzzy area, depending as it does on predictions of economic indicators, it is a relatively simple matter for most utilities to justify "no-build" decisions by adopting a forecast which concludes that there will be no need for new capacity.

This is where "engineering" models come in. An engineering model, also called an "end-use" model, attempts to determine future levels of demand by modeling electric appliance saturation rates and utility load curves. Engineering models are far more useful for predicting short-term electric power capacity requirements, and an engineering model which diverges from reality will much more quickly brought to light.

For purposes of engineering models, air conditioners are extremely important appliances, because it is their use which results in the sharp spikes in utility system demand in the middle of the summer, and which often is directly responsible for the need for new generating capacity.

Tax reform reduces the penalty for consumer spending (except for those who itemize, and who will lose their consumer interest credit deductions), by reducing, or at the low end, eliminating the marginal Federal tax rate. Today a New York City plumber may pay as much as $.70 out of the last dollar earned in taxes. Tax reform is meant to help him by allowing him to keep, and spend, more of that last dollar.

On a hot summer day, a cheap, inefficient air conditioner is going to start looking good. If the plumber is feeling a little less put-upon by taxes, he is more likely to buy it. If electric prices have been moderating due to the oil price drop, the extra cost of an efficient unit will not be as attractive. This will not only increase overall demand, but will increase it during the system peak, creating aneed for new generating capacity.

The bottom line is this. While the depreciation and tax credit aspects of the Tax Reform Act will be critical in determining what type of powerplant to build, the reduced marginal tax rate may be the most important aspect of the bill, due to the impact it is likely to have on increased power consumption.

---

## THE IMPACT OF THE DEFICIT REDUCTION ACT OF 1984 AND THE "WALLOP AMENDMENT" ON ENERGY SALE CONTRACTS

**by Lee M. Goodwin**

The Deficit Reduction Act of 1984, as approved by the House and Senate tax conferees, contains a significant provision, popularly known as the "Wallop Amendment", which establishes a new rule for distinguishing "service contracts" from "leases" for tax purposes. It is critical that energy project developers understand this rule and ensure that their energy sale contracts comply with its requirements. The tax consequences of all alternative energy and cogeneration projects will be governed in large part by the extent to which these requirements are satisfied.

## BACKGROUND

Even before the Deficit Reduction Act, a major area of concern in structuring an alternative energy project was whether the contract for the sale of the electrical or thermal output of the project should be treated as a legitimate service contract or as a lease of property for tax purposes. If the energy sale contract is treated as a legitimate service contract for tax purposes, the project owner is treated as being in the business of producing and selling energy, and the owner's receipts from the sale of energy are treated as income from the active conduct of a service related business. However, if the energy sale contract is treated as a lease of property for tax purposes, the project owner is treated as if it had leased the energy project to the energy customer, and the owner's receipts are treated as rental income for tax purposes.

The question of whether an energy sale contract is a service contract or a lease for tax purposes is significant in a number of areas. If an alternative energy project sells its output to a governmental unit or tax exempt organization, and if the energy sale contract is characterized as a lease rather than a service contract, the project owner will lose any investment and energy tax credits which would otherwise be available with respect to the project, and the depreciation period for the project will be substantially extended. Even if the output is sold to a taxable entity, if the contractual arrangement is characterized as a lease, the non-corporate lessor rules would apply, and any individual or Subchapter S corporations that own a portion of the project could lose their share of any investment or energy tax credits that are available for the project. If electricity produced by an alternative energy project is sold to a public utility and the power purchase contract is characterized as a lease, the project could also be classified as public utility property for tax purposes even if the project itself is a qualifying facility under PURPA. This, in turn could cause any investors in the project to lose the benefit of the energy tax credit and five-year depreciation. The distinction between a service contract and a lease is also critical in numerous other areas of the tax law, since leasing transactions and rental income are generally treated less favorably than other types of business transactions for tax purposes.

## ENERGY SALE CONTRACTS FOR ALTERNATIVE ENERGY PROJECTS

The tax treatment of an energy sale contract does not automatically follow the characterization of the contract by the parties involved. On the contrary, the IRS has long taken the position that if a contractual arrangement more closely resembles a lease than an actual service contract in substance, it should be classified as a lease of property for tax purposes regardless of how the contract is characterized by the parties. Prior to the Deficit Reduction Act, service contracts, including contracts for the sale of electrical or thermal energy produced by cogeneration or alternative energy projects, were classified as service contracts or as leases on the basis of several judicially and administratively established criteria. These criteria included the allocation of operating responsibility for the project and the risk of loss with respect to the project.

The Deficit Reduction Act of 1984 contains a special rule, the "Wallop Amendment", which replaces the rules of prior law with a detailed list of statutory criteria which an energy sale contract must satisfy in order to qualify as a service contract and avoid classification as a lease for tax purposes. The amendment applies to contracts for the sale of electrical or thermal energy produced by a cogeneration or alternative energy facility. For purposes of the amendment, an alternative energy facility is a facility which derives no more than fifty percent of its energy from oil, natural gas, coal or nuclear power.

Under the Wallop Amendment, an energy sale contract will be classified as a service contract if it satisfies four specific statutory requirements.

First, the service recipient cannot operate of the facility. Accordingly, where the project owner does not want to operate the facility directly, it should arrange for an unrelated third party to operate and maintain the facility, and should not delegate the operation or the maintenance of the project to the energy customer or to a party related to the energy customer.

The second requirements is that the service recipient cannot bear any significant financial burden if there is nonperformance under the contract. Accordingly, contracts with minimum payment clauses, such as "hell or high water" power sale contracts, will not qualify as service contracts.

However, the mere fact that the energy customer may be required to purchase substitute energy at a higher price if the project fails to operate does not impose a significant financial burden on the energy customer within the meaning of this provision.

The third requirement is that the energy customer cannot receive a significant financial benefit if the operating costs of the facility are less than the standards of performance or operation contemplated in the contract. Under this provision, any increase or decrease in the operation and maintenance expense of the project must be borne by the project owner, and arrangements whereby the power purchaser either pays the entire operation and maintenance cost or shares that cost with the project owner, directly or indirectly, will not be permitted. Accordingly, contracts which pass these costs through to the energy customer, such as "net" power sale contracts, will not qualify under this rule.

The final requirement is that the energy customer cannot have an option or an obligation to purchase all or part of the project at a fixed or determinable price other than fair market value. Under this provision, discount purchase options and fixed price options at any price may not be included in a power sale contract. However, the contract can include an option to purchase the property either at its fair market value, as determined by appraisal at the time of the exercise of the option, or at a value determined pursuant to a formula which is reasonably expected to yield fair market value when the option is exercised.

The new rules added by the Wallop Amendment apply to contracts with tax exempt and governmental entities entered into on or after May 23, 1983, and to other energy sale contracts entered into after November 4, 1983. Contracts entered into prior to those dates will continue to be governed by the rules of prior law.

## OTHER ENERGY SALE CONTRACTS

The Wallop Amendment Only applies to contracts for the sale of energy produced by cogeneration and alternative energy projects. Energy sale contracts which do not qualify under the Wallop Amendment, and other types of service contracts, such as energy management contracts, will be governed by the general rules contained in the Act. Under these general rules, the determination of whether a contract is a service contract or a lease must be made by considering "all of the relevant factors". Six specific factors are set forth in the Act:

1. Whether the service recipient controls the property;
2. Whether the service provider bears any substantial risk of loss due to nonperformance;

3. Whether the service recipient has a significant possessory or economic interest in the property. (One way in which such an interest is acquired is by granting the service recipient the right to purchase the property's output under a long-term contract);
4. Whether the service recipient has physical possession of the property;
5. Whether the service provider concurrently uses the property to provide significant service to more than one customer; and
6. Whether the total cost of the contract substantially exceeds the fair rental value of the property.

The first two factors are similar to the criteria considered under current law, and are not difficult for alternative energy projects to satisfy. However, some of the other factors could present problems for alternative energy projects which do not qualify under the Wallop Amendment. For example, many components of industrial and commercial energy management systems must be located on the customer's premises, thus giving the energy customer "physical possession" of the equipment. Similarly, many alternative energy projects are highly capital intensive, and while the cost of the peripheral services associated with such projects may be substantial in absolute terms, those costs often represent a relatively low percentage of the total cost of the output of the project. Because the price for the output of such projects largely reflects the cost of capital to the project developer, it would therefore not differ substantially from the typical return on capital from the rental of a similar item of property.

Alternative energy projects could also have trouble satisfying other criteria set forth in the Act. For example, industrial and commercial energy management systems generally provide their services to a single industrial or commercial user. Thus, contracts for such projects could also fail to satisfy the criterion which is based on the number of customers served by the project. Similarly, in order to obtain financing for an alternative energy project, it is often necessary to enter into a long-term contract for the sale of the services provided by the project. However, under the Act, this would violate yet another of the criteria because it would be deemed to give the power purchaser an economic interest in the project.

The more of the criteria set forth in the Act which a contract fails to satisfy, the greater the risk that it will be classified as a lease rather than a service contract. Indeed, statements by Treasury officials have suggested that a contract could be classified as a lease even if it only fails to satisfy one or two of these criteria. Accordingly, the new rules for the classification of contractual arrangements will make it much more difficult for contracts for the sale of the output of alternative energy projects which do not satisfy the requirements of the Wallop Amendment to qualify as service contracts.

Alternative energy projects could also be adversely affected by another provision in the Act, which would apply the same six criteria set forth above to characterizing other arrangements (such as partnerships) as leases of property for tax purposes. The legislative history of the Act contains an example in which a partnership which leased space from a tax-exempt hospital in order to install a CAT scanner to be used to provide services to patients of the hospital was actually treated as having leased the CAT scanner to the hospital. This rule could create particular problems for certain types of contractual arrangements which are becoming prevalent in the alternative energy industry. For example, an increasingly common practice in the small scale hydropower industry is for a private developer to agree to develop a municipal dam to produce power for sale to a third party. The developer agrees to own and operate the project for a period of time less than the full useful life of the project, after which the developer agrees to transfer the project to the municipal entity for no charge. Under the Act, the developer could actually be treated as having leased the project to the municipal entity for the entire period of its useful life. This in turn would cost the developer most of the tax benefits associated with the project. However, because the municipal entity is not the energy customer, the Wallop Amendment would not apply, and its "safe harbor" relief would not be available. Accordingly, project developers should carefully scrutinize all of their contractual arrangements, and not just their energy sale contracts, in light of this new Act.

## CONCLUSION

The Deficit Reduction Act of 1984 will have a major impact on the structuring of alternative energy projects in coming years. Nevertheless, by paying careful attention to the new requirements set forth in that Act, the tax benefits associated with alternative energy projects will continue to be available.

# Section E

## Power Contracting and Procurement

# Section E

## Power Contracting and Procurement

The process of contracting for and procuring power supplies has often been referred to as "competitive bidding". This is a misnomer. Competitive bidding is a process typically used for procuring well-specified commodities, with which the only critical means of differentiation is price. Thus, for example, one might use a "competitive bidding" process for procuring No. 2 pencils. However, this is the incorrect approach to procuring powerplants. The success of a power procurement is dependent on a great many variables, only one of which is price. In fact, price itself has virtually no meaning outside the context of the quality of the powerplant and the contractual guarantees accompanying it. The consideration of these variables is a complex project by itself, but, if handled correctly, will substantially increase the likelihood the power purchaser will be satisfied with its choice.

In the first part of this section we discuss the problems states initially had in procuring supplies under PURPA. In essence, they were determining what it would cost them to make pencils and then offering that price to all comers. As the states and utilities gained more experience, they learned that it would be better to let the market decide the best price.

They also learned something else: procuring power supplies is a multi-faceted process. Price is important, but so are many other factors. Hence, the second part of this section deals with what a prospective purchaser should look for in selecting a supplier.

While the language in the second part of this section is geared toward utility-sponsored procurements, it is equally applicable to any large industrial or commercial user who wishes to pursue the self-generation option, or even power marketers seeking to add a new power plant to their resource portfolios.

### PROCUREMENT UNDER PURPA

States had numerous difficulties in implementing administratively the avoided cost standard under PURPA. If they set the price too low, they ended up with no capacity. On the other hand, setting the price too high led to excess capacity with undesirable characteristics.

These difficulties led some states to experiment with letting the market decide by procuring supplies through a competitive bidding process. To date, there is a wide variation in the approaches adopted and the philosophies underlying the bidding process. It is probably fair to say that no two states are exactly alike in their approach, though there are some similarities.

In the following paragraphs we review the manner in which competitive bidding has evolved following a state-by-state approach. Our approach is illustrative rather than comprehensive, attempting to highlight the key differences between the various approaches.

## LEGISLATION AND IMPLEMENTING REGULATIONS

Section 210 of PURPA created a special regulatory regime for cogenerators and small power producers. Exempting these suppliers from regulation under both the Federal Power Act and the Public Utility Holding Company Act, it gave the generator the right to sell power at wholesale up to a rate not in excess of "the incremental cost to the electric utility of alternative electric energy." The "incremental cost ... of alternative electric energy" is defined in the Act as "the cost to the electric utility of electric energy which, but for the purchase from such cogenerator or small power producer, such utility would generate or purchase from another source."

The Congress gave regulators relatively little guidance in interpreting these provisions. PURPA was but one part of President Carter's National Energy Act. Even on this portion of the legislation, Congressional attention was focused primarily on Title I, which dealt with setting standards for state regulatory proceedings.

FERC issued its implementing regulations in 1980. Although it wasn't required to do so, these rules require that, absent a waiver, the rates for purchases from these so-called "qualifying facilities" (QFs) should be set at the utility's full "avoided cost" for new capacity. The Commission justified its use of full avoided costs as necessary to promote QFs. It explained the avoided cost standard was based on the economic concept of marginal cost. The utility's avoided cost at any point in time was to be determined by the combination of the marginal cost of production and the incremental cost for new capacity.

## STATES – AVOIDED COSTS AND THE ENTITLEMENT ISSUE

FERC's regulations left considerable discretion to the states to implement PURPA, including determining what constituted full avoided cost. As states grappled with this new legislation, two issues quickly arose: (a) What was the proper methodology to use in setting cost? And (b) what should a state do when supply exceeded demand?

### California

California was the first state to implement PURPA in a major way, foreseeing a need for a quick 6000 MW. The Public Utilities Commission (PUC) implemented a standard offer system, whereby the proposal was the offer of a legally binding contract which anyone could accept. In other words, the standard offer created a legal entitlement: anyone willing to sign the offer was entitled to the price contained in it. The Commission's justified the standard offer requirement on the unequal negotiating positions of utilities and QFs.

The most significant of these contracts was one which came to be known as Standard Offer 4. Standard Offer 4 had three variants:

- Option 1 allowed the QF to lock in a forecasted rate for a ten year period. This was useful for capital intensive projects.
- Option 2 provided for a levelized contract which was front-loaded in the early years. This was primarily used for hydro units.
- Option 3 provided for two components: (a) a capacity cost equal to that of a gas turbine, which was fixed at $110/kW/year, and (b) an incremental energy rate, based on the utility's marginal cost of gas.

This led to a huge number of acceptances, some 18,000 MW of capacity within 15 months.

Why?

California, relying on a negotiated price, set the "avoided cost" under Option 3 way too high. A cogenerator needed about \$70/kW/year to cover debt service, but California was requiring its utilities to pay 50% more than that amount. Moreover, the heat rate was based on the utility's most inefficient units, some of which had a heat rate of 15,000 btu/kWh. A cogenerator's heat rate often was below 10,000 btu/kWh because it used more efficient turbines.

This illustrates the difficulty in implementing the avoided cost approach. The California approach might have used a utility's marginal capacity cost and heat rate, but this administrative approach didn't reflect the new realities emerging in the marketplace. The result: California received signed contracts for substantially more capacity than it needed at a higher price than it otherwise would have had to pay.

California then compounded the problem by failing to place a limit on the capacity their utilities were forced to purchase.

Generally, one issues a nonbinding request for a proposal in procuring goods or services. There is no contract unless the parties can agree on the price and terms. This allows the purchaser to limit its purchases to those suppliers offering the most advantageous price and terms.

However, the California commission treated PURPA and avoided cost as an entitlement. It forced its utilities to make offers, leaving it up to the supplier to determine whether it wished to accept the specified price and conditions. Moreover, all suppliers were assured the same price and conditions since the commission interpreted the utility's obligations under PURPA as requiring it to buy all power generated by QFs.

To be fair, at the time regulators adopted Standard Offer 4 they didn't think independent producers could supply very much power. They have been dealing with the consequences of this miscalculation ever since.

### New York

If California illustrates the difficulties in determining administratively the proper figure for avoided costs, New York is a testament to what can occur if legislators get into the act.

A statute in New York sets a minimum rate of 6 cents/kWh for QF power. This was extremely attractive since the six cent floor applied in the early years of a project – when the QF might not ordinarily qualify for capacity payments. This statute wasn't repealed until the 1990's.

After oil and gas prices dived in 1985, New York was inundated with proposals for gas turbines, the marginal cost of which was well below six cents. New York regulators, which at the time didn't see the need for new capacity for at least a decade, had to deal with offers for thousands of unneeded megawatts.

The regulators ultimately solved this problem by deciding that PURPA was not an entitlement program. They then said that QF energy payments should be based on the cost of oil during peak only; at all other times the energy payment was based on the production costs of existing coal units. In addition, the regulators radically reduced demand forecasts and fuel price forecasts used for estimating avoided costs. These lowered avoided cost estimates became the basis for new contract offers, which have proven to be unfinanceable.

### New Jersey

Unlike New York, New Jersey by the mid-Eighties recognized it needed more capacity.

Its regulators concurred with California's view that a standard contract was an appropriate means of procuring that power. The problem they faced was how to procure it without setting themselves up for the oversupply problems already apparent in California.

The New Jersey Board of Public Utilities (BPU) came up with a novel approach: the first 600 MW offered would be accepted. To ensure that only qualified applicants were in this group, the BPU required a developer to obtain a QF certification from FERC, although PURPA does not require this.

There were two major problems with this approach.

The first come, first serve approach is economically inefficient. Single person developers won contracts, though they had no way of developing a project.

The requirement of FERC certification exacerbated this problem. It led to the construction of gas turbines since they could be easily certified. On the other hand, a waste facility couldn't qualify so quickly since FERC examines these on a case by case basis.

Recognizing this problem, the BPU decided upon a lottery system for the next solicitation. This solved the FERC certification issue, but, once again, small developers without the means to build a project could and did receive contracts.

## RISE OF COMPETITIVE BIDDING

### Texas

Faced with a need for new powerplants, Houston Lighting & Power in 1983 proposed a competitive procurement system. The Texas PUC rejected the proposal on the grounds it might discourage cogeneration and set an arbitrarily low ceiling on capacity payment levels, but it did permit its utilities to negotiate with potential suppliers.

This was the first significant deviation from the concept of avoided cost. Here avoided cost was used as a cap. It also was the first time that the seller's, rather than the utility's, cost of producing power was proposed to be the key factor.

Just as significantly, the Commission permitted the utilities significant discretion in negotiating contracts. They could pick their fuels, business partners and projects – so long as they did not exceed the theoretical avoided cost. For example, they can assume some of the fuel risk. Some of the companies have assumed a 70% availability for a project whose avoided cost is based on a coal or lignite plant and are willing to pay for energy above the 70% level at the average weighted cost of gas.

The Texas PUC does not approve cogeneration contracts and is bound by legislation which limits its flexibility to second-guessing negotiated prices at a later date.

### Maine

While Texas deviated from the avoided cost standard, Maine was the first state to look to the market to determine avoided cost.

In the early Eighties Central Maine Power (CMP) sold its Seabrook 2 interest and sought power from elsewhere based on Seabrook 2's avoided costs of 14 cents/kWh. It was bombarded with offers.

In 1984 CMP announced it would allow potential suppliers to bid against each other. In June 1984 the Maine PUC amended Chapter 36 of its regulations to establish criteria to guide this process.

Under the Maine approach, bidding is used to determine parties for negotiation, with the price being subject to bargaining even after negotiations begin.

CMP solicits bids via RFPs to fill blocks – known as decrements – of power. Bids must be below the ceiling price for that decrement. CMP then ranks the proposals and accepts enough bids to fill the current decremental block of capacity. CMP retains substantial discretion to modify the RFP or capacity selection process, including the ability to revise prices offered, if decremental ceiling costs change at any time prior to the execution of the contract.

After considerable experience, CMP has developed a five-part bid evaluation process:

1. A price index weighs bids according to prices.
2. The capacity index reflects the capacity factor of the facility.
3. An endurance credit is given if the facility posts a letter of credit guaranteeing performance.
4. The security index is based on pricing formulae and security guarantees.
5. The operating index is based on the project's dispatchability, maintenance scheduling and peak and off-peak power production.

The Maine process is notable in that the PUC has given the utilities a fair amount of latitude to examine both price and nonprice factors. As of early 1988, the PUC was involved in only 4 of 102 contracts (the last being in 1984). Since the PUC doesn't usually see the contracts, it doesn't review them.

The PUC doesn't get involved for several reasons.

First, the Commission has set a very clear policy. In 1983 the PSC reduced rates of return to show the utilities it was serious.

Second, bids are let for small increments of capacity.

In addition, administrative mechanisms have been established to encourage negotiations and keep parties from running to the Commission. In Maine a QF which files a petition gets the avoided cost at the end of the proceeding, absent bad faith by the utility. Since avoided costs have been declining, this has kept QFs away from the PSC. If the utility is found to have negotiated in bad faith, the cost at the beginning of the proceeding is determinative – and utility shareholders must pick up the difference.

Finally, utilities are prohibited from entering their own bidding contests so the potential for conflict is not present.

In mid-1989 CMP said that the utility had executed 84 long-term power purchase agreements with QFs since 1978, and only six contracts had been terminated or canceled. Only one contract had been breached, representing 0.5% of CMP's QF capacity currently on-line.

**Massachusetts**

While Maine regulators set a clear policy and let their utilities implement it, the Massachusetts Department of Public Utilities (DPU) sought to limit utility discretion by coming up with a "scientific" approach to bidding.

In 1986 the BPU issued detailed rules requiring utilities to purchase via competitive bidding a minimum of 5% of their annual maximum load.

A utility submits a 20 year resource plan to the DPU for review and approval. If this process reveals a need for capacity, a utility submits a proposed RFP to the DPU for review and approval. The RFP must set forth a scoring system which outlines the weight to be given to both price and nonprice selection criteria. Comments by potential bidders are encouraged. If the DPU approves, the utility may proceed.

Boston Edison was the first utility to follow this procedure – issuing an RFP in 1987. It developed a weighting scheme which, at least nominally, counted price for 50% of the total evaluation. Nonprice factors counted for the remaining 50% and included consideration of such items as whether the power would be available at peak, the project's stage of development at the time of the application, and the sponsor's financial strength. Each was given a weighting: e.g., a developer offering the maximum amount of financial security was given points which were equivalent to lowering the bid price by 5%.

The Boston Edison proposal was designed to be so precise that each bidder could score itself. Every factor – both quantitative and qualitative – was to be quantified. The utility was to have no discretion in choosing a winner, though it could return a proposal if it thought the bidder scored itself too highly.

The formulae used included price and quality factors, taking into account the present value of the bid, the type and amount of security offered, fuel type, plant characteristics and other matters.

As a practical matter, the price factor dominated since the other factors were easily met by most bidders.

The biggest "winner" was a 200 MW coal project to be located in near downtown Boston. This amounted to 58% of the capacity accepted (a total of 340 MW). Four of the nine winning QF bids were sponsored by a wholly-owned subsidiary of a neighboring utility, NEES. Every winning bid failed except one, which was able to leverage another contract. (All the NEES projects failed for lack of a site.)

Another reason this system failed was because of the commission's involvement: it was too active in setting rules. No security deposits were required. Boston Edison just followed its regulators' instructions to the letter, thinking it could blame the regulators if things didn't work.

Massachusetts Electric Company (MECO), a retail sub of NEES, obtained an exception from the formal competitive bidding process to engage in negotiation. Analysis of this approach showed that:

- Agreements reached by negotiation contain price levels that are statistically similar to those provided under negotiated bidding, but pricing formulas that are significantly different (reflecting the different types of plants);
- The nonprice features of the trades reached via negotiation provide greater benefits to the utility than those resulting from bidding (e.g. right to approve and oversee construction as well completion security);
- Although it is too early to draw firm conclusions, performance under negotiated agreement is an improvement over performance under bidding; and
- Developers (i.e. sellers) overwhelmingly prefer negotiation to bidding.

**Connecticut**

About the same time that Massachusetts created its competitive program, the Connecticut Department of Public Utility Control (DPUC) established a bidding system for that state. Recognizing the problems inherent in self-scoring, the DPUC sought to ensure the success of the program by involving itself in every aspect of the procurement process.

A utility submits its resource plan to the DPUC. After holding a public hearing, the DPUC, if it finds additional capacity is necessary, will tell the utility how much capacity is needed, the proposed factors to be considered, and the proposed weighting of these factors.

The utility then submits its RFP to the DPUC, which must approve the RFP.

The utility issues the approved RFP and then evaluates the responses.

The DPUC holds a public hearing on the utility evaluation and the DPUC notifies winning bidders. The utility must then reach agreement with these parties – at which time they must file with the DPUC the negotiated contracts.

The Connecticut regulation exclude QFs of under one MW, renewable resource QFs of less than five MW, municipal solid waste facilities under 10 MW, and QFs not seeking front-loaded rates.

As of early 1993, no Connecticut utility has issued an RFP utilizing these rules.

**Virginia**

Virginia took a different approach from those we have reviewed thus far: it approved utility sponsored bidding programs without adopting formal rules.

In January 1988 the State Corporation Commission authorized Virginia Power (Vepco) to issue an RFP for 1750 MW. This RFP contained numerous innovations.

First, there was no Commission involvement in the bidding program. Vepco did not seek Commission approval of the projects it selected.

Second, regulators gave Vepco considerable discretion in both setting the terms of the RFP and choosing winners.

Third, IPPs were allowed to participate.

Fourth, the utility required projects to be dispatchable.

Fifth, the pricing formula accommodated the needs of capital-intensive units by offering front-loaded prices.

Finally, the proposal was broadly advertised, was significantly larger than any other previous solicitation, and attracted many more proposals than any of the others we have discussed.

Virginia Power's first solicitation selected 19 projects representing 1974 MW from almost 14000 MW of proposals.

A second solicitation for 300 MW of peaking capacity led to Virginia Power declaring itself the winner, saying it could install the peaking capacity 15% below that offered in the competitive solicitation. Regulators have approved the construction of this unit.

**Florida**

Florida also has allowed its utilities to proceed with RFPs without adopting formal mechanisms.

A 1989 Florida Power & Light (FP&L) solicitation for 800 MW invited responses from all types of suppliers, including IPPs, small power producers, cogenerators and other utilities.

Responses were evaluated under a comprehensive framework balancing a number of price and nonprice factors. The latter included items such as the experience of the developer in similar projects; the level of development; the developer's financial resources; project viability; interconnection effects; fuel supply reliability; its fuels mix; its location; the extent to which it utilized Florida resources; its ability to be licensed; the project's planning and schedule flexibility; dispatchability and other operational considerations; and other factors that may affect the reliability and dependability of the project.

There were three major steps in the evaluation process. After initially screening the proposals to eliminate those failing to meet minimum requirements, FP&L would screen those remaining to choose those with the most desirable characteristics. (FP&L refused to release detailed criteria as to what those characteristics might be.) It would then enter into negotiations with the developers of the most promising projects.

Unlike Virginia Power, FP&L insisted upon regulatory approval of the contracts before they entered into force.

The solicitation was notable in that it excluded oil facilities and required all gas units to be coal-capable.

Florida Power & Light's solicitation also was notable for the response it drew: 107 preliminary bids totaling 27,634 MW. The response also was unusual in the size of individual projects offered: 73 were for plants over 100 MW and 27 were for over 300 MW. IPPs bid 34 plants with a capacity of 14,117 MW.

**New York**

The New York Public Service Commission (PSC) adopted a very formal bidding system in 1988 and 1989.

This system had five innovations.

First, the program is open to all sources – QFs, IPPs, DSM and utility subsidiaries. (CMP in 1987 was the first utility to open competitive bidding to DSM.)

Second, it adopted "wheeling in" and "wheeling out" provisions for utilities and their subsidiaries in the state. Under wheeling in, a utility participating in another utility's solicitation would be required to wheel power through its territory for any other bidder. Wheeling out requires a utility issuing an RFP to wheel power to its borders for any losing bidder which sought to locate a plant in its territory.

Third, in some cases, bids will be opened by an independent party who will rank bids by score.

Fourth, the price of competing utility life-extension projects will be raised by 10% because of the risk to ratepayers of cost overruns.

Finally, the RFP takes into account environmental factors.

The New York system requires utilities to give explicit details of an objective ranking system. For instance, Niagara Mohawk's 1989 RFP contained the following weightings, having a total score of 1310 points:

Price:
850 points (64.9%)

Success:
64 (4.9%)

Economic risk:
75 (5.7%)

Operational – supply side, or performance - DSM: 80 (6.1%)

Longevity:
21 (1.6%)

Environmental:
220 (16.8%)

Rankings differ widely in the various solicitations thus far in New York. For example, Long Island Lighting's RFP contained a total score of 550 points, with the following weightings:

Price:
175 points (32%).

Success (project development confidence: 50 (9%)

Breakeven period:
90 (16%)

System optimization:
115 (21%)

Price risk:
25 (5%)

Environmental:
80 (15%)

Operational longevity:
21 (3%)

Lilco said all bids must score at least 273.3 points, which is 90% of the score achieved by a Lilco-built 150 MW gas combined cycle plant.

In reviewing New York State Electric & Gas' proposal the PSC showed a willingness to get into minutiae. Here the PSC raised the importance of location from 30 to 85 points and significantly increased the importance of environmental protection.

### FERC NOPR

No overview of the evolution of competitive bidding would be complete without a discussion of FERC's 1988 Notice of Proposed Rulemaking on Competitive Bidding (NOPR).

In the mid-Eighties several state commissions questioned whether PURPA permitted competitive bidding. After much discussion, the Nevada commission asked FERC for its opinion.

Nevada got much more than it bargained for. FERC issued a proposed rulemaking which sought to mandate detailed rules for the conduct of bidding programs.

Under the proposed FERC rules states would have had to follow a number of conditions if they were to adopt competitive bidding programs:

- All sources – including purchased power and demand reduction – must be taken into account. If a state decides not to pursue all-source bidding, any source excluded from direct participation must be accounted for in a benchmark avoided cost calculation.
- An electric utility may not negotiate to acquire capacity and related energy from the time the final bids are submitted to the time the winning bids are selected.
- Nonprice factors which affect the operational integrity of the system must be addressed in the bid solicitation and be submitted to the state.
- The bid solicitation and selection process should be specified in advance of the solicitation. The RFP should specify the quantity and characteristics of capacity needed, the terms of the offer to purchase capacity, the criteria for participation, and the criteria for bid selection.
- The evaluation of all bids submitted in a bidding process must be in writing and must be based on criteria identified in the bid.
- The state must certify the price and selection of the winning bid.

In addition to these mandatory conditions, FERC also made suggestions as to further "improvements." These guidelines suggested using all-source bidding, binding prices without reopeners, an evaluation process which would permit an independent third party to replicate the winning selection, and limiting the use of front-loaded payments.

Comments on the proposed rules were almost uniformly adverse. Most commentators felt that, if FERC decided to issue regulations, significant revisions to the proposed rule were necessary if it were not to have the opposite result from what the Commission desired: a state aversion to bidding, discontinuities in existing state programs, and a hiatus in the signing of new power supply contracts while the courts seek to entangle the mess caused by unanticipated developments.

In 1993 FERC formally abandoned its effort to come up with a formal rule to guide state competitive bidding programs.

## ELEMENTS OF SUCCESSFUL POWER PROCUREMENTS

Until now we have discussed state-sponsored procurement programs and necessarily focused on the interaction between utilities and regulators. The experience we have thus far had with competitive bidding seems to indicate that the most effective state programs are those which give the utility flexibility in implementing a program and choosing winners. This is not surprising since putting together a well-designed power procurement system is an arduous and time consuming task. A power purchaser is unlikely to put forth the necessary effort if it thinks it will be second-guessed every step of the way.

At this point it is appropriate to examine the actual task of implementing a power procurement program. Successful procurements generally consist of a five-step process:

1. a solicitation process which identifies needs upfront and lets potential bidders know what these needs are;
2. an evaluation process designed to weed out weak competitors;
3. evaluation criteria aimed at ensuring successful bidders meet the power purchaser's needs and are viable projects;
4. the successful negotiation of a power sales contract which meets the need of all parties; and
5. post-contract monitoring to ensure the process stays on schedule.

The purpose of this section is to provide guidance in the implementation of requests for proposals for electric generating capacity by a utility. However, large industrial users seeking on-site generation will find this information equally useful, though they, of course, won't have the need to interact with the state regulatory commission.

## SOLICITATION PROCESS

For most companies the solicitation process will consist of three steps: a resource plan, a solicitation package, and adequate publication of the RFP to ensure a broad response.

The process should begin with a *resource plan* presented to utility commission setting forth the reasons why it believes it needs new capacity. This plan should state (1) how much new generating capacity is projected to be needed; (2) what type of capacity would be optimal (e.g., base or peak load, gas or coal fired); (3) when each increment of new capacity is required; and (4) where such capacity would optimally be located. This plan is subject to refinement and finalization based upon information garnered during the RFP process.

The power purchaser then designs a *solicitation package* and presents this package to the utility commission for comment prior to its release. This package will, as a minimum, include:

- A summary of the resource plan sufficient to communicate to bidders the power purchaser's perceived and expectations. It should include (a) the power purchaser's oil and gas price projections; (b) the power purchaser's projections of the general inflation rate and other appropriate price escalation indices; and (c) the power purchaser's projected discount rate, which will be used for purposes of comparing proposals. The Resource Plan itself may also be included.
- Solicitation procedures, including salient dates, contacts and addresses, instructions to bidders, schedules and contracting procedures.
- Evaluation criteria (evaluation criteria are discussed at length below).
- Bid preparation instructions. Bidders should be instructed to respond to the RFP in a fashion which will provide useful information in a form lending itself to evaluation by the power purchaser.
- A sample power sales agreement. This will provide a starting point for negotiations and give the power purchaser's contracting preferences. Explanations of contractual preferences should accompany the sample power sales agreement.

- A mechanism for bidders to obtain an estimate from the power purchaser of interconnection costs, including line studies, so that these costs may be included in the proposal.

After receiving comment from the Commission and others, the power purchaser should publish a *notice of the RFP* at least three times in a national publication and in the trade press. In addition, the power purchaser should issue press releases to appropriate members of trade press. Copies of the entire solicitation package should be made available to all who request it at either no cost or a price no greater than reproduction and mailing cost.

The RFP should provide bidders with a minimum of 120 days to respond to the RFP, although bidders may be required to declare their intent to bid within 60 days. Whenever feasible, the power purchaser should provide additional time: either by providing for a longer response time or by issuing a preliminary notice of intent to issue an RFP.

## EVALUATION PROCESS

The evaluation process should be divided into at least two stages.

The first stage is the qualifying stage, during which bidders will be requested to provide information regarding their qualifications to construct their proposed project, as well as a general description of their proposal. The purpose of this stage is to eliminate bidders failing to meet minimum criteria before they incur significant costs on a full response. A power purchaser seeking too many complete bids will be at a disadvantage since a bidder with only a small chance of winning is not very likely to expend the time and effort to optimize its bid.

Projects surviving the initial stage should all be capable of fulfilling at least some portion of the power purchaser's resource requirements. The bidders on this "short list" will then submit full proposals sufficient to enable the power purchaser to fully evaluate the bids and begin negotiating contracts with the parties it selects.

## EVALUATION CRITERIA

The Bidder in a competitive solicitation is proposing to take on a bundle of responsibilities, which traditionally been assumed by the procuring electric power utility. These responsibilities can be broken down into seven categories:

1. Construction and operation
2. Long-term operation and maintenance responsibility
3. Fuel supply and transportation
4. Site and permitting
5. Financing
6. System integration issues
7. Non-power-related responsibilities (proffers)

The power purchaser will need to request information for each of these categories. This information should include the party responsible and the specifications for performance. In addition, the power purchaser will wish to obtain guarantees or representations to assure that the responsible party will in fact perform to specification.

A. Assignment: to a responsible party

B. Specify: performance

C. Guarantee: through contractual enforcement provisions.

In evaluating projects the purchaser should remember that one gets what one pays for, like in most other endeavors in life. If a project comes in at a much lower price than its competitors there may be a reason for this other than financial or technical ingenuity. Hence, the purchaser must set forth a detailed plan to guide its evaluation to protect itself against surprises after it signs a contract.

The subheadings provided below are commonly used areas for inquiry. Within each are examples of minimum criteria and preferences which have been used by companies procuring power. The choice as to minimum criteria and/or preferences is one which the power purchaser must make based upon its needs and options.

**Construction plan**

This part of the inquiry should look at four different matters:

(a) the projected construction schedule;

(b) major project components;

(c) construction management; and

(d) cost estimates.

Bidders should be required to provide their *project schedule* for constructing the new facility. This schedule should include such matters as completion of design and engineering and projected dates for commencement, major intermediate milestones, and the reaching of commercial operation. Operation should be attained by a specified date.

In evaluating this schedule the purchaser should examine not only the likelihood that the bidder will be able to meet these dates, but the bidder's willingness to:

- incorporate this into a project tracking and following procedure;
- be bound by this schedule;
- keep to the schedule despite ordinary construction delays; and
- accelerate or delay commercial operation to reflect resource plan modifications.

In addition to providing information as to the schedule, the bidder should specify *major project components,* with notations as to the delivery schedule of the primary components included in the projected construction schedule. The bidder should be encouraged to discuss unusual or special aspects of the project having an impact on the evaluation.

The power purchaser may wish to specifically require information on significant components, such as major equipment items, mechanical support systems, electric support systems, environmental control systems, control systems, plant arrangement, and site layout.

The purchaser should insist that all equipment be new, utility grade, and that equipment of the type specified has been used for power generation previously. In addition, the power purchaser may specify a heat rate and/or fuel capability, with requirements for dual fuel firing, coal capability or coal conversion capability.

In many cases, suppliers of major project components will offer guarantees. Given the usual fragile financial structure of the bidder itself, the power purchaser should consider the potential for these guarantees running to its benefit, and request information as to the extent to which such guarantees apply or are available.

A third line of inquiry should focus on *the project's construction manager and the architect engineer.* What are their qualifications, including references? The power purchaser should also seek the resumes of the individuals who will bear project responsibility. The evaluation of experience should go beyond a mere inquiry into the corporate resume. The purchaser also needs to examine closely the resumes of the actual individuals who will be working on the construction project.

In the case of nonmanagerial personnel, purchasers should seek assurances that all employees will be members of the appropriate qualified union or are otherwise certified professionals in their specialty.

Finally, the purchaser should require *detailed cost estimates* for design and construction, as well as construction financing costs as of the time of completion and permanent financing. This information will not only be useful in identifying overpriced offers, but also in ferreting out problems with low-priced offers. If critical components are left out, or underpriced, or if unskilled and unqualified personnel are budgeted, the detailed cost estimate will aid the evaluator in discovering this, and thus help to avoid nonviable projects.

Bidders should be required to use the customer's discount and inflation rates, as well as fuel cost projections so that the offer may be designed around the power purchaser's view of the future.

A checklist of things to look for in evaluating the construction plan is contained in Appendix E-1.

### Long-term Operation and Maintenance Responsibility

The next area of inquiry should focus on whether a proposed plant will provide reliable and cost-effective generating capacity during the term of the agreement. This portion of the evaluation will focus on (a) the O&M plan and (b) proposed O&M practices.

The power purchaser should examine the *bidders' planned maintenance program and organization,* including, but not limited to planned training, schedule, recordkeeping, procedures, spare parts inventory and quality control. The purchaser should evaluate any contracts signed or under negotiation with an operation and maintenance organization and review that organization's qualifications and references.

In the case of nonmanagerial personnel, the purchaser should ensure the bidder provides assurances that all employees will be members of the appropriate qualified union or are otherwise certified professionals in their specialty.

If available, the bidder should provide actual costs of the operating and maintenance plan, or estimates if actual costs are not available. The plan should require the power purchaser to obtain the full range of information regarding personnel. This can include information on the number of O&M personnel and their positions, staffing levels, training and experience criteria for each position. The plan should include reasonable contingency funds.

Particularly in the absence of an acceptable third party O&M contractor, the bidder should be required to provide specifics as to *O&M practices.* In addition the bidder should be requested to state whether its facility will be able to cooperate with the customer in scheduling maintenance for optimal periods. A checklist is provided in Appendix E-2.

Depending upon the contractual relationship envisioned, this can be an ongoing process, as in cases in which an operating committee which includes customer representatives makes decisions regarding O&M practices.

### Fuel Supply and Fuel Transportation Plan

The purchaser will use different criteria for judging the adequacy of fuel supply depending on the type of fuel, the area of the country, and the level of reliability it prefers. For example, gas transportation will be a critical issue in New Jersey, while it may be unimportant in Texas. Spot purchases may make sense for one purchaser, while another might insist upon firm contracts covering all of the plant's requirements for its entire useful life. Despite the type of fuel, however, the purchaser will always want to evaluate such matters as the supply requirements of the plant, the project's transport plan, its fuel supplier and the bidder's fuel management experience.

In order to assess the adequacy of the *fuel supply plan,* the power purchaser will need to obtain a description of the specific fuel characteristics that will be used to produce steam, including secondary and backup fuels, and the annual fuel requirements of the plant. As part of this evaluation, the purchaser should review the fuel inventory management plan including average and minimum fuel inventories for primary and secondary fuel sources, and how long that inventory would last in the event of a fuel supply interruption. Appendix E-3 provides a checklist of points to be covered in a fuel supply plan for coal or gas-fired facilities.

The next item for analysis should be the *fuel transportation plan.* The purchaser will review the reasonableness of the mode of transport which the bidder suggested and the proposed or potential transporters. The power purchaser also may wish to look at evidence of pipeline reservations (for gas), or reasonable commitments of rail transportation availability for coal. In light of rapid developments in the natural gas industry, Appendix E-4 includes a primer on natural gas transportation.

The purchaser will need to review the list of *fuel supply contractors* who will be relied upon, letters of commitment from these suppliers stating quantity, price terms and length of commitment, and the bidder's plan for obtaining new supplies of fuel. If available, fuel supply contracts should be examined.

The bidder should submit information regarding who in the project team will be responsible for *fuel supply management and procurement* and their experience and qualifications.

**Siting and Permitting**

Siting and permitting encompasses several interrelated project components. First, the project developer needs to have a site. This is particularly important in the case of qualifying cogeneration facilities, which need to be sited so that they may simultaneously provide thermal output for a useful purpose. Thus site control becomes a critical and qualifying issue. Second, location is important from a transmission perspective. Finally, the bidders site must be comply with applicable permitting and zoning requirements if it is to be a viable project.

The power purchaser will need to acquire the *address and a general description of the proposed project site,* as well as contact information for the current owner(s). The power purchaser also will need to obtain information regarding the current status of bidder's plans to obtain rights to the site. This could be outright ownership, an executed contract to purchase, an option to purchase, or other right to use the site in question.

The purchaser needs to consider items relating to *permitting, licensing and zoning.* Often permitting is the critical path for project development. The power purchaser must therefore request a schedule for applying for, and projected dates of issuance of, all approvals, licenses, permits or variances. As bidders often are not locally based, it is useful for the power purchaser to include a list of permits and permitting agencies in the RFP as an aid to bidders. As the ability to negotiate the complex permitting process is key to successful and timely project development, the power purchaser will also need to obtain from the bidder the status of its applications, if any are in process, and its experience in obtaining licenses and permits for similar projects, including the experience of any consultant or team member employed for the purpose of obtaining such licenses and permits. Appendix E-5 provides a basic checklist of approvals which will be required for power plant construction.

**Financial Plan**

In order for the power purchaser to assess the financial viability of the bidder, it must obtain the bidder's financial plan. This plan should cover and distinguish between the two primary stages of financing – construction financing and permanent financing.

Construction financing is subject to a unique set of risks, the most critical of which are discussed in Appendix E-6(A). The power purchaser will want to obtain information on who is responsible for the costs associated with these risks in doing the financial evaluation.

Information required of bidders regarding permanent financing should include, at a minimum, the projected capital and financial structure of the project over the term of the pow sales agreement, the sources of equity and debt financing, and projected annual cash flow statements for the facility over the term of the power sales agreement, with supporting assumptions.

As most bidders are companies set up for the purpose of this particular project, the power purchaser also will need to obtain financial information regarding the project's principal participants. Principal participants may be defined as parties who are providing guarantees which support the financing of the project. This is important because often the guarantees of project participants act as a substitute for equity. Insurers may play a similar role, using some of the more innovative insurance products available. A list of insurance products used by bidders appears in Appendix E-6(B). Audited financial statements for the past two to three years should be sufficient.

The power purchaser also will need to know about the company(ies) providing debt financing and guarantees. For these companies, references on previous financings, and a list of previously financed projects and their current status would be useful to gauge the likelihood the project will be able to enter construction.

The bidder's financial plan is constrained by any announced preferences the power purchaser's company may have regarding payment streams. For example, a preference for low costs in the early years of the contract may yield higher contract-cycle costs. Linkage of payment streams to an avoided cost measure can greatly increase the complexity of financing. A brief discussion of the relationship between avoided cost payments and finance is included in Appendix E-6(C). The subject is also treated in the discussion of the evaluation of price offers, infra.

**System Optimization**

The power purchaser, assuming it is a utility, needs to examine the manner in which the plan fits into its existing system.

It will need to examine the site and its impact on the transmission system. The power purchaser's service territory has unique transmission characteristics which will militate in favor of some project locations. Therefore, the power purchaser should include in the power purchaser's RFP a map of the service territory and a description of the power purchaser's needs and preferences from a transmission perspective. This will permit the bidders to plan their project locations accordingly. Section G discusses the issue of who pays the costs of any necessary upgrades.

The power purchaser also must determine the extent to which integration of a bid project into the utility's normal economic dispatch is feasible. Bidders should be encouraged to state what those parameters are for their plant, in as much detail as possible, so that the utility may compare plants based upon their likely operation on the system. Appendix E-7 includes a checklist of performance characteristics which may be useful in designing performance parameters for the power purchaser's RFP.

### Non-power-related Criteria

There are a variety of non-power-related commitments which have been made by bidders, and which the power purchaser may wish to consider in assessing the bids. If so, the purchaser should communicate its interest in these factors to prospective bidders as early as possible. Some of the more commonly examined proffers include:

- *Steam sales:* Qualifying cogenerators under PURPA are required to have a useful thermal output. This is often provided at low cost, to the benefit of local industry. The power purchaser may wish to consider a preference for projects which provide such output. Some contracts require continued qualifying status under PURPA as a condition to continuing the contractual relationship. If so, the power purchaser will need to gather information on the thermal application, and the terms and conditions of the thermal energy agreement.
- *Employment:* A commitment to continue employment, or to enhance the likelihood of the survival of the thermal host, or to increase employment may be a part of the bid considered of value, even though the benefits do not flow directly to the utility-customer. Similarly, a commitment to use local labor in construction or operation, or union labor, may be a factor for consideration. Employment projections, guarantees, or contracts may be sought to support and document claimed employment benefits of a bid.
- *Environmental commitments beyond those required by law:* All reasonable environmental control measures should be considered. Bidders should therefore be instructed to include in their proposals alternative environmental control measures, and their associated costs and benefits.
- *Community enhancements:* While not required, the power purchaser might review bids based upon proffered funding of road improvements, beautification programs, or other community enhancements.
- *Conservation efforts:* In recognition of the differences in procurement techniques applicable to conservation efforts, the power purchaser should not be required to consider conservation or other demand-side management approaches as a part of the RFP process. However, a utility may be required under other provisions of law to consider such alternatives. As a result, the utility commission will provide guidance on the consideration of conservation alternatives on a case-by-case basis.

## NEGOTIATING THE POWER SALES CONTRACT

Once a short list of bidders has been chosen and their bids evaluated, the power purchaser will commence the final selection process by entering into power sales contract negotiations.

The power purchaser may wish to include one or more sample contract formats as a part of its RFP package. Contract language is a convenient means of telling prospective bidders what the power purchaser is looking for. Also, the sample contract can accelerate the negotiation process by providing a starting point.

From the perspective of the bidder, therefore, the power purchaser's preferences as to contract language, as expressed in the standard contract included in the bid package, is an alternative mechanism for stating evaluation criteria which do not lend themselves to simple description. Particularly in the complex areas of risk allocation, precise contract language is often the only way to delineate the responsibilities of the members of the bidder team.

Appendix E-8 contains a list of provisions to consider for inclusion in the power sales contract.

### Integrating Contracting and Evaluation

The power purchaser should review each project component and make sure that (a) the performance contracted for has been adequately *specified* to reflect bargained-for performance, (b) all responsibilities have been *assigned* to an accountable party, and (c) there are provisions assuring that performance is *enforceable.*

The essence of, or even the entirety of the agreements between the developer and the members of the development team, should be incorporated into the omnibus power sales agreement, so that the power purchaser will have a means for directly addressing the accountable party in the event of difficulty.

Only once the project has been adequately specified, and the allocation of risk between the power purchaser and the various accountable parties has been established, will the power purchaser be in a position to establish pricing terms.

### Negotiating Pricing Terms

In designing price bid evaluation criteria, an analytically useful delineation is that between fixed and variable price components of the bid. This is useful because there is perceived to be greater risks for the customer associated with variable components of the bid, and less with fixed.

A fixed price may have components which change by a fixed percentage each year, or by a set formula, such as the GNP price deflator. The risk that costs will be subject to dramatic change is still well-controlled, so that these are still considered to be fixed price terms.

*Fixed Price Bid Component*

The fixed price component of a bid may be further divided into the portion committed to recovery of investment (debt service and return on equity) and fixed operating and maintenance costs.

The power purchaser may have preferences as to how much of the price bid should be fixed, and what specific components should be included in fixed costs. These may either be expressed as mandatory minimum qualifications, or as mere preferences. For example, gas-fired projects may be required to obtain firm, long-term contracts for gas transport. Coal-fired projects may be required to have long-term coal-supply contracts for a percentage of their requirements. The variations are many, and there is no right or wrong answer. However, to the extent the power purchaser has such proclivities, they should be communicated to the bidders.

Much of the information presented by the bidder in making its fixed price component offer will have already been obtained in the form of cost information in the various substantive nonprice criteria (primarily in the engineering, construction and procurement contracts). The fixed price bid translates those costs into a payment stream to be required of the power purchaser. The bidder should be requested to show how the cost information translates into the fixed price component of the bid.

The fixed payment may be linked to performance as a means of encouraging excellent performance. Typical means for providing these benefits is through the linkage of fixed payments to performance indices such as the equivalent availability factor. These may be year round, only during peak periods, or during such period as is relevant.

*Variable Price Components*

The most important variable cost generally is fuel cost. Particularly with gas-fired facilities, intractability has often been recognized by the request for bids based upon formulas which take into account market rates for fuels. The actual variable cost bid is the product of the market fuel price and a bid heat rate.

There are other potential variable cost provisions. A common one is for variable operating and maintenance costs. These may be handled in any number of ways, from an index, to a ceiling, with incentives for remaining below the ceiling, to an operating committee which includes representatives of the plant operator and the utility, which jointly makes decisions as to required O&M and costs to be incurred therefor.

Unlike fixed costs, however, variable costs are subject to practical modification by dispatchability provisions. If the power purchaser is requesting bids for fully dispatchable plants, then any variable cost becomes little more than a price ceiling. In practice, should the variable price provided for in the contract not be attractive, dispatchability will provide the opportunity to renegotiate.

In the absence of dispatchability, information regarding ceiling prices and the guarantees supporting them will be needed in order to assess the price stability of the fuel supply.

*Responsibility for Other Potential Costs*

Unless bid documents make clear that expenses are the responsibility of the bidder, they may well end up being the power purchaser's responsibility. Two areas which lend themselves to this concern are costs of transmission upgrades and costs relating to changes in environmental law. However, the general category of force majeure can effectively transfer contingent costs to the power purchaser. It may not be possible, or even desirable, to prevent this. However, the power purchaser may wish to assess the likelihood of these events and their potential cost.

The cost of *transmission upgrades* can include everything from interconnection costs to major system upgrades required by the addition of generating capacity to the system. While under PURPA, qualifying cogenerators and small power producers are responsible for the cost of interconnection, their responsibility for transmission upgrades is problematic. Nevertheless, if such upgrades are not specified in the bid, and accounted for by the bidder, as a practical matter the bidder will lack the financial capability to provide the upgrades. This will require the power purchaser to either do without the power from the facility or to pay for the transmission upgrades itself. Consequently, the power purchaser may wish to perform an interconnection study which assesses these costs as a part of the evaluation process, and perhaps provide an opportunity to reformulate projects based upon these interconnection studies.

Over the decades during which your company will be purchasing power from a successful bidder, there likely will be changes in environmental law, requiring additional investment. Unless a grave error was made in the procurement process, the bidder (now supplier) will lack the funds for major retrofit. This means that the risk of *changes in environmental law* is effectively on the power purchaser, regardless of the terms of the solicitation.

Finally, *unforeseeable and uncontrollable events* may arise which affect the cost of operation of the project. Given the relatively fragile economic condition of project-financed power projects, the power purchaser could well be faced with the choice of renegotiating payments or doing without contracted-for power. Litigation, strikes, fuel supply upheavals, and political interference, all could impose costs which the bidder may be unwilling or even unable to bear. Should the project become insolvent, power may still be available, but at a higher than foreseen cost.

### Security

In the case of penalties for poor performance, the power purchaser should be secured against damage in the event of bidder's failure to perform. One obvious source of security is the project itself. However, often the project has been heavily obligated for financing purposes, leaving little apparent equity to be attached as security in the event of default.

The power purchaser will need to go into the contracts negotiated between the bidder and the team members to determine who is actually taking the risks of poor performance and obtaining the rewards of excellent performance. The team member which is to benefit from excellent performance would in most cases be the appropriate source of security for damages in the event of poor performance. The power purchaser should request such security, in the form of corporate guarantees from the appropriate team member or the equivalent (such as a bond purchased by that team member).

As fuel is by definition available at market rates, and failure to maintain a low heat rate provides its own penalties, the power purchaser's major concern on the fuel side will be to secure the continued operation of the plant in order to protect its customers should generating capacity come to be in short supply.

In such circumstances, a poorly operating plant would be poor security. The power purchaser will therefore need to obtain information on the underlying guarantees supporting efficient operation to support the effective cap on variable costs of operation. For nondispatchable plants, it needs to obtain information on the underlying guarantees supporting continued availability of fuel at a rate equal to or less than the ceiling price.

#### *Final Evaluation of Negotiated Contract Offer*

In evaluating the price offered by bidder, the power purchaser must look to (a) the relationship between value offered and price, (b) the effectiveness of any penalties and bonuses offered as a means of encouraging excellence, and (c) the overall price level.

The benefit to be gained from competitive power sales contract negotiations is primarily to be found in the reduction of risk taken on by the power purchaser. All controllable risks should be placed on those most able to control them. All project responsibilities should be assigned to those best qualified to handle them and willing to do so for reasonable compensation.

## POST CONTRACT PROJECT MONITORING

Meaningful post contract project monitoring can be a complex and personnel-intensive process. Appendix E-9 includes a list of key project stages.

Project development is still very much an art. No two development teams are likely to bring their project to fruition in precisely the same way. However, in all cases, the power purchaser should require that the bidder provide a schedule which shows how it intends to move the project through its key stages and into operation as scheduled. During the project evaluation stage, a preliminary version of this schedule should be made available. A final version should be incorporated by reference into the power sales agreement. The bidder-provided schedule will be the basis for monitoring efforts. Appendix E-10 is an example of key stages which might be denoted in such a project schedule.

The personnel required for project tracking and monitoring efforts are directly analogous to those used in the initial project evaluation. In fact, this can be considered an extension of the project evaluation process, with similar requirements for substantive experts who can assure fulfillment of the pledges made during the procurement process. The same people who performed the project evaluations would likely be the most effective at tracking those projects, and a similar organizational structure might be employed.

A project manager is assigned to each project. Project managers are responsible for overall progress of their assigned projects. They should be assisted by experts who take specific responsibility for tracking and analyzing that component of projects within their areas of competence. Thus, a project manager might have supporting him an environmental compliance specialist group, which would be responsible for providing reports on the permitting progress of all projects being tracked, and similar specialist groups for construction, fuel procurement, financing, etc.

Post-contract monitoring begins with the drafting of contractual provisions providing the power purchaser with the right to act and with the right to information needed to act intelligently. Those provisions will differ depending upon the project requirements, the purchaser's capabilities, and the state of the marketplace. Once in place, their implementation will require a continuing, intelligent understanding of the realities of project construction and operation, a purchaser's alternative suppliers, and the risks and costs involved in pursuing those alternatives.

Project monitoring doesn't end until the power sales contract ends. Once construction is complete, that component of the monitoring terminates. But fuel supply, operation and maintenance, and perhaps other components of project implementation will continue to require careful follow-up in order to assure reliability.

This last stage of the procurement process blends back into the first. That is, in the process of resource planning, the power purchaser will need to make judgments as to the likelihood and level of contractual compliance by previously selected contractors based on the monitoring reports. Reports of the project managers will be fed back into the resource planning process on a continuing basis in order to assure reliable, cost-effective power suppliers.

# APPENDIX E-1

## CHECKLIST OF THINGS TO WATCH FOR IN EVALUATING THE CONSTRUCTION PLAN

One can shave capital costs by either leaving something out, or installing used or substandard equipment. If this occurs, there is an increased risk of failure, particularly as the plant ages. This is important because while investors provide a modicum of protection against these practices during the early years of operation, during later years, when the investors' role diminishes, the only one looking out for assured performance is the procuring utility. The utility must therefore assure itself that the bidder has not shaved costs in an undesirable way, unless the utility is specifically aware of it, and the price bid has provided for it.

Some of the things which a utility would ordinarily include in its planned powerplants, but which have been proposed for exclusion by some competitive power producers as a means of cutting costs, are:

- proper weather protection
- redundant auxiliary equipment
- adequate instruments and controls
- maintenance space and machinery
- tools and spare parts
- adequate water treatment equipment
- turbine lube oil purification equipment
- air dryer equipment
- laboratory space and equipment
- backup fuel source and delivery system
- plant fire protection

In addition to these hardware items, several intangible items often are added to the exclusion list:

- boiler envelope size margin
- boiler fan capacity margin
- scrubber gas flow margin
- boiler feed pump capacity margin
- condensate pump capacity margin
- operator training programs
- startup and checkout program
- performance testing program
- environmental testing program.

Every item which is excluded from the project presents a further risk of reduced capacity, or availability, or both. Each threatens to reduce the potential for the project to produce reliable power.

Buying used equipment, on the other hand, is just like buying a used car. The power purchaser trades the lower cost of the equipment for the increased risk of an early failure. The magnitude of the risk depends on such factors as the original quality of the equipment, how old it is, how it has been used, and how well it has been maintained. There are many possible variations of these factors, so there is a wide range of risk exposures associated with projects which include used equipment.

# APPENDIX E-2

## CHECKLIST OF THINGS TO WATCH FOR IN EVALUATING THE O&M PLAN

Some common problems to look out for in bidder's O&M plans are:

- plans to hire the minimum number of personnel.
- plans to use inexperienced operators
- exclusive reliance on "on the job training"
- failure to stock spare parts
- failure to set aside funds for supplies and expendables
- assumption that no contract maintenance will be required
- failure to plan for any outside laboratory and testing services
- failure to provide contingency funds for replacements and renewals

The following checklist should be used to ensure that an operating plan is in place:

A. Definition of operating staff
   1. Organization chart
   2. Qualifications outline for each position
   3. Salary levels
   4. Hiring schedule
   5. Training schedule

B. Spare parts to be stocked
   1. For each major equipment item
   2. For mechanical support systems
   3. For electrical support systems
   4. For control and instrument systems

C. List and estimate of expendable supplies
   1. Chemicals
   2. Welding gases
   3. Gaskets
   4. Pipe and tubing materials
   5. Conduit and cable materials

D. Definition of outside services required
   1. Boiler maintenance
   2. Turbine maintenance
   3. Water analyses
   4. Fuel analyses
   5. Environmental testing

E. Five year operating budget

# APPENDIX E-3

## FUEL SUPPLY CHECKLISTS

The specific questions which must be answered to determine the adequacy of fuel supply arrangements will vary with the type of fuel. Below are checklists for the two primary fuels expected to be used by bidders.

### Coal

(a) Delivered cost (overall level and match with payment stream expected from power sales, whether there are market or operating cost adjusters).

(b) Heating value of coal

(c) Sulfur and ash content

(d) Transportation and operation costs

(e) Expected performance of the supplier

(f) Does coal supply and current boiler specifications?

- i btu/lb
- ii % sulfur
- iii % ash
- iv % moisture
- v % volatile
- vi grindability
- vii % sodium oxide in ash
- viii base/acid ratio in ash

(g) Does coal supply meet ash fusion requirements (reducing atmosphere, oxidizing atmosphere)?

(h) Sampling and testing approaches planned

(i) Length of coal supply contracts and whether there are renewal options

(j) Overall purchasing strategy: is there a preference for long-term supplies or spot market purchases, a portfolio of suppliers or a single supplier?

(k) Transportation (method and cost)

(l) Who is responsible for coal procurement?

(m) Does the coal come from in-state or out-of-state sources?

(n) Is the supplier financially sound? What is the amount of mineable reserves at their disposal, number and thickness of coal seams and other characteristics, status of mine permits and environmental matters affecting the mine in question and the chemical characteristics of the coal as delivered? What are the Vendor's commitments to other customers? Is there any litigation ongoing in which coal vendor is involved? What is the availability of substitute coal if problems arise at one mine?

(o) Inventory of coal planned to be on hand

(p) Mine site visit and production operation review plans

(q) Penalty and cancellation provisions in fuel contract

### Natural Gas

There are a number of strategic issues pertaining to gas supply contracting which are not generally controversial for coal-fired plants. Should gas supplies be purchased under long term contract, spot market, or some combination of the two? To date, cogenerators have virtually been required to enter into long-term agreements, albeit with variable or indexed price provisions, by the Scylla of avoided-cost based power sales agreements and the Charibdas of project financing. This may not make sense for a particular utility, or for any utility, given the realities of today's gas markets. Gas suppliers who have been disappointed by the enforceability of long-term agreements, are likely to require price reopened provisions in return for long-term commitments.

In determining the ideal fuel arrangements for gas-fired plants, the purchasing utility should take care to assess how important it is to have security of price and supply. Both are attainable, but may be too costly. The RFP should communicate a particular utility's desire for less secure supplies.

*Gas Supply Options*

Before setting criteria for the optimal gas supply contract, the purchasing utility should consider the real-world options of bidders. Today these include numerous possibility, including the ability to:

- Purchase daily gas from the Local Distributor (LDC)
- Purchase daily gas from the pipeline
- Purchase monthly spot gas from a marketer
- Purchase monthly spot gas from the pipeline
- Contract for long-term delivery at spot
- Contract for long-term supply at fixed price with escalators and market openers every 3-5 years
- Contract for long-term supply with corporate warranty
- Prepay for the gas reserves to the gas producer
- Purchase a gas producing company outright
- Purchase proven producing reserves with reliable operator
- Invest in drilling programs with reliable Exploration and Development firm
- Purchase Leases, and become an Exploration, Development and Producing firm

Today, the two most realistic and viable options for assuring the supply of natural gas for projects are: 1) long-term, fixed escalator contracts and 2) outright reserve acquisitions.

In the case of long-term fixed escalator contracts, a great deal depends upon the guarantees the purchasing utility insists upon in comparing different projects. It will be rare that absolute guarantees will be available from a responsible party. The better the guarantee, the more carefully the purchasing utility must investigate its enforceability.

In the case of outright reserve acquisitions, risks are incurred in that the purchaser depends on reservoir engineering analysis of the formation to determine future decline rates and useful life of the field. In addition, the utility/customer must also assess the efficiency of the operator who manages the wells. Most firms that perform this function are usually very well qualified and should in the long run give comfort to the purchaser.

Just how long this window of opportunity will remain open is uncertain. Customers must continually apprise themselves of market conditions in order to decide on realistic and optimal criteria for comparing bidders' supply plans. Gas supply markets, and not avoided cost methodologies, will dictate the terms a customer may realistically offer should it expect to obtain viable bids for gas-fired capacity.

*Gas Quality*

While less of a factor than with coal, there are still a number of quality criteria applicable to the gas supply contract which can make or break the project. These include:

- BTU content of the delivered gas
- Delivery pressure and temperature
- Moisture content and other quality factors

# APPENDIX E-4

## GAS TRANSPORTATION PRIMER

Due to the rapidly changing nature of the gas supply industries, this Appendix provides a look at today's options for transportation of natural gas from wellhead to burnertip. There are myriad combinations available to the prospective supplier – each with their own risks and costs. Only by understanding these risks and costs can the utility-customer provide guidance as to the type of transportation arrangements it prefers and/or evaluate the offers made by bidders.

Any transportation package will be comprised of one or any combination of three segments: gathering, transportation and distribution. Supply and transportation patterns have traditionally been well established in the gas industry, but the advent of contract carriage has unsettled the normal mode of operation and allowed for creative transport routing. The relative location of the gas production to the cogeneration facility will determine the optimum plan. In fact, the location of the cogeneration facility will probably determine which production basin will be utilized. For example, Appalachian gas would probably not be utilized in California but it may be an optimum choice for New England. Each segment will require a separate service agreement negotiated in parallel with the other segments and simultaneously with the gas purchase agreement. Each link in the transportation chain should fit hand in glove with the next.

Below is an outline of the major factors which will influence costs and the quality of service of each.

### Gathering Segment

A gathering system takes gas in the production field and transports it to a transmission pipeline. It is usually the responsibility of the gatherer to compress the gas to a pressure which will allow injection into the transmission pipeline and to treat the gas to meet pipeline quality specifications. It is likely that the gathering agreement will most closely resemble an arms length commercial transaction. Gathering system rates may be unregulated, though, in some circumstances they can be regulated by the states in which they operate or the federal government.

Generally speaking, the terms and conditions of the service are negotiable. The most important features of the arrangement are to be sure that the gatherer has adequate pipeline, compression and treatment facilities to move the wellhead production to the transmission pipeline in the quantities needed. Transmission pipeline pressure and quality requirements should be incorporated into the gathering agreement.

### Transmission Segment

Transmission pipelines are typically large diameter, high pressure pipelines which transmit gas for long distances. A transmission pipeline which operates entirely within one state is an intrastate pipeline and it is regulated by that state. A pipeline which has facilities crossing a border between states is an interstate pipeline. It is regulated by the Federal Energy Regulatory Commission under the Natural Gas Act.

Interstate pipelines transport gas under rate schedules approved by the FERC. Probably the most crucial decision to be made about transmission service relates to whether the service chosen will be firm or interruptible. A shipper receiving firm service will receive preference to pipeline capacity during periods of capacity curtailments. The trade-off is that the shipper will be obligated to pay a capacity reservation to demand charge each month whether he uses the capacity he has reserved or not.

The load factor characteristics of many cogeneration projects make them ideal candidates for firm pipeline capacity. In fact, on many pipelines as the load factor approaches 100%, the total cost of transportation (demand plus commodity charges) may actually approach or be lower than that of interruptible transport service. Firm transportation rates usually incorporate a commodity charge as well as the capacity component. The monthly commodity charge will be equal to the actual throughput of gas transported times a specified unit charge. Each pipeline has unique rate characteristics and the relative costs of firm and interruptible service must be analyzed carefully.

Interruptible transportation is a lower priority service in which transportation capacity is provided on an "as available" basis by the pipeline. Because interruptible shippers have no preference to capacity, they do not pay a demand charge. Normally, shippers pay a commodity rate based upon actual flow from month-to-month. Interruptible capacity is generally rationed on a first-in, first-out basis so the chronological order of an interruptible transportation agreement may be very important in determining the potential reliability of the agreement. On underutilized pipelines where capacity is readily available old interruptible transportation agreements may provide what is tantamount to firm service without the demand charge.

Current ratemaking design allows for some variation in the rate charges. Most interstate pipelines post both maximum and minimum rates for firm and interruptible service giving them the flexibility to price their services competitively. There is good potential for negotiating rates within this range. Cogenerators, with their strong potential for underwriting pipeline fixed costs, have a great deal of leverage with pipelines when negotiating rates.

Intrastate pipelines may operate under rates which fall under the jurisdiction of a state. These regulations vary from state to state, however, the underlying rate issues are the same and opportunity for negotiation is manifold.

**Distribution Segment**

The final segment of the transportation chain is the local distribution utility. LDC's operate entirely within one state and are regulated by that state's utility commission. Most utilities are willing transporters today and have unbundled their rates to reflect the cost of both firm and interruptible transportation and storage services. Rates may not be negotiable on the surface, but utilities have demonstrated a willingness to design new rate schedules that are custom tailored for major load centers.

Cogenerators and the utility/customer may question the necessity of a gas distribution utility in the transport chain. Bypass is currently a very controversial topic and even some states which take a free market attitude toward regulation have enacted anti-bypass measures. The distribution segment must be analyzed in perspective with the other components of the transportation plan. To the extent that an LDC complements the entire transportation chain, it adds value. Services such as storage, balancing and emergency gas advance may actually enhance reliability and reduce the cost of the entire transportation chain.

**The Transportation Plan**

A successful transportation plan will be as inexpensive as possible when taking into consideration an acceptable level of risk. A plan which incorporates firm transportation from wellhead to burnertip will undoubtedly be relatively expensive. It is likely that the optimum plan will incorporate both firm and interruptible transportation service on perhaps multiple transmission systems and possibly one of the transportation segments may be eliminated altogether. The combinations and permutations of potential transportation paths are virtually unlimited.

Operating flexibility will determine how reliable the transportation will be. Certainly, all transportation services are not the same and operational flexibility varies from transporter to transporter. The ability to utilize multiple receipt and delivery points is very important Firm service limited to one receipt point cannot be reliable because of the operational characteristics of a gas well. For example, if the gas supply is located in the Gulf of Mexico, the power purchaser can count on losing significant production each year because of hurricanes and tropical storms. If the transporters do not have liberal gas advance or balancing provisions there may not be gas available for the plant even though the cogenerator is obligated to pay demand charges. Another concern is whether the cogenerator will have to shut in production during scheduled plant maintenance periods or will the pipeline store for the duration of the shutdown?

It would probably be appropriate to have several redundant interruptible transportation agreements with other combinations of shippers to allow emergency alternatives and the ability to reduce costs as opportunities arise. By using interruptible agreements, the fixed costs of demand charges can be avoided.

The marketplace provides the options. Each utility must determine which options it prefers, so it can send the appropriate signals to its prospective suppliers as they enter preliminary negotiations for their fuel supplies.

# APPENDIX E-5

## PERMITS AND LICENSES CHECKLIST

It would be useful to identify the appropriate agency, and even the name, address and telephone number of a contact at each agency, and a description of its mandate in issuing permits and licenses. Of course, not all of the following will apply to each project, and there may be requirements which haven't been listed. This Appendix may nevertheless serve as a starting point for constructing the power purchaser's own permitting and licensing listing.

### Federal

- Federal Aviation Administration Notification of Proposed Construction
- NEPA Certification
- U.S. Army Corps of Engineers Section 10 Permit
- U.S. Army Corps of Engineers Section 404 Permit
- U.S. EPA NPDES Permit

### State

- Coastal zone management certificate of consistency
- Cross connection permit
- Environmental impact statement
- Floodplain development
- Gas pipeline approval
- Groundwater discharge permit
- Historical Commission approval
- Industrial user discharge permit
- Oil storage tank construction permit
- PSD/air plans review
- Sewer extension/connection permit
- Siting approval
- Solid waste facility operating permit
- Solid waste facility site assignment
- Surface water discharge permit (with NPDES)
- Water quality certification
- Water withdrawal permit
- Wetlands approval of local order of conditions

### Local Approvals

- Board of Health
- Building Inspector
- Conservation Commission
- Department of Public Works
- Fire Department
- Historical Society
- Planning Board
- Water Department/Sewer Commission
- Zoning Board of Appeals

# APPENDIX E-6(A)

## MAJOR CONSTRUCTION FINANCING RISKS

### Delays in Construction

As any utility which has been involved in the nuclear industry can attest, delay can be expensive. Interest expenses can easily exceed the noninterest costs of construction of a powerplant. These risks are often assigned to the turnkey contractor. Typical causes of delay include:

- Permitting and other governmental approvals
- Litigation
- Equipment unavailability
- Personnel unavailability
- Transportation problems
- Cost overruns

Cost overruns may be due to any number of factors, some of which may be controllable, others not. Typical reasons for overruns will include:

- *Utility requests for project modifications.* In this case, the utility will be required to cover the cost overruns. As the developer has a strong interest in keeping costs as low as possible, unless there is adequate specification of the project, the utility may find itself subject to charges for modifications to achieve a level of quality it thought it had already paid for.
- *Incorrect estimates.* This is generally due to a failure to accurately account for all costs or contingencies. Original estimates may be off even with the best of analysis. Particularly to the extent that there are provisions for passthroughs in the event an estimate proves incorrect, independent evaluation of cost and contingency estimates is needed.

### Design and Operating Risks

The plant may not operate adequately without unbudgeted work. Heat rate, availability, and shifting emissions targets may not be met without additional expenditures. Additional personnel, or more training than originally budgeted, may be required to achieve optimal operation.

In each case, the question must be answered as to who is responsible, and to what extent, for costs associated with each type of overrun. The risks are relatively open-ended and may be handled, to some extent, through liquidated damage provisions.

# APPENDIX E-6(B)

## INSURANCE AND OTHER FORMS OF CREDIT ENHANCEMENT

There are basic insurance requirements which should be fulfilled by any electric power supplier. These include casualty, property damage and bodily injury, automotive and workman's compensation insurance. None of these are particularly complex or controversial although periodically there are runups in the cost of ordinary liability insurance, and it may even prove difficult to obtain in certain markets.

Rather, here we are concerned with the use of some innovative insurance products as credit enhancements. The best of these products can be helpful in assuring operations, on-time construction, availability, or other factors already discussed. For purposes of this factor, the purpose of insurance is to provide credit support for the project. However, if the insurance is "smart" insurance, i.e., is sold by a company which has the ability to oversee construction and operations, and to help assure these are done properly, then it can be a valuable service to the utility also. Requiring that such insurance be carried beyond the debt service period could provide valuable comfort to the utility relying on continued availability of independently-supplied capacity. These include:

- Insurance of financing costs associated with force majeure caused delays and penalties associated with those delays;
- Insurance to repay the construction lender in the event of physical loss, damage or destruction of the project or any physical components thereof;
- Insurance to guarantee a predetermined level of project production and the cash flows associated with that production (systems performance or efficacy insurance);
- Insurance against named perils, unexpected business interruption due to physical damage;
- Environmental impairment liability insurance;
- Errors and omissions insurance; and
- Performance bonds.

These forms of insurance can help assure that the project will continue providing power, or at least the wherewithal for the supplier to pay the utility/customer for purchasing replacement power, under a broad range of circumstances.

Yet, as always, offloading risk has a cost. The premiums demanded, and the terms of coverage will vary with market conditions, and may not be worth obtaining in all situations. In some circumstances, the utility/purchaser may itself be the most cost-effective "insurer," due to its ability to diversify its generation portfolio. Of course, in such circumstances, the price charged for energy and capacity should reflect the additional service provided by the utility/customer.

# APPENDIX E-6(C)

## FINANCING AND AVOIDED COSTS UNDER PURPA

Generally, independent power projects are financed using project finance techniques, under which most of the money going into the project is debt, which is supported through a combination of the right to be paid by the utility/customer, which guarantees a market for the power, and a variety of commitments to take on responsibilities by the members of the project development team. The commitments, their form and strength, are so important that they often take the role of equity in the project.

These commitments are thus the key to evaluation of the bid, as their terms will specify who is taking on the responsibilities for each component of the project.

### The "Matching" Problem

Many successful bidders to supply new generating capacity will construct plants which are qualifying facilities under PURPA. As we discussed earlier in Section E, there are numerous methodologies for calculating a utility's avoided costs. One, generally called the "proxy method," is even based upon the costs of constructing and operating a particular plant. However, all of these methods, except if one were planning a facility which in all particulars was like the proxy, produced rates having nothing to do with the real options available to cogeneration and small power project developers.

Simply put, the payments offered to the cogenerators and small power producers and the costs incurred in building and operating their facilities, don't "match". They have nothing to do with each other, except by happenstance.

This problem place both prospective supplier of capacity and customer in a quandary. How can the legal avoided cost ceiling be reconciled with the cash flow requirements of projects?

The answer has generally been the use of "reconciliation" or "recapture" accounts.

Under this scheme, two payment streams are calculated: one based upon the utility's avoided cost, the other on the real requirements of the offeror. To the extent that the offeror requires payments in excess of avoided costs in order to meet debt service requirements, a "reconciliation" or "recapture" account is kept, which tracks the "overpayments" and accrued interest. This is to be paid back after debt service has been completed.

In theory, when the plant completes its contractual obligation, all overpayments and interest have been repaid to the utility-customer. Sometimes the early overpayment is even characterized as a loan.

Few projects have reached the stage at which the repayment obligation comes into play. Yet a great deal of effort has been expended in attempts to "secure" this repayment.

Typically, repayment is secured through a junior lien on the plant itself. As this is generally the cogenerator/small power producer's only asset, it is the only form of security available. On occasion, the "avoided cost" payment is high enough to permit a cash, or cash equivalent escrow account to be set up from cash flow, which will partially or totally secure repayment.

Neither type of security is likely to be very effective. To the extent that the cogeneration/small power facility has a value, it is linked to the cash flow to be generated from its power sales agreement. If the cash flow becomes inadequate to support payback of the reconciliation account, then the junior lien on the plant is inadequate security for the "overpayment". If there is a cash escrow supporting repayment, then the contract was probably too rich in the first place, otherwise no funds would be available for the cash escrow.

At bottom, the recapture approach is a sham, designed to reconcile what is generally irreconcilable – utility avoided costs, which are based on one plant or a mix of plants, and cogenerator/small power producer financial requirements, which are project specific and completely unrelated to the utility's avoided costs.

When PURPA was first implemented, the right to sell power was deemed an entitlement for qualifying facilities. The avoided cost measure was the level of the entitlement. Competitive procurements are not entitlements. In order for them to work, pricing must be wholly related to the product to be delivered, and not to an arbitrary avoided cost standard. Each different plant is likely to have different financing requirements, depending upon the project's component parts.

Direct comparison of prices bid by different suppliers is therefore not likely to be fruitful on its own. Pricing must instead be judged to be appropriate relative to the project bid, and the project bid appropriate to the needs of the utility-customer. Significant differences in prices bid by similar plants might best be taken as a warning that something is amiss in the plans of the low bidder. In the absence of mitigating circumstances, low bids are likely not to be financeable.

This is particularly applicable in the common circumstance in which there is an attempt to require a financial structure which will not result in a rate increase. Power from new generating capacity often is more expensive than power from existing capacity. While financial structures may be arranged which temporarily mitigate the effects of this, that mitigation will generally cause higher life-cycle costs.

The utility-customer should be aware of its options (in terms of its prospective suppliers' real constraints), make a conscious decision as to its preferences, and communicate those preferences to prospective bidders so that they may configure their bids accordingly.

However, while the concept retains some importance from a regulatory perspective, comparisons of price offers to an avoided cost standard will generally not yield analytically useful information.

### Getting to the Reality of Project Finance

It is often pointed out that cogenerators and other independents are far more heavily leveraged than utilities are permitted to be by law, and that this leverage, while the source of apparent cost reductions, may also be the source of additional ratepayer risk, as the enterprise becomes less reliable.

This is not necessarily so. Commonly, guarantees provided by project participants take the place of traditional equity in cogeneration and independent power projects.

The utility/customer's guarantee that it will purchase the output of the facility is but one example of the substitute-for-equity in what would otherwise be unacceptably leveraged transactions. Utilities which are unable to uphold their traditional part of the risk equation, i.e., the guarantee of a market under known terms, may be unable to obtain new supplies. While once thought unlikely, events in recent years have made the possibility of the utility/customer's bankruptcy a real consideration during financing. A utility against whom the obligation to purchase is inadequately enforceable will find it increasingly difficult and expensive to purchase new supplies, as someone else will be required to absorb the demand-related risks traditionally absorbed by the utility/customer. Available terms will thus vary not only due to market conditions, but also due to the condition of the utility/customer.

Ideally, financing is the mechanism by which risks are assigned to those who can best understand, and control those risks, in return for associated rewards. Uncontrollable and unforeseeable risks, if not insurable will be assigned to those willing to bear them at lowest cost.

The ideal is rarely achieved, however. The major reason for this failure is a lack of full understanding of the risks associated with electric power projects by one or more parties. It is not sufficient to simply off-load all risks possible. This can only be achieved at a price. It is more cost effective to retain risks which one can control best. The premium required to off-load such risks will necessarily be too high.

It is therefore necessary, when analyzing the capital structure of a project, to look beyond the face of the numbers to determine who is really taking which risks, and garnering the rewards associated with a successful project.

As noted above, every project responsibility should be specified, assigned, and enforceable. If this is done properly, a project may be 100% debt financed, and still present a perfectly acceptable risk to the utility and its ratepayers.

For example, if an uninterruptible fuel supply is guaranteed by a major natural gas company, which is subject to damages for failure to deliver, this guarantee will do more to provide project reliability than even substantial project equity. The value of the fuel throughput will be greater than the value of the powerplant. Similarly, guaranteed availability and heat rate, with enforcement provisions running to the benefit of the utility-customer, will be more valuable than a few percentage points of equity should there be problems.

But the question of appropriate risk and reward will also be dictated by market circumstances, and not by any absolute values.

For example, in a buyer's market, the utility-customer may be successful in laying off virtually all risks upon a mix of contractors, investors and insurers. A hungry supplier may offer options not available during normal times. Perceived risks may also be affected by market circumstances.

This has sometimes seemed to be the case in recent years, as such apparently intractable risks as those relating to fuel costs, inflation, unforeseen regulatory changes and environmental requirements, have been absorbed by prospective cogeneration and independent power projects.

In the early Eighties it was possible for utilities to purchase power at costs linked to their marginal costs of production, as the market perceived little risk that those costs would fall. For example, a utility purchaser which had oil or gas fired generating capacity on the margin would be able to purchase solid-fuel fired generating capacity at a price linked to the cost of operating oil and gas units.

The drop in oil and gas prices in 1985, and the reduction in generating costs which followed, rendered the debt on many of these solid fuel projects unserviceable. In response, today that type of risk would be difficult and expensive to off-load.

In a seller's market, which may quickly develop as power supplies tighten, experienced and reputable project development team members will be less willing to take on risks which they cannot readily control, simply because there are other customers who won't require them to do so, or because they are unwilling to risk an unprofitable contract.

Financial risks are in large measure reflective of the other risks associated with power projects. In the early years of PURPA implementation, many utilities even considered investors' financial review of projects to be adequate to protect the utility's interest. This clearly is not true: investors are neutral as to whether risks are assigned to the utility/customer, or to others involved in the project, so long as the risk does not fall upon their interests.

---

## APPENDIX E-7

### AVAILABILITY, RELIABILITY AND INTEGRATION INTO THE SYSTEM

Questions which should be answered by bidders to allow the power purchaser to evaluate availability, reliability, and ability to integrate the facility into the system include:

A. Facility generating and outage information. (NERC definitions may be employed.)

B. Proposed method of operation (base load, cycling, peak load, other) and limitations of such operations.

C. Startup time from cold conditions to synchronization.

D. Ramp rate in MW/minute for automatic and manual control for normal and maximum sustained rates.

E. Minimum and maximum continuous capability in MW for automatic and manual control.

F. Net reactive capability, including details of excitation system capabilities.

G. Host dependency if facility is a cogenerator.

H. Net heat rate under expected operating conditions.

I. Regulated voltage range.

J. Startup and shutdown operating restrictions, such as soaking times, cooling times and MW output limitations.

K. Participation in automatic generation control systems.

L. Extent to which facility is subject to commitment/decommitment on a daily and/or weekly basis.

M. Availability guarantees.

# APPENDIX E-8

## PROVISIONS TO CONSIDER FOR INCLUSION IN THE BASE POWER SALES AGREEMENT

The actual drafting of the agreement should be a multidisciplinary effort, in which members of the various technical specialties explain their requirements and preferences to qualified legal counsel who can accurately and effectively reflect those requirements and preferences in the power sales agreement.

1. Definitions
2. Term of agreement
3. Deferral or cancellation
4. Sale of power
5. Performance guarantees
6. Operation
7. Maintenance
8. Facility and equipment design and construction
9. Delivery and metering
10. Billing and payment
11. Land rights; facility and records access
12. Force majeure
13. Insurance
14. Liability
15. Indemnity
16. Representations, warranties and covenants
17. Events of default
18. Liquidated damages
19. Assignment
20. Regulatory approval
21. Notices
22. Amendment and modification
23. Joint/several obligations
24. Waiver
25. Captions
26. Disputes
27. Waiver
28. Choice of law
29. Entirety, severability
30. Notices

**Contract Appendices**

a. System guidelines
b. Interconnection requirements
c. Security
d. Project critical path schedule
e. Equivalent availability factor
f. Pricing formulae
g. EPC contracts
h. Fuel supply contracts
i. Fuel transportation contracts
j. Financial agreements
k. Dispatchability
l. Licenses and permits

# APPENDIX E-9

## KEY PROJECT STAGES

**Project Initiation**

1. Finalize Development team
2. Execute contracts

**Planning**

3. Project program
4. Project budget
5. Environmental compliance plan
6. Site evaluation
7. Coordinate with permitting agencies
8. Conceptual estimate
9. Study financial feasibility
10. Preliminary project master schedule
11. Secure financing commitments

**Design Schematic**

12. Refine program budget estimate
13. Prepare detailed design schedule
14. Select primary components and systems
15. Prepare functional flow diagrams
16. Prepare schematic drawings
17. Prepare outline specifications
18. Conduct environmental study
19. Prepare alternative design schemes
20. Conduct economic operations studies
21. Conduct value engineering analysis
22. Establish project execution/management procedures
23. Develop bid package format
24. Identify long-lead purchase items
25. Develop construction schedule
26. Initiate preliminary permitting agency reviews
27. Initiate preliminary insurance company review
28. Update conceptual estimate
29. Update master schedule
30. Update project budget

**Design Development Phase**

31. Refine program
32. Update design schedule
33. Evaluate component and system alternatives
34. Refine outline specifications
35. Complete design development drawings
36. Establish generation conditions
37. Prepare preliminary construction estimate
38. Prepare preliminary construction schedule
39. Secure permitting agency approvals
40. Evaluate labor and trade contractor market
41. Prepare trade contractor bid lists
42. Prepare vendor bid lists
43. Update master schedule
44. Prepare long-lead purchase and construction documents
45. Complete project financing

**Contract Document Phase**

46. Finalize project budget
47. Prepare cash flow schedule
48. Review contract plans and specifications
49. Finalize trade contractor bid lists
50. Final permitting agency approval
51. Final insurance company review
52. Update construction schedule

**Contracting**

53. Advertise for trade contractor bids
54. Conduct trade contractor pre-bid conference
55. Receive and tabulate bids
56. Update construction and master schedule
57. Issue notice to proceed
58. Develop detailed master schedule

**Construction/Startup**

59. Mobilization for construction
60. Verify insurance requirements
61. Approve trade contractor progress payments
62. Maintain shop drawing and sample control
63. Provide project cost control
64. Evaluate progress and update construction scheduling
65. Change orders
66. Inspect and monitor work quality
67. Inspect for conformance to design
68. Coordinate startup schedule
69. Conduct final inspections
70. Startup
71. As-build drawings, manuals and warranties finalized
72. Perform final accounting
73. Secure operating permits

# APPENDIX E-10

## SAMPLE PROJECT CONSTRUCTION SCHEDULE

**(Hypothetical Project Mass-burn coal-fired facility)**

| Activity Description | Start | Finish |
|---|---|---|
| 1. Permitting | Dec. 90 | Mar. 92 |
| 2. Award of contract | April 91 | April 91 |
| 3. Engineering | April 91 | Dec. 93 |
| 4. Manufacture of equipment | Sept. 91 | Oct. 94 |
| 5. Start construction | April 92 | April 92 |
| 6. Site work | April 92 | Aug. 95 |
| 7. Foundations | June 92 | Sept. 93 |
| 8. Electrical | July 92 | June 95 |
| 9. Boiler | Sept. 92 | July 95 |
| 10. Buildings | Oct. 92 | Feb. 95 |
| 11. Condenser | May 93 | Oct. 94 |
| 12. Scrubber and baghouse | March 93 | April 95 |
| 13. Chimney | May 93 | April 94 |
| 14. Coal handling facilities | June 93 | Jan. 95 |
| 15. Switchyard | Aug. 93 | Dec. 94 |
| 16. Turbine generator | Feb. 94 | June 95 |
| 17. Checkout | June 94 | Nov. 94 |
| 18. Startup | Sept. 94 | Dec. 95 |
| 19. Synchronization | Aug. 95 | Aug. 95 |
| 20. Commercial Operation | Dec. 95 | Dec. 95 |
| 21. Performance testing | Oct. 95 | May 96 |

# Section F

## New Entrants into the Competitive Power Marketplace

# Section F

## New Entrants into the Competitive Power Marketplace

In recent years a new breed of independent generator has arisen, the so-called independent power producer (IPP). Like QFs, this type of generator is generally permitted to price its power at market-based, rather than cost-justified, rates and is exempted from PUHCA. At the same time utilities and their affiliates have also been granted, in some cases, greater flexibility to compete in the competitive marketplace.

This section discusses historical obstacles to the rise of independent power, when FERC will allow competitors in the competitive power marketplace to charge market-based rates for their power, and those type of facilities qualifying for relaxed treatment under the Public Utility Holding Company Act (PUHCA).

### HISTORICAL OBSTACLES TO THE RISE OF INDEPENDENT POWER

Historically, there were several significant obstacles to the rise of a competitive electric generating sector.

First, the electric industry was a declining cost industry until the 1970s. It was widely perceived that economics favored vertical integration and the economies of scale which were present at this time seemingly supported this notion. Moreover, with rates declining, customers felt little need to explore alternatives. Not surprisingly, they decided to focus on the their core business and left the electric generation business to the local utility.

Second, utilities and their regulators guarded their franchise zealously. Backup power rates were kept extremely high so as to deter on-site generation. Moreover, anyone who built their own on-site generation capacity had no market for their power. Utilities either refused to buy that power or paid next to nothing for it.

Federal regulation reinforced this trend. Any owner of a plant selling power into the bulk power markets was subject to rate of return regulation under the Federal Power Act. If a company (other than a utility) owned more than one plant, it would become subject to the Public Utility Holding Company Act, which would have placed all its corporate activities under strict regulatory scrutiny.

The first break in the utility monopoly over power generation occurred with the passage of PURPA. As noted in Section B, this legislation provided a market for power sold by cogenerators: the local utility had to buy it at that company's "avoided cost". Moreover, any facility qualifying for PURPA treatment was guaranteed backup power at nondiscriminatory rates and exempted from the requirements of the Federal Power Act and PUHCA.

At the same time qualifying facilities from PURPA began generating large amounts of power, there were significant changes in attitudes occurring within the utility industry and at FERC.

Many within the utility industry began looking to independents for large amounts of power. Burned by large disallowances in the cost of plants coming on-line in the mid-1980s, many utility executives began realizing that the decision to build a large, baseload plant was equivalent to a "bet the company" proposition. Believing the regulatory compact broken, they decided purchasing power was a less risky alternative. Moreover, a new generation of utility executives were gaining power within the industry. Many believed new approaches had to be found to serving their customers needs.

Meanwhile, FERC became concerned about the potential for shortages. The Commission recognized that cogenerators, with their need for a steam host, might not be able to satisfy this need entirely. Consequently, FERC began promoting independent power producers.

Proponents of IPPs found three significant obstacles that impeded the concept from becoming a reality. First, the traditional means of administering the Federal Power Act, rate of return regulation, deterred investment in new facilities. Another deterrent was PUHCA. Finally, there was a need for greater access to the transmission system. Otherwise, IPPs would be at the mercy of the host utility.

The issue of transmission access is the subject of Section G. The remainder of this section is devoted to the effort to overcome the first two obstacles.

## ENTITIES ALLOWED TO CHARGE MARKET-BASED RATES

As FERC sought to promote IPPs, it recognized that there were problems with using rate-of-return regulation to govern new entrants. Rate of return regulation was designed as a means of dealing with monopoly providers; its use simply is inconsistent with a competitive marketplace. Its continued use would likely thwart the agency's goal of attracting capital to build independent plants. Moreover, traditional regulation had high transactional cost and sent incorrect price signals. Consequently, FERC increasingly began examining proposals to allow various entities to charge market-based rates.

In general, FERC decisions on market-based prices divide sellers into three categories, distinguished by their perceived market power (which in FERC-speak has meant the seller's control over transmission to the prospective purchaser).

The first category consists of independent power producers (IPPs). These have been defined as sellers who don't own or control transmission and have no affiliation with a traditional electric utility having a franchised service area.

A second group is affiliated power producers (APPs). These producers are affiliated with a traditional utility, though they may be a separate corporate entity.

Finally, there are utility producers (UPs, to coin a new acronym). These are traditional utilities seeking to sell power to their neighbors at noncost-based rates.

## KEY FACTORS IN FERC CASES

FERC's decisions in cases involving market-based pricing for long-term power sales tend to focus on several key issues.

First, does the seller have market power over the purchaser? FERC analyzes a seller's market power by examining evidence as to the following:

- whether the seller or any of its affiliates is a dominant firm in the sale of generation services in the relevant market;
- whether the seller or any of its affiliates owns or controls transmission facilities which could be used by the buyer in reaching alternative generation suppliers; and
- whether the seller or any of its affiliates is able to erect or otherwise control any other barrier to entry.

FERC has defined market power as the ability of a seller to significantly influence the price of power, either by restricting supply or by denying the purchaser access to other sellers.

In the PSI case and elsewhere FERC has said its primary concern in a market power analysis is that customers have genuine alternatives to buying the seller's product. Opinions mention various means of analyzing market power: HHI determinations, market shares, concentration ratios, shares of total generation capacity in the region, and potential entry by utility and non-utility generators. However, the Commissioners consistently have said they do not believe any one type of evidence is sufficient for this analysis, and that they would not rely on any mechanical analysis in this regard.

Indeed, in reading the FERC decisions, one sometimes gets the feeling FERC makes up its mind what to do in a particular case and then writes up a market power analysis to support it. Perhaps the best that can be said is that, in FERC's view, market power is like obscenity: one knows it when one sees it.

If the seller or its affiliates have market power, FERC looks to see whether they have adequately mitigated this power.

FERC then looks at the relationship between the parties, the so-called "affiliation" issue. This involves concerns of self-dealing and preferential pricing. That is, a utility might sell its power to an affiliated broker at less than cost so as to divert money from ratepayers to shareholders. FERC also is concerned about the potential for reciprocal dealing, as might occur when utility A buys power from utility B at market-based prices in exchange for an implicit promise on B's behalf that it will pay higher prices for A's power.

Finally, in reviewing these cases FERC often seeks to ensure that a "market process" is at work. This has meant the agency will examine a specific transaction to assure itself that the process by which the power was selected was designed to yield a reasonable result.

### IPPS

After a rough start, FERC has now made its approval of market-based rates for IPPs almost routine. FERC now approves market-based rates for true IPPs without even discussing them at Commission meetings.

Past cases reveal that FERC will look at three factors before approving market-based rates for IPP.

First, it will check for the presence of a substantial number of alternative suppliers.

In Enron the applicant's offer represented 4% of the total capacity bid. Suppliers proposed 23 times the amount of capacity sought; more than 90% of the responses included proposals for non-peaking projects which competed directly with the Enron Power bids. Moreover, bids were received from cogenerators, small power producers, IPPs and a traditional utility, representing existing or proposed facilities in seven states. Therefore, the Commission concluded NEPCO had additional alternatives to choose from if it had not selected Enron Power.

Likewise, in Doswell the IPP traced its ancestry through a QF which won a solicitation where 27 developers offered seven times the amount of power sought in the RFP.

FERC will then examine the purchaser's ability to satisfy its needs from multiple bidders. In the Enron case NEPCO did not rely on Enron Power's bid to satisfy its entire 200 MW need, even though the supplier offered to sell the full 200 MW. The Commission noted NEPCO executed power agreements with three other developers for 124 MW and declined to execute contracts with 18 bidders that passed the first cut in the solicitation.

Finally, the agency will look at the record to determine whether the purchaser had bargaining power – i.e. whether the contract reflects the give-and-take of negotiation. For example, in Doswell it was satisfied this condition was met since the contract the purchaser was able to negotiate contained favorable nonprice terms and conditions.

In Enron the Commission noted that NEPCO was able to negotiate several favorable provisions in the agreement: NEPCO has the right to suspend its obligation to accept deliveries and pay monthly kilowatt-hour charges; NEPCO has the unrestricted right to terminate the agreement after fifteen years; and NEPCO has the right to receive capacity charge reductions if Enron Power's total unit price exceeds NEPCO's wholesale revenues on a per unit basis.

However, FERC has not required an IPP to win a competitive solicitation to obtain a market-based rate. The Enron contract resulted from a process combining elements of a competitive bid and negotiation. In Dartmouth Power the Commission approved market-based rates even though the contract resulted solely from negotiation.

As the analysis of these cases indicate, FERC has not been all that tough in reviewing filings by true independents. As we shall see, this contrasts with its treatment of cases involving APPs and utilities. FERC has approved IPP contracts resulting from a market process which would have been unacceptable if it had involved an APP or traditional utility. Part of this likely results from the agency's clear desire to promote IPPs. Another reason is FERC's recent fixation on transmission.

Will this different treatment continue? Is it justified?

The answer is that is probably too early to tell. Market power with regard to IPPs involves different considerations than it does for other producers. FERC has not yet had to deal with these issues in any significant fashion.

As the independent industry consolidates, one is likely to be confronted with many cases where a traditional equipment or fuel supplier to a utility now proposes to build an independent plant to sell power to it. Here, as in the APP cases, there is the potential for cross-subsidization and preferential pricing.

The one case where FERC discussed this issue did not provide a good test of the agency's feelings toward this issue. The Commission found a very clean situation in Enron Power, noting the applicant will purchase gas from an unaffiliated supplier and that the gas will be transported by an open-access pipeline and an unaffiliated local distribution company. Thus, Enron Power did not compete on more favorable terms than other bidders in the NEPCO solicitation that used natural gas as a fuel. Further, there is no evidence of reciprocal dealing since neither NEPCO nor its affiliates are involved in any other arrangements with Enron Power or its affiliates.

Future cases are unlikely to be so clean. If Enron or GE propose to sell power to one of their traditional utility customers, will FERC examine all other arrangements between the companies to ensure that these contracts aren't tied in some way to the independent project? That, to date, is a major unanswered question.

Looking further into the future, one also must consider the possibility, particularly now that PUHCA has been reformed, that certain independents will begin dominating certain regions. Again, the present status of FERC case law gives us no guidance on this issue.

One further issue should be raised. In some early cases FERC looked at a factor called "other barriers". FERC has defined "other barriers" as the ownership or control by the applicant or its affiliate of "resources which could have been used to create any barriers to other suppliers who sought to enter the market in response to [a] solicitation". Such barriers could result from control of key inputs to electricity production or the transportation of those inputs. In Doswell the Commission said such a barrier might include sole rights to the last or one of the last viable sites for new generating capacity located in or near the buyer's service territory.

To date, however, the Commission has not found such a barrier in any of the cases it has considered and it has not explicitly examined this factor in recent cases.

## APPS

In contrast to the relative ease with which IPPs have gotten their rates approved, APPs have had a much more difficult time getting permission to charge market-based rates. An obvious concern is that the affiliate's parent will use its control of the grid to frustrate competitors. In addition, FERC also is concerned about any potential for self- or reciprocal-dealing.

Self dealing issues arise in any transaction where an entity may sell to or buy from an affiliate on terms more favorable than would be available to other market participants. The concern is the purchaser won't have an incentive to negotiate the best package of price and other terms. The reason for this is that the affiliates' interest in generating profits for shareholders owning both entities creates an incentive to maximize profits for the affiliate at the expense of the utility's ratepayers.

The concern about reciprocal dealing is that utility X's APP will receive more favorable treatment from utility Y in exchange for utility Y receiving additional benefits from utility X.

The APP cases break down into two types: (a) an APP selling power to an unaffiliated utility and (b) an APP selling power to its parent.

### APP Selling Power to Unaffiliated Etility

FERC approved market-based rates for an APP in a case involving Commonwealth Atlantic, a partnership building a powerplant in Virginia. The partnership included Mission Energy. Mission is a wholly-owned subsidiary of SCEcorp, a holding company which also owns Southern California Edison, a utility with its owns service territory.

In FERC's view Commonwealth's affiliation with a franchised utility that has captive customers "raises a warning flag". FERC went on to say:

> However, such affiliation only becomes a problem when there are transactions or locational considerations that can give rise to favoritism. The Commission's concern is not with the affiliation per se but with the abuses that may arise when the affiliates do business together or when the affiliated utility uses its control over transmission or some other entry barrier to block others from competing against its affiliate.
>
> There is no evidence that these abuses were present here. First, Commonwealth is not purchasing power or transmission services from Mission, Southern California

Edison or any other subsidiary of SCEcorp. Second, Commonwealth is not selling to any affiliated entity. Third, Virginia Power could not use Southern California Edison's transmission facilities to reach any of Commonwealth's competitors. Fourth, neither Commonwealth nor any of its affiliates is affiliated with Virginia Power or any of its affiliates.

FERC made special note of the fact Southern California Edison's 7,000 circuit miles of high voltage transmission lines are so distant from the geographic market in which Virginia Power sought new capacity that the lines could not have been used to prevent Virginia Power from reaching competing suppliers, and thus could not have been a source of market power to Commonwealth.

Apart from the Commonwealth project, FERC found that the only business relationships between Commonwealth and its owners and affiliates, and Virginia Power and its owners and affiliates, are other contractual agreements for power sales. Mission owns three QFs in Virginia Power's service territory, but each had already signed a power sales agreement with Vepco by the time this contract was signed. Given this factual situation, Mission could not have exercised market power over Virginia Power in the negotiation of these QF contracts.

The Commission next addressed potential abuses of reciprocal dealing. A Virginia Power affiliate, Dominion, is a part-owner of three QFs that sell power to Southern California Edison. This raises the potential for reciprocal dealing in that it is possible Vepco could have agreed to pay more for power to Commonwealth in return for Southern California Edison paying more for power from the QFs in which Dominion, Virginia Power's affiliate, has an interest.

The Commission relied on the several findings to dispel its concerns about reciprocal dealing. First, the rates Southern California Edison has agreed to pay the QFs in which Dominion has an ownership interest result from standard offer rates approved by the California Commission. Moreover, there is no evidence to suggest Virginia Power gave special treatment to the bid here or a bid by the same parties in a subsequent solicitation. Lastly, neither Commonwealth nor any of its owners or affiliates has undertaken any joint ventures with Virginia Power or its affiliates. A turnkey contract between a Mission subsidiary, Mission Power Engineering Company, and a project in which Dominion has an ownership interest is not in FERC's view a joint venture because Mission will not participate in the risks/benefits of the project.

While Commonwealth Atlantic involved an affiliated producer whose parent's service territory was several thousand miles away, FERC has taken a hard look where APPs have been used to sell power to utilities located near the parent's franchise.

In one of the first decisions on market based pricing, FERC approved market-based rates in a decision involving Ocean State Power. In this case three of the four utilities purchasing power from the facility had an ownership interest in it. FERC concluded self-dealing was not a problem because (a) the rate charged the owner-purchasers was the same as that charged the nonowner-buyer and (b) the local regulatory commissions had found that this power was priced below each purchasers' avoided costs.

However, the continuing validity of Ocean State Power has been called into question by a series of cases between 1990 and 1992.

In Portland General Exchange the Commission said the utility, if it desired to sell power to an affiliated broker which would then sell the power to third parties, should be required to offer to others similar terms and conditions of service as those extended to its affiliate. The fact that PGE was not asked to provide similar service is irrelevant: the purchasers may have wanted this service if it had been available.

In TECO FERC applied a similar condition where a utility permitted its affiliate to bundle existing capacity in a proposed power sale (even though the affiliate was selling that capacity at no markup). It said TECO should have allowed non-affiliated suppliers to bundle that capacity with their own offers. It also required TECO to wheel power for any potential competitors if it wished to obtain market-based rates.

In Terra Comfort the Commission came down hard on another utility affiliate. Terra Comfort, an affiliate of Iowa Southern, sought market-based pricing for a sale of power to Iowa Electric. FERC imposed similar conditions to those contained in the TECO order.

This case is also interesting for the tone it conveys toward utility affiliates. Though FERC had ample grounds to reject the proposal, the Commission substantially expanded the burden of proof on the market power issue for a utility affiliate. It said an "independent" reason for making this decision was the lack of evidence that Iowa Southern did not exercise market power over Iowa Electric.

To reach this conclusion, the majority had to do some reaching: Iowa Electric is interconnected with six utilities other than Iowa Southern and has transmission access to two other utilities, through its part ownership of a line. The majority, however, "found" that because Iowa Southern also owned a portion of this line it could thwart any attempt Iowa Electric might have made to purchase power from others. This analysis ignores two other important facts: (a) that there were six other owners of the line and (b) Iowa Electric never requested the use of this line for this purpose.

The Commission found that the existence of an interconnection does not imply the availability of transmission. The very fact that the applicants in this case could deny Iowa Electric access to any seller could foreclose more economical alternatives. Moreover, other utilities with which Iowa Electric was interconnected did not have a generally available tariff.

The majority then made the following statement without citing any evidence: "Iowa Electric and its potential suppliers knew there was no point in asking for transmission. Iowa Southern's position would be substantially strengthened by evidence of a positive offer to provide transmission service rather than a showing that no requests were made."

In another case involving Iowa Southern FERC required the utility to charge one of its municipal customers, the City of Pella, the same transmission rate it offers its affiliate, Terra Comfort. In this filing, Iowa Southern proposed a rate of \$1.10/kW/month, reflecting its full cost-of-service rate. It offered the same service to Terra Comfort at \$0.69/kW/month. The Commission stated that, while it normally does not view rate discounts as being in violation of the Federal Power Act, in looking at dealings between affiliates the Commission is presented with a different set of concerns because of the incentive for preferential transactions. The Commission explained, therefore, that it only authorizes rate differentials involving an affiliate where there are cost or service-related differences. In this case, there were none, and, thus, no legal justification for the rate differential.

Thus, while Ocean State Power has never been explicitly overruled (FERC never seems to admit it has changed its mind), its value as a precedent clearly has been eroded.

A distinguishing feature of these cases is that FERC has, to date, paid much greater attention to the process by which an APP obtains a contract than it has in its decisions on IPPs.

In TECO the Commission carefully reviewed the response to the procurement process. Here, in rejecting the proposal, FERC said it was troubled by what it called the "sparse response" to the purchaser's bid and the fact that the arrangement agreed to was not the same as that initially proposed. (The solicitation in this case yielded only nine bids for about 4.5 times the amount of capacity sought.)

In Sun-Peak the Commission rejected market-based rates for an IPP where the purchaser did not undertake a "well-organized and comprehensive market search, either by competitive bidding or negotiation, which can provide evidence that the seller did not have market power over the affiliate". Though the Commission did not directly address the APP issue (and the ruling seemingly would apply to IPPs as well), various background information makes it clear the FERC staff was especially wary because of the affiliate issue.

Even in Commonwealth the Commission carefully reviewed the procurement process. Here, FERC noted that Commonwealth's offer represented 1.6 percent of the total capacity bid and that the bid was less than 4 percent of the cumulative capacity offered to Virginia Power through 1992. FERC noted that the applicant's bid was compared with all other bids and Virginia Power's own expansion plan. Even after being chosen for negotiation, Commonwealth still could not exercise market power: 11 out of 30 finalists were not selected.

One further matter should be mentioned: other than in Ocean State Power, FERC has given no credence to a state's findings that the deal is in the best interest of its ratepayers. In both TECO and Sun-Peak the Commission struck down deals which state regulators had already approved.

In Sun-Peak, for example, FERC dismissed Nevada's support for the project, saying that its inquiry had a different focus. It claimed Nevada's review centered on comparing the cost of the project, to a single alternative, self-construction. This "avoided cost" determination was insufficient to permit noncost-based pricing, which requires a finding that the market disciplined the rate.

In other words, where an APP is concerned, the process, not the price at the busbar, is the determining factor.

Where does this leave us? FERC will give extra scrutiny to cases involving utility affiliates and market-based rates, saying that in such cases the mere opportunity for undue preference will lead to rejection of the agreement. An applicant can overcome this problem by a timely offer to sell the service to all bidders at the same price offered to the affiliate, or by providing the Commission with a benchmark of the market value of similar services based on contemporaneous data. FERC also will require the affiliate's parent to provide access to competing bidders whenever an affiliate seeks to sell power to a nearby purchaser. The specific nature of these relationships are important in these cases; each case will have to be looked at on its own merits.

Based on the Commonwealth case, it appears that FERC seems to be moving toward adoption of the "zone of economic influence" test first proposed in the 1988 IPP NOPR. At that time some suggested FERC should preclude IPPs affiliated with utilities from participating in projects located within the same power pool or reliability council. The manner in which this line is drawn will be critical in determining the ability of APPs to compete in future procurements.

### APPs Selling Power to a Parent

As noted previously, FERC approved market-based rates in a decision involving Ocean State Power. In this case three of the four utilities purchasing power from the facility had an ownership interest in it. FERC concluded self-dealing was not a problem since the rate charged the owner-purchasers was the same as that charged the nonowner-buyer and the local regulatory commissions had found that this power was priced below each purchasers' avoided costs.

In the Edgar case, however, FERC placed substantial hurdles in the way of utility affiliates seeking market-based rates for the sale of power to their parent. While going out of their way to say that the decision should not be construed as prohibiting such transactions, the Commissioners seemingly have raised the barriers so high that few, if any, utilities, will go this route.

Here, as when an APP seeks to sell to a purchaser located near its parent, FERC said it would rely on a market test. Under this standard there may be several ways in which a utility could demonstrate a lack of affiliate abuse.

One alternative would be for Boston Edison to submit evidence of direct head-to-head competition between Edgar and competing unaffiliated suppliers either in a formal solicitation or in an informal negotiation process. If the affiliate is not the lowest priced option, the applicant must provide sufficient justification for why the affiliate was chosen over alternative non-affiliated sellers.

Another alternative would be for Boston Edison to show that nonaffiliated buyers were willing to pay similar prices for similar services from the Edgar project.

Boston Edison also could offer benchmark evidence which shows the prices, and terms and conditions of sales made by nonaffiliated sellers. This evidence could include purchases made by Boston Edison itself, or by other buyers in the relevant market. Two major considerations with respect to the credibility of the benchmark evidence would be whether the benchmark sales are contemporaneous and whether they are for similar services when compared to the instant transaction.

Boston Edison followed the benchmark alternative in this case, comparing Edgar to four different historical benchmark groups: (1) 15 different suppliers with which Boston Edison negotiated contracts over a three-year period ending 1990; (2) Boston Edison's RFP-2, a 1989 QF solicitation process under Massachusetts DPU procedures where Boston Edison received proposals from 48 QFs; (3) a group of 34 QF and IPP projects negotiated by other Massachusetts utilities between December of 1984 and December of 1989; and (4) two IPPs, Dartmouth and Enron, the rates for which had been approved by the Commission.

The Commission had problems with this approach, saying the utility's comparative analysis raised numerous questions about underlying assumptions because Boston Edison included few details or explanations of the assumptions concerning variables (e.g., fuel cost, plant factor, and indices) used to compare the rates for various projects. In addition, almost all the benchmark data included in Boston Edison's comparative analysis reflected projects that are not contemporaneous and that do not provide similar services as compared to the Edgar facility.

FERC made clear that this was only one of the hurdles involved in getting an affiliate transaction approved. Even if it had found the rates supportable on the basis of market value, it still said it would need to analyze market power issues involving transmission or other barriers to entry that could have been used to exclude competitors (e.g., power plant sites). It also would then consider the issue of how cross subsidies, if any, between Boston Edison and Edgar may affect the reasonableness of Edgar's rate, i.e., a comparison between Edgar's rate and rates offered by non-affiliated suppliers.

After taking a hard-line against affiliated transactions involving sales to a parent, FERC, in a second case involving Ocean State Power, then allowed that APP to sell power at market rates to three utilities which together own 49.9% of the plant.

This marks the second time – and first time in four years – that FERC has allowed a so-called affiliated power producer (APP) to use market rates. Many commentators, as noted previously, thought the first decision, also involving Ocean State, was a relic of a bygone era, that more recent decisions showed a skepticism toward APPs that might be impossible to overcome. Here, however, FERC went beyond the earlier decision. This case, unlike the earlier one, did not involve an unaffiliated purchaser whose willingness to accept the rates could be used to validate them.

Does this case involve a new FERC approach toward APPs? Probably not. A key portion of this order says that older projects should not be held to the same standard as newer ones. Nevertheless, the next few APP cases to come before FERC should be examined carefully to determine whether this decision is an aberration or whether the Commission is rethinking its past tough stance toward this type of producer.

Finally, the Energy Policy Act provides, as a general rule, a utility may not enter into a contract with an EWG which is an affiliate or associate company of the electric utility company. State commissions may allow affiliate purchases, however, if they conclude the transaction (a) will benefit consumers, (b) does not violate any applicable state law – including any Least Cost Planning provisions, (c) would not provide the EWG any unfair competitive advantage, or (d) is in the public interest.

## UTILITY SELLS AT MARKET-BASED RATES

As the APP cases make clear, market power over transmission is the key factor in determining whether an UP gets to charge market-based prices. At the risk of oversimplification, FERC believes transmission line ownership by its very nature conveys market power. If the applicant or its affiliate owns or is affiliated with entities owning transmission lines, FERC will examine whether the applicant could block competing suppliers by denying transmission access to the purchaser.

As a general rule, FERC will require any line-owner to take mitigating measures before permitting market-based pricing. At a minimum this will include opening the grid to competing suppliers under terms and conditions comparable to those enjoyed by the lineowner. This is true even where the purchaser is interconnected with several other utilities.

In two early cases, involving captive customers of Pacific Gas & Electric, FERC allowed the utility to charge market-based rates after it gave the municipals access to several other suppliers and permitted them to return to the system with several years notice.

In a series of cases over the past few years, including Public Service of Indiana, PG&E's interconnection agreement with Sacramento, and the Western Systems Power Pool, FERC policy has evolved to where it now essentially requires utilities seeking market-based prices to file an open access tariff. The tariff must contain the following terms and conditions:

1. The utility must be willing to provide long-term firm transmission service at cost-based rates.
2. It must build new capacity when needed within a reasonable period (usually five years). It must forego market-based pricing if long-term firm service is unavailable for any reason.
3. Firm service contracts must be able to be resold.
4. Service must be made available to IPPs. It also must be made available to QFs provided they are willing to forego their rights under PURPA.
5. The purchasing utility must be able to both import and export power.

It is unclear whether FERC will seek to extend its finding in Northeast Utilities with regard to native load preference to this type of case. (In that merger case FERC would require NU to give firm service for others preference over economy sales for native load customers.) If past experience is any guide, look for it to appear soon in these types of cases.

One might also expect FERC to add some additional conditions in these cases. One likely candidate is requiring systemwide, rather than point-to-point, service in appropriate cases.

Utilities seeking market-based prices face at best a suspicious FERC which will emphasize academic theory at the expense of reality. The following passage from FERC's decision in the PG&E (Sacramento) case is a classic. Regarding PG&E's argument that export rights were meaningless in this case, FERC said: "while such limitations may not have a direct impact on PG&E's ability to exercise market power..., they limit development of efficient markets in general".

In reality, except for cases involving captive customers, it could be argued these cases require less, rather than more, oversight. These cases involve big boys. For example, in PSI all of the eligible customers under the FERC-approved power sales tariff are sophisticated buyers of bulk power, able to recognize and take advantage of their generating alternatives. PSI's plan will add to the number of alternatives those customers have. Moreover, the state regulatory bodies of both the selling and purchasing utilities have can effectively block the deal.

FERC, however, takes the opposite view.

There is one exception to this line of cases: a 1992 deal between Cleveland Electric and General Public Utilities. GPU issued an RFP for power from sources located within PJM. Given the provisions of the pool agreement and transmission constraints, non-PJM resources were not considered as viable alternatives. CEI, which was able to supply power from a jointly-owned facility within the pool, was permitted to charge market-based prices without any conditions because (a) it didn't own lines within PJM and (2) competing bids by PP&L and Philadelphia Electric provided market discipline.

## PUHCA EXEMPTION FOR EWGS

The Energy Policy Act of 1992 removed one of the most significant legal obstacles to independent power producers by creating a new legal entity called an exempt wholesale generator (EWG). An exempt wholesale generator is any person FERC determines to be engaged directly, or indirectly through affiliates, which is exclusively engaged in the business of owning and/or operating all or part of one or more eligible facilities and selling electric energy at wholesale.

The primary benefit of EWG status is that one is not considered an electric utility company under section 2(a)(3) of PUHCA. Nor will ownership of an EWG result in such person being considered as primarily engaged in the generation or sale of electricity within the meanings of sections 3(17)(C)(ii) and 3(18)(B)(ii) of the Federal Power Act. This means that an EWG is not limited to a 50% ownership interest in QFs.

Under EPAct, an eligible facility is a powerplant that is either used for the generation of electric energy exclusively for sale at wholesale or used for the generation of electric energy and leased to one or more public utilities. The term includes interconnecting transmission facilities necessary to effect a sale of electric energy at wholesale. An EWG may make retail sales only if the facility is located in a foreign country and none of the electric energy generated by the facility is sold to customers in the U.S.

Existing rate-based facilities may become an EWG provided that every state commission having jurisdiction over the facility makes a determination that allowing the facility to become an EWG (1) will benefit consumers, (2) is in the public interest, and (3) does not violate state law. Special provisions apply to an affiliate of a registered holding company.

Hybrid facilities are allowed with limitations. A utility or its affiliate may own or operate a hybrid facility if it was in the utility's ratebase on the date of enactment. However, a utility or its affiliate may not own a new hybrid plant. For example, Commonwealth Edison may convert 250 MW of an existing 500 MW plant into an EWG. However, it could not build and own a new 500 MW unit, seeking to place only 250 MW of the unit in the ratebase. In that case it must find an unaffiliated partner to own that portion of the plant which is regarded as an EWG.

As a general rule, a utility may not enter into a contract with an EWG which is an affiliate or associate company of the electric utility company. State commissions may allow affiliates, however, if they conclude the transaction (a) will benefit consumers, (b) does not violate any applicable state law – including any Least Cost Planning provisions, (c) would not provide the EWG any unfair competitive advantage, or (d) is in the public interest.

Other significant provisions of the Energy Policy Act insofar as it related to EWGs include:

- Reciprocal arrangements among companies that are not affiliates or associate companies of each other that are entered into in order to circumvent the provisions of this legislation are prohibited.
- State commissions may examine the books and records of an EWG and any affiliated electric utility. District courts have jurisdiction.
- There is an exemption from PUHCA for investment in foreign utilities.

# Section G

## Transmission – Access, Price and Terms and Conditions

# Section G

## Transmission – Access, Price and Terms and Conditions

The subject of wheeling has long been one of the most hotly contested issues in the electric debate. For years captive customers and independent generators have argued that the failure of lineowners to wheel power for others was at best discriminatory, and, at worst, violative of the antitrust laws. Lineowners countered with a broad range of technical and operational obstacles to increased wheeling. FERC generally took a hands-off approach, saying that the Federal Power Act (FPA) did not give it authority to order wheeling, and the courts upheld this view.

This issue is extremely important for cogenerators because it involves the ability of cogenerators to maximize the number of potential purchasers of their power. Wheeling access is particularly critical for projects where economics dictate location of the generating facility at or near the fuel source, but where the utility in whose service territory the plant is located does not need the additional generating capacity.

### ACCESS

As recently as 1987, one could safely say that wheeling was largely a matter within a utility's discretion. A few utilities were obligated to wheel as a result of conditions attached to their nuclear powerplant licenses, but other than that they generally transmitted power only for neighboring utilities.

FERC began reassessing its traditional view in a 1988 case involving the merger of PacifiCorp with Utah Power & Light. Here, the Commission held steadfast to its view that it did not have authority to order wheeling, but said that it could not find the merger met the appropriate statutory criteria under section 203 of the FPA unless the parties agreed to conditions requiring them to wheel power for others. When the parties acceded to these conditions, the movement to open the lines gained momentum.

Between 1988 and 1992, FERC added wheeling conditions to a number of requests by utilities and their affiliates to sell power at market-based rates under section 205 of the FPA. In each one of these cases the Commission acknowledged that it didn't have authority to impose these conditions directly, but maintained that it could approve the application only if the parties voluntarily agreed to wheel power under specified conditions.

### EPACT: THE ACCESS DEBATE ENDS

The passage of the Energy Policy Act of 1992 (EPAct) ended the great access debate. This legislation, which substantially opens the grid to independents and other wholesale sellers of power, is likely to have as significant an impact on the electricity industry as the passage of PURPA more than a decade ago. For all practical purposes, the transmission network will now be treated as a common carrier.

EPAct, for the first time, grants FERC authority to order wheeling. Section 211 authorizes any electric utility, Federal power marketing agency, or any other person generating electric energy for resale to request a wheeling order. The application may include a request that FERC require the lineowner to increase its transmission capacity if the lineowner claims a lack of capacity.

The Commission shall grant the order unless it finds that such order would (a) unreasonably impair the continued reliability of electric systems affected by the order, (b) replace existing contractual obligations, or (c) amount to a retail transaction or a sham wholesale transaction. This criteria essentially allows FERC to do what it pleases. While FERC can deny an order for reliability reasons, the burden of proof with regard to reliability is placed on the party opposing the order. FERC has been unsympathetic to such claims in the past.

The only positive finding that FERC must make is that the order is in the public interest. FERC cases over the past few years indicate this will not be a difficult standard to meet.

While an application may request that FERC require the lineowner to increase its transmission capacity, this obligation to build is not absolute. A utility will be excused from a FERC order to enlarge its transmission system if, after making a good faith effort, it fails to obtain the necessary approvals or property rights under applicable Federal, State and local laws.

Significantly, the provisions of EPAct are in addition to whatever authority FERC has under existing legislation. As noted above, FERC has used sections 203 and 205 of the Federal Power Act to condition the approval of mergers and market-based pricing proposals on the willingness of the applicants to accept an open access regime.

As a result, parties who seek to wheel and are located in utility service territories which have accepted conditions as part of their efforts to merge or sell power at market-based rates may have additional rights not available in other cases.

Perhaps the two most significant of these rights involve native load priority and line upgrades.

For example, in a case involving the merger of Northeast Utilities and Public Service of New Hampshire (the NU case) FERC said NU must back its economy energy transfers that benefit its native load customers when a third party seeks to use that capacity for a firm sale of power. In this and other cases, FERC essentially has placed an absolute obligation to build on the lineowner.

These issues are discussed below under terms and conditions of service.

## GOOD FAITH REQUESTS AND REPLIES

In a policy statement issued in July, 1993 FERC established standards for "good faith" wheeling requests and responses sufficient to satisfy the requirements of EPAct.

This issue is important for two reasons. First, FERC may not order transmission services under section 211(a) unless the applicant has made a request for such services to the transmitting utility at least 60 days prior to filing an application with the Commission. Second, under section 213(a), unless the transmitting utility agrees to provide transmission services pursuant to a good faith request, at rates, charges, terms and conditions acceptable to the requester, the transmitting utility is required to respond in writing to the requester within 60 days of receipt of such request, or within some other mutually agreed upon period.

A threshold issue for FERC was the relationship between Section 211 and Section 213(a). FERC believes that Congress, in enacting section 213(a) of the FPA, intended that potential applicants for transmission services under section 211 of the FPA and transmitting utilities provide one another with as much information as reasonably available concerning requests for, and ability to provide, transmission services, before a person seeking transmission services avails itself of section 211. FERC further believes that the exchange of detailed information by potential applicants and transmitting utilities will encourage constructive business transactions through negotiated agreements. Accordingly, it adopted standards which provide for a broad exchange of information.

The primary advantage of this, according to the Commission, was that a broad exchange of information will permit transmission requesters to file focused and more detailed applications under section 211(a) and will allow FERC to expedite section 211(a) applications. It may also further the goal of encouraging negotiated agreements.

### Good Faith Requests

FERC's policy statement outlines twelve components of a buyer's "good faith" request for transmission services:

1. the identity of the purchaser;
2. assurances that the prospective purchaser of the transmission services is eligible to request such service;
3. assurances that the requested transmission services are a type of transmission that FERC is authorized to order *(i.e. non-retail wheeling);*
4. specifics on the type of services requested, including specifics on point-to-point service or network service;
5. names of other parties expected to be delivering and/or receiving power from the transmitting utility;
6. the proposed dates for initiating and terminating the requested service;
7. the total amount of transmission capacity requested;
8. the description of the flow of power and energy into the transmitting utility's system *(i.e. a load profile and points of interconnection);*
9. the degree of firmness of the requested service;
10. whether the request is made in response to a solicitation;
11. terms and conditions of the requested services, specifying the rates; and
12. any additional information that would aid in evaluating the request.

### Good Faith Replies

The five components of a seller's reply to a "good faith" request as set forth by the commission are:

1. an acknowledgment within ten days of the receipt of the request;
2. requests by the transmitting utility for clarification of information;
3. the transmitting utility must respond within 60 days or other mutually agreed upon time;
4. the transmitting utility must offer the requester an executable service agreement if it believes it can provide the requested service from existing capacity; and
5. the transmitting utility must provide specified information to the requester if it determines that it must construct additional facilities or modify existing facilities in order to provide all or part of the request.

While a policy statement does not have the binding nature of a rulemaking, it is important as an expression of Commission sentiment on an issue. The guidelines set forth seek to get the parties moving toward settlement of their differences, and, if this is not possible, to provide a record for Commission decisions.

## REGIONAL TRANSMISSION GROUPS

In another policy statement FERC affirmed its support of regional transmission groups (RTGs) and outlined the basic components that should be included in RTG agreements. According to the Commission, "RTGs promise efficient and expeditious solutions to problems that may stem from expanded transmission access".

FERC explained that it does not have the authority to certify RTGs. However, under section 205(c) of the FPA, public utilities must file agreements affecting rates and charges for any transmission or sale subject to the Commission's jurisdiction. All contracts which in any way affect or relate to jurisdictional rates, charges, classifications, and services must be filed with the Commission. An RTG-related agreement that affects or relates to jurisdictional transmission rates or services must be accepted by the Commission under the FPA. Therefore, FERC's policy statement is intended to provide guidance on the basic components that should be included in RTG agreements.

FERC defines an RTG as a voluntary organization of transmission owners, transmission users, and other entities interested in coordinating transmission planning (and expansion), operation and use on a regional (and inter-regional) basis. According to FERC, RTG agreements, at a minimum, should contain the following seven basic components:

1. An RTG agreement should provide for broad membership and, at a minimum, allow any entity that is subject to, or eligible to apply for, an order under section 211 of the FPA to be a member. An RTG agreement should encompass an area of sufficient size and contiguity to enable members to provide transmission services in a reliable, efficient, and competitive manner.

   Component number 1 allows for the broadest possible membership for RTGs, including foreign utilities that are interconnected with the national grid. Broad membership will extend the benefits of RTGs to the greatest number of market participants, thereby leading to greater efficiency.

   Component number 1 also provides that the geographic area covered by an RTG agreement should be sufficiently large and contiguous. It is implicit in section 202(a) (which concerns "regional districts" for voluntary coordination and interconnection) that there should be coordinated operation in areas large enough and contiguous enough for economic efficiency.

2. An RTG agreement should provide a means of adequate consultation and coordination with relevant state regulatory, siting, and other authorities.

   Component number 2 provides for adequate consultation and coordination with states. FERC agrees with the comments of transmission-owning utilities and transmission-dependent entities that consultation and coordination with the states are critical to the successful implementation of RTGs. FERC feels that state involvement in RTGs can allow state agencies to improve communications with utilities and with each other in dealing with transmission concerns, and can facilitate coordinated treatment of siting issues among the states.

3. An RTG agreement should impose on member transmitting utilities an obligation to provide transmission services for other members, including the obligation to enlarge facilities, on a basis that is consis-

tent with sections 205, 206, 211, 212 and 213 of the FPA. To the extent practicable and known, the RTG agreement should specify the terms and conditions under which transmission services will be offered.

Component number 3 provides for an affirmative obligation to provide transmission services. An inability to obtain service on reasonable terms and conditions will likely result in filings with the Commission under sections 211 and 212 of the FPA. In addition, FERC believes the service obligation of RTG members should extend to all types of transmission services and should include a commitment to expand or upgrade facilities when needed to meet service requirements.

4. An RTG agreement should require, at a minimum, the development of a coordinated transmission plan on a regional basis and the sharing of transmission planning information, with the goal of efficient use, expansion, and coordination of the interconnected electric system on a grid-wide basis. An RTG agreement should provide mechanisms to incorporate the transmission needs of non-members into regional plans. An RTG agreement should include as much detail as possible with regard to operational and planning procedures.

   Component Number 4 provides for coordinated transmission planning and sharing of transmission planning information. The coordinated planning process should be open to participation by all members and should address the transmission needs of members as well as non-members. The term "coordinated planning" is a broad term that should encompass the goal of efficient use and expansion of the nation's transmission system. It also includes planning to make expansions that are economically justified from a regional perspective. This component assures that the economic trade-offs between generation and transmission expansion will be weighed appropriately.

5. An RTG agreement should include fair and non-discriminatory governance and decisionmaking procedures, including voting procedures.

   Component Number 5 provides for fair and non-discriminatory governance and decisionmaking procedures. In general, FERC thinks an RTG should have rules or procedures to protect the rights of entities that are more susceptible to the exercise of market power, such as transmission dependent utilities (TDUs).

6. An RTG agreement should include voluntary dispute resolution procedures that provide a fair alternative to resorting in the first instance to section 206 complaints or section 211 proceedings.

   Component Number 6 provides for voluntary dispute resolution procedures. FERC particularly encourages RTGs to develop high quality alternative dispute resolution procedures for resolving technical and reliability issues. The Commission encourages proposals under which FERC would afford substantial deference to outcomes resulting from appropriate alternative dispute resolution (ADR) procedures that are specified in the RTG agreement.

7. An RTG agreement should include an exit provision for RTG members that leave the RTG, specifying the obligations of a departing member.

   Component Number 7 provides for an exit provision for RTG members who wish to leave the RTG. If a party has accepted a responsibility under an RTG agreement and then decides to leave the RTG, the obligation of such departing party to comply with its prior commitments should be set forth in the RTG agreement.

FERC clearly sees RTGs as a means of reducing its workload in implementing the Energy Policy Act. While FERC isn't willing to cede too much authority up front, don't be surprised if RTGs become a de facto means of resolving disputes in an increasingly deregulated environment.

## PRICING

Given the monopoly nature of transmission lines, FERC traditionally has permitted utilities to charge only embedded cost rates when they wheel power for others. In the past these decisions were relatively easy given the prevailing view among utilities that transmission was largely incidental to generation. As more players have gained access to the grid and utilities have begun perceiving transmission as a key strategic resource, lineowners have sought compensation for what they believe to be the full economic cost of using the lines.

EPAct gives FERC great latitude in deciding a proper pricing regime. Any section 211 order shall allow the transmitting utility to recover "all" costs incurred in connection with the transmission services and necessary associated services. These include an appropriate share, if any, of legitimate, verifiable and economic costs. The charge should recognize any benefits to the transmission system of providing the transmission service, as well as the costs of any enlargement of transmission facilities.

The legislation further provides these rates should promote the economically efficient transmission and generation of electricity and shall be just and reasonable, and not unduly discriminatory or preferential. To the extent practicable, the applicant for the order, not the transmitting utility's existing customers, should pay the costs incurred in providing the wholesale transmission services.

FERC has spent a fair amount of time in recent years looking at some of the tough issues involved in setting wheeling prices in a competitive environment. These issues included charges for opportunity costs, the proper allocation of costs for system upgrades, and stranded investment.

Nothing in the 1992 legislation is likely to change the way FERC looks at these cases, though one should recognize that the courts will be the ultimate arbiter in determining whether this view is consistent with the nebulous language contained in the Energy Policy Act.

## OPPORTUNITY COSTS

Opportunity costs, in the context of electricity transmission, are the revenues lost or costs incurred by a utility in providing third-party transmission service when capacity is insufficient to satisfy both a third-party wheeling request and the utility's own use. For example, opportunity costs might include the revenues lost or costs incurred because a utility must reduce its own off-system purchases or sales in order to overcome a constraint on the grid. In FERC's view legitimate lost opportunity costs do not include the loss of revenues that result from losing a wholesale sale because of competition and do not include foregone revenues from purchase and resale (brokering) transactions such as those condemned in PacifiCorp.

In the NU rehearing case the Commission said it would allow utilities to charge opportunity costs in certain cases. It announced three principles would govern its decisions on the permissibility of such costs in individual rate cases. First, the native load customers of the utility providing transmission service should be held harmless. Second, transmission customers should be charged the lowest reasonable cost-based rate for third party firm service. Third, the pricing mechanism should prevent the collection of monopoly rents by the transmission owner and promote efficient transmission decisions. In ruling on specific proposed rates, it said it would balance these three goals in light of the facts and circumstances presented at that time.

The Commission said that, if NU filed a proposal to recover opportunity costs for firm transmission service, the utility should address the following issues: (1) whether opportunity costs should be capped by incremental expansion costs or any other cap; (2) whether current wheeling and wholesale requirements customers should be treated differently from future wheeling and wholesale requirements customers (e.g., by receiving "grandfather" rights to embedded cost rates for the amount of transmission capacity they already use); (3) how NU will identify those customers responsible for growth on its system and what particular new facilities are necessary to accommodate that growth; (4) whether and how third parties should be protected from uncertainty regarding fluctuations in opportunity costs; (5) how the proposed rates will prevent the collection of monopoly rents; and (6) how the proposed opportunity costs will be verified.

In the Pennsylvania Electric case, however, the Commission seemingly restricted when utilities may charge opportunity costs by ruling a lineowner may charge embedded or incremental costs, but not both. This decision involved a transmission service agreement between Pennsylvania Electric and Penntech Papers. The agreement required Penntech to design, construct and operate a 115 kV transmission line extending approximately 7.5 miles from the facility site to a Penelec substation. Penelec was to design, construct and install – at Penntech's expense – all interconnection facilities necessary to connect the Penntech transmission line to Penelec's transmission system. The parties agreed the charge would include a 100 percent contribution to the embedded costs of Penelec's system plus an "increased energy cost component" to reflect the costs associated with actual reduction in economic energy transfers to Penelec's native load customers resulting from the use of Penelec's "limited" transmission system to accommodate the transmission of Penntech capacity and energy. Penelec proposed capping Penntech's "increased energy cost payments" at Penelec's estimated cost of expanding its transmission system.

FERC concluded that Penelec's proposal failed to adequately satisfy the principles enumerated in the NU rehearing order. In effect, the utility would be charging twice for the same transmission capacity (i.e., charging the basic non-firm rate it would receive if it used the capacity to serve the third party and charging the foregone benefits it would receive if it used that same capacity for a different purpose). Since the utility cannot use the same capacity at the same time for two different purposes, it would be unreasonable to allow it to charge rates reflecting dual use of the same capacity.

Consequently, FERC conditioned its acceptance of Penelec's proposed transmission service agreement on Penelec's agreement to eliminate the opportunity cost "adder" and to charge Penntech no more than the greater of: (1) the basic firm transmission rate or (2) validated opportunity costs, capped at Penelec's incremental cost of expanding its system in order to alleviate any transmission constraints that may occur during the 20-year contract period as a result of providing the Penntech service.

The cap in this contract did not ease the Commission's concerns for potential "double dipping". Nevertheless, FERC found the cap appropriate in the context of the "or" condition discussed above because it provides an incentive for the utility to expand its system when it is efficient to do so. Specifically, the cap at estimated or actual expansion costs gives Penelec an economic incentive to build additional transmission capacity if opportunity costs exceed expansion costs. If Penelec does not build, it will be unable to recover all of its opportunity costs and, therefore, cannot satisfy its stated goal of keeping its native load customers harmless.

In a decision involving NU and NEPCO FERC rejected hourly calculations of opportunity costs for long-term firm service, saying these charges should be based on total costs over the lifetime of the contract. In their proposed tariffs the utilities sought to recover the higher of two rates, as determined hourly: (1) an embedded cost rate; or (2) the out-of-rate charges billed to the utility by NEPOOL. FERC disagreed, contending this approach failed to provide for the lowest reasonable cost-based rate for third party firm service. Instead the commissioners offered a "simple calculation" that would allow utilities to bill for opportunity costs exceeding their embedded cost rates as they incur these costs, but which uses "a continuous true-up" to ensure that the customer pays no more than the higher of embedded or opportunity costs for the transaction period. At the end of the contract term the utilities would be compensated for out-of-rate charges only if the total incurred during the life of the contract exceeds total fixed costs.

FERC took a more lenient approach toward opportunity costs in the context of nonfirm wheeling in still another case involving Northeast Utilities. The agreements involved in this case covered situations when NU forecasts or requires the use of its transmission facilities to reliably and economically meet the needs of its requirements and/or retail customers, and such needs cannot be met simultaneously with the supply of the transmission service called for under the agreements. In such cases the transmission customer may choose either to accept the reduced deliveries, or to continue receiving service (in essence, upgrading its service to firm) by paying Northeast a "Lost Opportunity Charge". If the buyer opts to continue service, the charge is based on the estimated economic savings foregone by the NU Companies.

According to FERC, the firm, cost-based transmission tariff, to be filed pursuant to the merger conditions, will provide a viable alternative service option for the nonfirm transmission customer, and it will serve as an effective ceiling on what amounts Northeast may seek to obtain through the lost opportunity charge. With firm transmission service available at a cost-based rate, the nonfirm customer will have the option of converting to that firm service if, over time, the lost opportunity charges associated with the nonfirm service exceed the firm transmission rate. The cost-based firm service tariff backstop, and the implicit price cap it places on opportunity costs, will ensure that opportunity cost pricing is not a mechanism for Northeast to collect monopoly rents, and the nonfirm transmission rate will provide an efficient method to price short-term imbalances between supply and demand for transmission service, based upon the savings that Northeast foregoes.

However, FERC did not permit Northeast to recover opportunity costs for a particular hour as an "adder" to the basic nonfirm rate. The Commission said this would go beyond holding the native load customer harmless and would overcompensate the utility to the detriment of the third- party transmission customer.

The Commission required Northeast to maintain sufficient records to enable the Commission independently to verify and monitor any lost opportunity charge Northeast proposes to collect, including:

- a statement of the factual basis for the conclusion that an economy energy purchase would have been made or was suspended;
- the cost factors (e.g., number of kWh of economy energy purchases foregone, and the generation cost differential – in mills per kWh – of the parties involved) which went into the calculation of the savings foregone;
- the identity of the selling party; and
- evidence that the relevant transmission interface was constrained and that the customer's curtailment would have relieved the constraint.

The validity of the firm-nonfirm distinction is questionable. As NU pointed out in a petition for rehearing, all energy is scheduled on an hourly basis, and the price signal that determines how constrained transmission will be used applies for that hour. This could lead to the emergence of third party brokers who can force NU to sell its transmission capacity to them on a firm basis and who can then resell it under the more favorable terms allowed for non-firm service.

## COST OF TRANSMISSION UPGRADES

In its August 1991 ruling on the merger case the Commission said NU in certain circumstances may charge its transmission customers the incremental costs of the specific facilities used to provide service. It said it would apply a "but for" test – i.e., incremental cost pricing would be appropriate when firm wheeling across a particular interface would degrade reliability absent upgrades. The Commission declined to decide certain other pricing issues, including the issue of whether NU may charge a wheeling customer both a pro-rata charge for particular facilities and a rolled-in rate for the rest of NU's transmission system. The Commission stated that the proper forum to decide the details of cost responsibility issues is a section 205 rate case.

On rehearing, FERC clarified what this meant, saying NU may not propose to collect from any transmission customer at any time an amount exceeding the greater of: (1) the incremental cost of new network facilities required at the time the customer's new transmission load is added or (2) the rolled-in cost of all network facilities required to serve the combined transmission loads of the NU Companies, including any required transmission additions.

This pricing rule does not preclude particularized cost allocations to specific customers where appropriate, e.g., distribution facilities, generation leads. But where transmission facilities are fully integrated and support the entire transmission system, NU will not be permitted to charge both an embedded cost rate and an incremental cost rate for firm wheeling service, since charging both rates would unjustifiably require a wheeling customer to pay rates for part of NU's costs based on cost causation, while paying rates for other costs based on use of the system. If the upgrade would have been needed and built eventually to satisfy native load requirements and the third-party wheeling request merely accelerates the construction schedule, the incremental cost rate should reflect only the cost differential associated with building the upgrade sooner.

NU will bear the burden of proof to justify specifically assigning the incremental cost of upgrades to a third party requesting wheeling service. This will require NU to justify any direct assignments of costs and support any arguments that reliability is degraded by a particular firm transmission service. No presumption is created by NU's "but for" criterion that firm wheeling customers always cause the need for upgrades.

In Public Service Electric & Gas (PSE&G) the utility sought approval of an interconnection and transmission agreement between the utility and a cogenerator, EEA I, which sought to charge EEA I an embedded cost transmission rate of $0.85/kW/month plus an additional "system facilities reinforcement advancement charge" of $224,400/month (approximately $1.50/kW/month).

The Commission recognized that the situation in this case was different from that presented in Penelec. There, like here, the system is constrained. But, unlike Penelec, the utility in this case chose to expand its system to remove the constraint. Moreover, PSE&G sought to recover expansion costs, not opportunity costs. Public Service maintained that another distinction between the EEA I filing and the Penelec filing is that Public Service used a distance-based rate design, while the utility in Penelec used a postage stamp rate design. Public Service stated that, if it were to use a postage stamp rate design, its embedded cost rate would be higher.

Nevertheless, the Commission did not believe these differences were sufficient to allow additive pricing – i.e., embedded plus incremental cost pricing. The fact that the expansion costs are incurred because they are necessary to provide the quality of service requested (firm transmission service) simply demonstrates that the facilities meet the "but for" test, i.e., the test which the Commission uses to determine when integrated grid service may be priced using expansion costs (instead of embedded costs). As noted above, once that test is met, the transmitting utility is allowed to charge the higher of: (1) its standard embedded cost rate (including the costs of the upgrade rolled-in with the costs of the existing system) or (2) the upgrade costs that it would not have incurred "but for" this transmission service.

In Public Service of Colorado (Colorado) the utility requested the Commission permit it to amend the transmission service tariff to provide that when an upgrade attributable to a specific service request does not immediately contribute to the system as a whole, the costs of the upgrade may be assigned to the customer requesting service until such time that other customers benefit from the service.

Colorado asked the Commission to recognize certain exceptions to the general policy that a transmitting utility may charge either embedded cost rates or incremental cost rates, but not both. If a wheeling customer does not pay both embedded costs and appropriate incremental costs, Colorado argued, native load customers and existing wheeling customers will subsidize new wheeling customers. Colorado recognized that the Commission permits the recovery of incremental costs for radial feed lines and certain other discrete transmission facilities. However, Colorado believed that the recovery of both the incremental costs of expansion and embedded costs is also appropriate in certain other circumstances. Colorado said such circumstances may arise when an upgrade is added to a geographically remote part of the transmission system where no appreciable load growth is expected and where the upgrade is used only to provide service to the prospective wheeling customer.

For instance, Colorado stated a transmission customer's request for service may require upgrades that are only needed to serve the requesting customer and do not provide any benefit to the system as a whole (i.e., because the upgrade does not provide for increased economic transfer capability or reliability to Colorado, the transmission system owner). Colorado argued that its native load customers should not pay for these upgrades and that the wheeling customer should pay the entire incremental cost.

Once again, FERC rejected the additive approach, saying Colorado "simply misunderstands" the Commission's transmission pricing model and the model's "but for" test. An integrated transmission grid is a cohesive network moving energy in bulk. Because the grid operates as a single piece of equipment, the Commission has consistently priced transmission service based on the cost of the grid as a whole. The Commission has rejected the direct cost assignment of grid facilities, even if the grid facilities would not be installed but for a particular customer's service. The Commission has reasoned that, even if a customer can be said to have caused the addition of a grid facility, the addition represents a system expansion used by and benefiting all users due to the integrated nature of the grid. Believing that the grid is a cohesive network in a dynamic state of development, the Commission has even included remote facilities in the grid on the ground that they were merely the first segment of what would eventually be a network loop. The Commission has reserved direct assignments for only those transmission facilities which fall into what we have referred to as an "exceptional category" consisting of radials which are so isolated from the grid that they are and will remain non-integrated. In a footnote, it referred to a 1978 Idaho Power case, which concerned the direct assignment of unusually long lines ranging from 67-110 miles from the utility's main system.

### STRANDED INVESTMENT

In Entergy Services the Commission allowed Entergy to recover stranded investment costs in its supply contracts. Though the utilities involved in Entergy had failed to design their power supply contracts in a manner to protect themselves against increased stranded investment risks, the Commission permitted the recovery of these costs on the grounds the company was seeking an open-access transmission tariff, which could exacerbate Entergy's risk of stranded investment far beyond what was contemplated when it negotiated its current power supply contracts. Any utility would be reluctant to open its transmission system voluntarily if it meant being subjected to self-inflicted stranded investment.

On rehearing in Entergy, the Commission clarified that, in general, issues regarding stranded investment costs must be decided based on the facts of the specific case in question. However, the Commission provided some "general guidance" on the issue by stating that in order to recover stranded investment costs from a particular customer, Entergy: (1) must be able to demonstrate that it has, for a particular customer, incurred generation investments or other obligations on the customer's behalf based on a reasonable expectation at that time that the customer's power contract would be renewed; (2) cannot levy stranded investment cost liability on a customer which is more than what the customer would have contributed to fixed costs under its existing rate had the customer remained on Entergy's system; and (3) must mitigate a customer's stranded investment obligation when the customer leaves the Entergy System.

The Commission seemingly further limited the application of its Entergy decision in Maine Public Service. Here the utility proposed the addition of a termination charge to its wholesale tariff designed to recover costs associated with a loss it incurred on its investment in Seabrook. Unlike the factual situation in Entergy, the Maine Public Service was not proposing the termination charge in the context of filing an open access tariff, so the Commission relied on a strict reading of the contract between the parties.

While FERC explicitly stated its decision here was not meant to limit its ruling in Entergy, one must wonder when, if ever, the circumstances of its earlier ruling will be replicated now that the Energy Policy Act of 1992 essentially makes all utilities "open access" companies.

## PRICING NOI

In June 1993 FERC issued its long-awaited notice of inquiry (NOI) into transmission pricing services (RM93-19). The inquiry requests comments concerning whether it is appropriate to revise the Commission's present pricing policy for transmission services provided by public utilities under the Federal Power Act.

Given the changes now taking place in the industry and the passage of the Energy Policy Act, the Commission let it be known that it would reexamine its transmission pricing policy to determine the impact of that policy on efficiency and competition in the wholesale power markets. In addition, FERC Chair Moler has been concerned that Commission decisions may not fully protect a utility's native load customers.

In its notice of inquiry, FERC noted any transmission pricing model adopted by the Commission must meet the substantive standards contained in the Federal Power Act. However, it invited comments on the scope of the Commission's discretion on transmission pricing under newly added section 212 of the Act, and on whether the Commission should apply the same pricing under section 205 of the Act as it applies under section 212.

Attached to the NOI was a staff paper asking for comments on 36 different matters relating to the proper means of pricing transmission services. These questions included such fundamental issues as whether FERC's current pricing regime allows adequate compensation for all costs, the need for financial incentives to provide service, the continued validity of the contract path method of pricing, the need for distance-sensitive pricing, whether providers of short-term wheeling services should be allowed to continue adding a capital component, and the proper pricing of network services.

The tenor of these questions makes it clear that FERC is contemplating a thorough revamping of its transmission pricing policy. Many believe the outcome of this NOI will be the electric equivalent of Order 636, which required natural gas pipelines to unbundle their services.

## TERMS AND CONDITIONS OF SERVICE

### Native Load Priority

In its original decision in the NU case FERC said that under no circumstances would NU be required to provide firm wheeling service out of existing transmission capacity when doing so would impair or degrade reliability of service to native load customers. It defined "native load customers" as those customers on whose behalf NU, by statute, franchise or contract, has undertaken the obligation to plan, construct, and operate its system to provide reliable service. This includes NU's retail native load customers as well as its wholesale full and partial requirements customers.

On rehearing, the Commissioners recognized the difficulty in precisely demarcating transmission needed for "reliability" purposes from that needed for "economic" purposes. They generally agreed with NU that: (1) some reservation of transmission capacity to back-up NU's large generating units is legitimate and reliability-related, (2) transmission capacity that is used to provide generation reserves cannot be fully preempted by third parties for their own wheeling transactions, and (3) "reliability" encompasses capacity needs for some reasonable planning horizon into the future.

With regard to the third point, FERC required that any capacity needed for reliability purposes within a reasonable planning horizon must be offered for wheeling use until NU expects to need the capacity for reliability reasons. Also, in these circumstances, native load customers must bear their pro rata share of the incremental cost of expansion. That is, if an upgrade is planned to meet the needs of both native load and wheeling customers, both sets of customers must bear part of the incremental cost.

While the Energy Policy Act reaffirms the importance of reliability, it implicitly regards native load priority as primarily an economic issue (see below). Moreover, it should be noted that NU involved a section 203 merger approval and it is unclear whether this authority exists under section 211 of the Energy Policy Act.

**Immutable Constraints**

In its initial NU decision involving section 203 of the FPA the Commission specified that, if NU's defense for not providing firm wheeling service was that a transmission constraint was "immutable," the Commission would convene a technical conference at which NU, all affected customers, and all affected state commissions could address the issues of whether the constraint was immutable and, if so, how to allocate NU's existing capacity most efficiently and equitably.

Not surprisingly, the state commissions went ballistic over this provision. For example, New Hampshire asserted that the Commission's "immutable constraint" procedures are unprecedented, lack any record support, and are unrelated to the merger's effect on competition. And that was one of the milder complaints.

In the face of this opposition the Commission backed down, eliminating the concept of a technical conference. Its justification was that the term "immutable constraint" was unfortunate because it suggested that a transmission constraint cannot be eliminated in any manner or at any cost. Most, if not all, constraints that arise can be eliminated at some cost of expansion, although the cost of such expansion may be so high as to render the proposed wheeling transaction uneconomical. Thus, the Commission should focus on determining what is the incremental cost of expansion.

FERC recognized that NU's identification of the facilities that need to be expanded as well as its initial estimate of the cost of expansion may be controversial and may be the subject of a proceeding before the Commission. NU, any affected customer, and all affected state commissions (including any regional transmission association or relevant multistate resource planning organization) will be afforded the opportunity to participate in such a proceeding. Such a proceeding will provide a reasonable basis (and a fully developed record) upon which the Commission can derive an appropriate incremental cost.

The Energy Policy Act relieves a utility from constructing new capacity if, after making a good faith effort, it fails to obtain the necessary approvals or property rights under applicable Federal, State and local laws. It, however, explicitly provides that FERC may use whatever authorities it has under other provisions of law to deal with such situations. In NU the Commission used its section 203 merger authority to force the issue. Presumably, FERC will continue to use this authority in similar situations in the future.

**Network Service**

In its first order mandating transmission service under the provisions of the Energy Policy Act FERC ordered Florida Power & Light (Florida Power or FP&L) to offer network service to its long-time nemesis, the Florida Municipal Power Agency (FMPA).

Although network service has proven to be one of the most controversial issues arising from the passage of EPAct, there is no industry-wide definition of the term. Here FMPA defined the term as meaning transmission service under which FP&L would transmit power without assessing multiple charges for each receipt/delivery point combination.

For example, under FMPA's definition of network service, if a power supplier had two resources (Unit A and Unit B, each with a capability of 100 MW) and two customers (Customer X and Customer Y, each with a load of 80 MW), the power supplier would pay for no more than 160 MW of transmission service, the maximum combined load of both customers. FMPA argues that, under the Florida Power's consistent practice of offering point-to-point service, Florida Power might agree to transmit power from Units A and B to Customer X and Y, but Florida Power would make the power supplier pay for 320 MW of transmission service, i.e., 80 MW of transmission service from Unit A to Customer X, 80 MW of transmission service from Unit B to Customer X, and 80 MW of transmission service from Unit A to Customer Y and 80 MW of transmission service from Unit B to Customer Y.

FMPA has long sought network service. FMPA and its members own interests in generating units which are used to meet a portion of the members' power supply needs. FMPA and its members also purchase partial requirements power and various interchange services (economy, short-term firm and emergency services) from Florida Power, Florida Power Corporation (FPC) and Tampa Electric Company. Florida Power and FPC also provide FMPA and its members with certain transmission services.

In the late 1980's, FMPA began investigating the feasibility of integrating and coordinating its resources. As a result of FMPA's unsuccessful efforts to form a Florida-wide power pool, FMPA designed its Integrated Dispatch and Operations (IDO) Project through which FMPA hopes to dispatch, operate and plan power supply resources for participating members on an integrated basis. FMPA estimates implementation of the IDO project will save it about $10 million annually.

In September, 1989, FMPA sent a proposed Transmission Service Agreement and, concomitantly, a formal request for network transmission service for the IDO Project to Florida Power. FMPA and Florida Power negotiated and exchanged information over approximately two years, but the parties were unable to reach agreement concerning the terms of such transmission service. After seeking network service in a number of different regulatory and judicial proceedings, FMPA filed its requests under sections 206 and 211 that the Commission direct Florida Power to provide network transmission service.

FERC justified its decision in extremely broad terms, saying that "[a]s a general matter, the availability of transmission service (or increased flexibility to use transmission) will enhance competition in the market for power supplies over the long run because it will increase both the power supply options available to transmission customers (thereby benefiting their customers) and the sales options available to sellers. This should result in lower costs to consumers. In addition, if a transmission customer determines that flexible service, such as network service, will allow it to serve its customers more efficiently, FERC believes the public interest will be served by requiring that service to be provided so long as the transmitting utility is fully and fairly compensated and there is no unreasonable impairment of reliability.

FERC stated that "[i]n this case, network transmission service will allow FMPA and its members greater flexibility to economically dispatch their generation resources. This flexibility should give them the opportunity to benefit ratepayers by using their generation resources more efficiently. Thus, it is in the public interest for FMPA and its members to have the opportunity to achieve greater efficiency and ratepayer savings."

Saying FP&L had not raised any reliability concerns, the Commission required FP&L to provide network service to support FMPA's IDO project, without multiple point-to-point pricing.

The Commission gave FMPA and FP&L sixty days to agree to the specific rates, terms and conditions under which this order is to be carried out.

FERC gave the parties little guidance as how such service should be priced. It did say it recognizes that the cost of providing network service may exceed the cost of point-to-point service because the transmitting utility may incur additional transmission capacity costs to accommodate the increased flexibility of network usage of the system. However, it added that Florida Power would bear the burden of substantiating higher charges if it seeks to charge a rate for network service that exceeds what it charges for point-to-point service.

Moreover, the Commission said that "Florida Power's rates, charges, terms and conditions for this network service must not be unduly discriminatory when compared to rates, charges, terms and conditions of transmission used to serve Florida Power's other customers". It is difficult to tell what this sentence means. Many believe this is the same comparability of service standard that FERC adopted in Order 636 for the gas pipelines. If so, this could be the first step in requiring utilities to unbundle its native load and off-system sales.

Finally, one should note that FERC was not troubled by the existence of a pre-existing transmission service agreement between the parties. Although Section 211(c)(2) precludes using the authority of Section 211 to break contracts existing at the time EPAct was enacted, FERC said it applies only to contracts relating to the provision of electric energy, not to those relating to the provision of transmission service.

# Section H

## Utility Standby Rates

# Section H
## Utility Standby Rates

### INTRODUCTION

Traditionally, standby service was designed to provide electric power to customers which have their own on-site generation during unscheduled outages. However, with the advent of retail wheeling, standby service takes on a new and critical role – the provision of service when the primary supplier of electric power fails to provide that service.

Standby service is critical, in that it permits customers to seek supplies on the open market, while protecting them against the possibility of interruption.

Standby service, while it serves the customer, will likely be paid for by retail power marketers seeking to assure their customer that no matter what, they will receive service. It will have a role similar to a homeowners insurance policy – the seller will buy it to protect the buyer, but also to induce the buyer to become a customer.

Thus, *reasonably priced* standby service is a necessity for the competitive retail market. Yet reasonableness for this service cannot be assumed.

Standby rates were designed early in this century to discourage interconnection by industrial firms and others that generated all or part of their electrical energy. Electric utilities purposely priced the service at a level which made it uneconomical for the customer to implement an on-site generation project.

PURPA (the Public Utility Regulatory Policies Act of 1978) made this illegal, and required the states order their jurisdictional utilities to implement the right to reasonably-priced standby service.

As this report makes clear, implementation has been mixed. While enforcement of the requirement that reasonably priced standby service be made available is important for companies considering on-site generation, it is even more important for other retail customers.

Prospective on-site generators are the largest and most sophisticated retail customers. They can often negotiate specialized electric contracts tailored to their needs, using the leverage of their ability to build a power plant to displace the local utility.

Other retail customers do not have this leverage, and generally have considerably less sophistication than the larger customers do regarding retail rates.

Therefore, as these customers gain access to alternative suppliers of power through retail wheeling, it will be essential that they have available to them reasonably priced standby service. Without such service, a competitive retail market is unlikely to be formed.

#### Methodology

This report was developed as the results of a study and survey on the diversity of standby rates and methods currently being used to set prices for standby service for utility customers having their own generating facilities.

The basis for the current study involved three principal activities:

- A review of the standby tariffs of over 100 investor-owned utilities,
- a literature search of publications and other appropriate sources
- and a survey of Public Utility Commissions and utility companies to gather comprehensive information on standby rates, terms, and conditions offered by the investor-owned utilities to customers otherwise considered firm retail customers.

This study observes that, in general, the area of standby rate design and methodology has yet to become as intensively studied as other areas of electricity costing and pricing. Some utility companies and utility commissions have undertaken complex analyses of the cost of supplying standby service, however, this study finds that the most frequently employed method of pricing standby service involves the application of rate design modifications to other general service utility tariffs.

This study does not provide a comprehensive examination of the methodology of standby rates: the objective was to demonstrate the similarity or differences between standby rates in various parts of the country and show the range of standby charges.

The standby rates schedules included in the appendix of this study reveal the applications of utility costing and pricing principles to circumstances unique to the tariff structure of each utility. These rates reflect the long evolution of utility generating resources, fuel mix, load shapes, customer mix and regulatory influence on the cost of service. The recent growth of customer-owned generation in the resource mix of many utilities requires the consideration of the saturation of customer-owned generation facilities, the number of these facilities, their size, location, fuel type, and reliability, in determining the cost of providing this increasingly important service for utility customers.

**General Findings**

The standby rate schedules included in the appendix of this report vary widely in the extent to which they address the following terms, conditions, and requirements:

- Customer generating unit nameplate capacity
- Unit outage verification and reporting procedures
- Contract versus actual demand and highest on-peak billing
- Customer load characteristics and load demand
- Energy and demand metering
- Customer and utility system protection requirements
- Interconnection expenses

Tariff components often have different combinations of the attributes listed above embedded in the customer, capacity and energy components of the standby rates.

Estimates of the performance of customer-owned generation equipment represents a key consideration in pricing standby service. The nature of standby load, irregular and unique when compared to firm service customer classes' load, challenges cost analysts trying to make capacity cost responsibility estimates. More detailed information on customer-owned generating equipment performance could be used to refine standby rate design.

Many utilities tailor the provision of standby service to the technical operating condition of the service territory, reflecting reliability concerns for areas within a given service territory.

Few standby tariffs include explicit provision for transfers of service between full and partial requirements service. Transfers are often negotiated between the utility and customer based on the terms and conditions of the general service schedule. Some tariffs for standby service require customers to give notice to the utility of intention to end standby service.

### Demand Charges and Ratchets

Practicability of standby tariffs generally turns on (1) the level of the demand charge and (2) the applicable ratchet.

The demand charge is a "reservation fee," expressed in terms of dollars per kW per month, paid by the customer regardless of the use of the standby service.

"Ratchets" are tariff provisions which require the customer to pay the reservation fee for some extended period, generally either a year or for the life of the standby service contract, generally based upon peak usage during any 30 minute period.

Some utilities charge the customer a demand charge similar to that charged their ordinary, full requirements customers. However, they only charge it during the months during which the customer actually uses power. Thus, a customer which relies on on-site generation for eleven months out of twelve would only pay one-twelfth the demand charge of the full requirements customer.

Other utilities charge the customer a demand charge which is deeply discounted from that charged to the ordinary customer – sometimes as little as a tenth of the ordinary rate. However, these customers are charged every month, based upon their peak requirement during any three month period, generally for a one year period – that is, they are "ratcheted" for the year, based upon the highest usage during that period.

A few utilities do both, charging both high demand charges, and ratcheting them for a full year. Such customers receive little savings from using alternative suppliers. The utilities charging such rates are violating PURPA.

The primary purpose of this study is provide a quick reference to the evaluated utilities, to permit one to determine whether a utility has practicable standby rates, for purposes of either on-site generation or retail competition.

## BACKGROUND

### Definition of Standby Rates

Self generating and other "partial requirements" utility customers such as those that engage in retail wheeling have a need for utility standby service. This need arises when customers' primary supply source becomes unavailable; typically in forced outage situations. The availability of standby-type services was made universal by the Public Utilities Regulatory Policies Act of 1978 (PURPA), requiring utilities to provide such serviced to a qualifying facility (QF) under that statute. Not all on-site generators are QFs, and customers engaging in retail wheeling are not QFs. Nevertheless, many of the partial requirements tariffs are written in a neutral fashion, available to all customers equally, and the principles used to create the tariffs will generally apply regardless of the partial requirements customers' primary source of electric power.

It is important to distinguish among different types of partial requirements services. These are:

- Supplementary Service – Energy or energy and capacity supplied by the utility which is regularly used by a facility in addition to what the facility generates itself or obtains elsewhere.
- Maintenance Service – Energy or energy and capacity supplied by the utility during *scheduled* outages.

- Standby or Backup Service – Energy or capacity supplied by the utility during *unscheduled* outages.

Back-up or maintenance service provided by an electric utility replaces energy or capacity which a qualifying facility ordinarily supplies to itself. FERC rules authorize certain facilities to purchase and sell simultaneously.

Supplementary power is electric energy or capacity used by a facility in addition to that which it ordinarily generates on its own. Thus, a cogeneration facility with a capacity of ten megawatts might require five more megawatts from a utility on a continuing basis to meet its electric load of fifteen megawatts. The five megawatts supplied by the electric utility would normally be provided as supplementary power.

Standby or backup power is electric energy or capacity available to replace energy generated by a facility's own generation equipment during an unscheduled outage. In the example provided above, a cogeneration facility might contract with an electric utility for the utility to have available ten megawatts, should the cogenerator's units experience an outage.

Maintenance power is electric energy or capacity supplied during scheduled outages of the qualifying facility. By pre-arrangement, a utility can agree to provide such energy during periods when the utility's other load requirements are low, thereby avoiding the imposition of large demands on the utility during peak periods.

Interruptible power is electric energy or capacity supplied to a qualifying facility subject to interruption by the electric utility under specified conditions. Many utilities have utilized interruptible service to avoid expensive investment in new capacity that would otherwise be necessary to assure adequate reserves at times of peak demand. Under this approach utilities assure the adequacy of reserves by arranging to reduce peak demand, rather than by adding capacity. Interruptible service is therefore normally provided at a lower rate than non-interruptible service. Supplementary, backup and maintenance power must be provided on an interruptible and on an uninterruptible basis.

There are qualifications to the right of cogenerators and small power producers to demand these services. For example:

- If a utility is not adding capacity (whether by construction or purchase) to meet anticipated increases in peak demand, the rates charged for interruptible service might appropriately by the same as for non-interruptible services. If interruptible customers provide no savings to the electric utility, the rate for interruptible service need not be lower than the rate for firm service.
- Certain electric utilities do not have any generating capacity, and to require the listed services might place an undue burden on the electric utility. In light of this, State regulatory authorities or the FERC will allow a waiver of these requirements upon a finding that provision of these services will impair the utility's ability to render adequate service to its customers or place an undue burden on the electric utility.
- FERC rules provide that rates for sales of backup or maintenance power shall not be based, without factual data, on the assumption that forced outages or other reductions in output by each qualifying facility on an electric utility's system will occur either simultaneously or during the system peak. Like other customers, qualifying facilities may well have intra-class diversity. In addition, because of the variations in size and load requirements among various types of qualifying facilities, such facilities may well have interclass diversity.

The effect of such diversity is that an electric utility supplying backup or maintenance power to qualifying facilities will not have to plan for reserve capacity to serve such facilities on the assumption that every facility will use power at the same moment. Probabilistic analyses of the demand of qualifying facilities show that a utility will probably not need to reserve capacity on a one-to-one basis to meet backup requirements. Utilities are prohibited from basing rates on the assumption that qualifying facilities will impose demands simultaneously and at system peak unless supported by factual data.

Rates for sales do take into account the extent to which a qualifying facility can usefully coordinate periods of scheduled maintenance with an electric utility. If a qualifying facility stays on line when the utility will need its capacity, and schedules maintenance when the utility's other units are operative, the qualifying facility is more valuable to the utility, as it can reduce its capacity requirements.

Many of the following issues have not yet been treated in regulatory proceedings. However, as of yet there has been little analysis by the commissions of the components of tariffs for this type of service. If you are planning a cogeneration project, or if you are planning on becoming a retail wheeling customer, it is likely that you will be among the first to broach these issues in your state.

## ISSUES IN PARTIAL REQUIREMENTS SERVICE TARIFFS

**Rights of the returning customer.** Should alternative supplies become relatively expensive, should a customer:

(a) have a right to return to utility service? Might there be mandatory waiting periods, or even an outright refusal to serve? What if the utility is capacity short?

(b) At what cost? What if the utility's marginal fuel source is oil or gas? Should the returning cogenerator be assured service at rolled in average costs, or only at marginal cost?

**What is the appropriate demand charge?** Under PURPA, assumptions that the cogenerator will experience outages on-peak or simultaneously with other cogenerators is illegal, unless supported by evidence. Is this relevant for the utility with excess capacity? If a utility has been recovering demand-related costs in its energy charges, might it actually charge cogenerators a higher demand charge than the ordinary customer? Under what circumstances? Would this discourage reporting of on-site generating capacity?

**Can rate stability for partial requirements customers be achieved?** Cost-based rates can change, but customers need stability in order to make investment and contracting decisions. Should there be vintaging of rates for partial requirements customers? Contracts rather than tariffs?

**Should there be multiple partial requirements customer classes?** Different customers are likely to have different cost of service characteristics – due to reliability, fuel choice, operating economics, etc. Should these customers receive separate treatment? Might they be covered by formula in a single tariff which accounts for operating characteristics?

**Is cost of service a useful concept in the context of competitive power markets?** If a prospective customer refuses to pay cost-based rates, is this grounds to abandon them? Does cost of service even have a definite meaning in situations in which there is surplus generating capacity? What if there is a competitive market for partial requirements service – should it continue to be regulated?

## THE STATE COMMISSIONS AND PARTIAL REQUIREMENTS SERVICE

Supplementary, backup and maintenance rates are similar to any of the service tariffs for which a utility has the primary responsibility to design rates. However, while utilities generally have little incentive to discriminate between classes of customers, so long as they receive enough income to fulfill their total revenue requirement, in serving customers with their own on-site generation, they are aiding their competitors. Thus, more than in other rate design cases, the role of the utility commission in reviewing the equity of utility rate proposals regarding service to cogenerators and small power producers is crucial. Below are several articles which describe the process of establishing different states partial requirements service. These are provided courtesy of *The Cogeneration & Power Marketing Monthly* newsletter.

Prior to PURPA, utilities would set rates for cogenerators and other on-site generators on the double presumption that backup power would be required simultaneously by all backup customers, at the time of the system peak. This approach was justified by the utility's claim that it was under an obligation to guarantee service to all such customers, and could only do so if it had the capacity to serve them all simultaneously. PURPA prohibits these assumptions from being used. The Niagara Mohawk case (below) shows an application of the prohibition, in which the commission ordered the arbitrary use of a 10% multiplier to reduce the capacity component of the standby charge by 90%. Another arbitrary multiplier which has been suggested is the utility's reserve margin. No studies have yet been completed on the actual capacity requirements of backup customers.

## NEW YORK PSC ISSUES ORDER ON NIAGARA MOHAWK'S TARIFF FOR SUPPLEMENTARY, BACKUP AND MAINTENANCE POWER (PARTIAL REQUIREMENTS SALES) FOR ON-SITE GENERATORS

The Public Service Commission has ordered substantial modifications to Niagara Mohawk's submitted tariff schedule S.C. 12, governing supplementary, backup and maintenance service for on-site generators. Backup rates may be chosen by on-site generators with low load factors who believe they might benefit from using those rates rather than the usual, "firm" rates. Supplementary power under the proposal is the service provided a customer who chose the backup rate, and who exceeds a 10% load factor.

The tariff, established in response to Commission Order 82-10, is a time of use format patterned after existing time differentiated rates offered by Niagara Mohawk.

All three options include a customer charge, a two block declining demand rate for distribution costs with a 50% ratchet, and a time differentiated energy charge. Production and transmission demand costs for backup and maintenance service are recovered through an increase in the on-peak energy charge. For supplementary service, these costs are recovered through an additional demand charge, which only applies during the peak period, but which includes a 75% ratchet for all usage over 40 kW.

There are special provisions in the tariff.

*Special Provision A* permits a customer to segregate any portion of his total requirement so that such portion may be served under a different service classification. Thus a portion of the additional power needed by an on-site generator may be purchased as normal "firm" power, and the rest as backup power under S.C. 12.

*Special Provision B* allows for a 3% reduction in registered demand when service is metered at the primary voltage level, in recognition of the increased efficiencies of service at that level.

*Special Provision C* provides penalties for exceeding contracted for demand. If the registered demand exceeds contracted for demand by 10% or less, there is a penalty in the form of a surcharge equal to six times the monthly distribution demand rate. If the excess demand is more than 10%, then the penalty is twelve times the monthly distribution demand rate.

*Special Provisions D and E* respectively prohibit instantaneous reactive power from exceeding measured demand, and states that operating mode, interconnection requirements, metering requirements and payments for additional interconnection costs shall be agreed to by the customer and Niagara Mohawk.

***Supplementary, backup service far too expensive.***

The Public Service Commission analyzed the rate schedule for backup power, and discovered that under the Niagara Mohawk proposal, customers requiring backup service are almost always far better off taking service under firm rates (S.C. 3), despite the fact that the firm rate tariff includes a ratchet of 75%, designed to penalize such low load factor customers as a backup customer might be presumed to be.

***Backup customers overcharged by presumption that all backup customers will take full contract demand at system peak.***

Supplementary service customers are charged .0484 under the proposed tariff for on-peak energy, while backup customers are charged a punitive .32 per kWh, for a difference of 27.2 cents per kWh. The Commission rejected this difference as improper because it assumes that all backup and maintenance customers take their full contracted demand at the time of system peak, despite the fact that the utility expects the class load factor to be only 5%. As a result of this approach, a customer using his contracted demand for 5% of the 3600 designated peak hours would have contributed, for each kW of contracted demand, the full "revenue reconciled" marginal cost per system peak kW for production and transmission.

The Commission concluded that, in the absence of actual operating data, backup customers would have average availability factors of 90%, and, therefore, the ratio of class demand to individual maximum contract demand is 10% (this figure having been offered by Commission staff and accepted during the ConEd case).

As the backup customer is only to be held responsible for 10%, and not the full, marginal cost per system peak kW, the "energy adder" to the backup rate is only one tenth of the proposed difference, or 2.72 cents per kWh. Thus, the on-peak energy charge for backup customers will only be 7.56 cents per kWh.

***Methodology to reconcile marginal based costs and embedded revenue requirements distorts charges***

The methodology used to reconcile marginal costs to actual utility revenue requirements results in distortions which penalize and subsidize on-site generators.

In this case the Commission found that the reconciliation factor used by Niagara Mohawk resulted in demand charges which are only 30% of marginal demand costs.

Due to this, the Commission has ordered a change in the methodology used to reconcile marginal cost based rates with embedded revenue requirements. A "system-wide reconciliation factor" will be applied to both energy and demand costs, resulting in an increase in demand charges and a drop in energy rates, in circumstances in which the embedded revenue requirement is exceeded by marginal costs (such as is true in this case), so that demand intrusive customers such as backup customers and some supplemental power customers will not be subsidized by other classes of customers.

(Note: the system wide reconciliation factor is the device used in New York to allow marginal cost pricing to be reconciled with actual utility revenue requirements. After marginal costs are calculated, and rates calculated on the basis of those costs, they are multiplied by the reconciliation factor to yield a projected revenue stream equal to total revenue requirements. If marginal costs exceed embedded costs, then the reconciliation factor is less than 1. If marginal costs are less than embedded costs, then the reconciliation factor exceeds one.)

Thus, should marginal costs come to exceed marginal revenue requirements, demand charges would rise and energy rates drop to the detriment of low load factor backup and supplemental power customers.

***Other determinations made in this ruling:***

On-site generators which choose the simultaneous buy/sell option should be treated under firm power tariffs, unless evidence is submitted showing that they belong in the partial requirements tariff. By choosing simultaneous buy/sell, they have opted not to accept the offer of backup and supplementary rates.

Rates should be revised to reflect latest approved marginal costs and embedded revenue requirements approved in Case 28225. This will tend to lower the rates for partial requirements sales, as the Commission approved less than the full rate request.

Rates for customers above 15,000 volts should be established.

*Special provision D,* prohibiting instantaneous reactive power from ever exceeding maximum demand, should be written to permit such demand when it does no harm to the system, but also accounts for situations in which the condition is not violated, yet Niagara Mohawk is forced to bear substantial costs because of the reactive demand. A charge of fifty cents per KVAR per month will be assessed of customers with contracted demand for over 500 kW per month.

Regardless of lead characteristics, any customer may choose to be subject to the "firm" power schedule, rather than the partial requirements schedules.

**NIAGARA MOHAWK POWER CORPORATION COMPARISON OF ANNUAL CHARGES**

S.C. 2,3 vs. S.C. 12

**Back-up and Maintenance Service**

Low Tension

| *Maximum Annual Connected Load* | *Annual Load Factor* | *kWh* | *S.C. 2* | *S.C. 12* | *Change Amount* | *%* |
|---|---|---|---|---|---|---|
| | | #2 | | | | |
| 10 | 0 | 0 | $123.12 | $374.40 | 251.28 | 204 |
| | 10 | 8640 | 780.20 | 1,835.80 | 1,055.60 | 135 |
| | | #3 | | | | |
| 100 | 0 | 0 | 4,680 | 1,723.20 | -2,956 | -63 |
| | 10 | 86,400 | 11,517 | 16,181.20 | 4,663 | 40 |
| 1000 | 0 | 0 | 4,680 | 8,743.20 | 4,063 | 87 |
| | 10 | 864,000 | 96,039 | 153,323.20 | 57,284 | 60 |
| 10000 | 0 | 0 | 4,680 | 78,943.20 | 74,263 | 1,587 |
| | 10 | 8,640,000 | 941,254 | 1,524,743 | 583,488 | 62 |

Compliance Tariffs:

Originally proposed rates by Niagara Mohawk:

On-Peak Off-Peak
Energy32 cents per kWh4 cents
(includes production
and transmission costs)

Plus a demand charge for distribution costs of between $.65 per kW to $2.84 per kW.

Supplemental service was apparently limited to purchases in excess of a 10% load factor (excluding scheduled outages) with purchases of energy up to 10% of contract demand billed at the backup rate.

The 32 cent per kWh peak period rate for backup service included 27.2 cents for generation and transmission costs. This was based on the assumption that all backup customers would take their full contracted demand at the time of the system peak. NIMO assumed that backup customers would have a load factor of 10%.

### Commission Decision

Following public comment, the Commission drastically reduced NIMO's backup rate because it assumed, contrary to the FERC rules, that all backup customers would take their full contract demand at the time of system peak. It also found that the 5% load factor assumed by Niagara Mohawk was incorrect. It directed that the capacity adder be reduced from 27.2 cents to 2.72 cents per kWh, based on a class coincidence factor of 10%.

It established a separate supplemental service definition which allowed QFs to contract for this service without purchasing backup power.

### Current NIMO Tariff

**ENERGY**
$.07579-$.08622/kWh on-peak
(Note: this resulted from taking the NIMO proposal of $.32/kWh and subtracting the disallowed capacity/transmission charge of 24.48 (27.2 minus 2.72), to get to $.072, and some other relatively minor adjustments.)

$.03352-$.03656/kWh off-peak
(depending on service voltage)

**DISTRIBUTION** (Generation and Transmission costs are recovered in the energy component during peak periods pursuant to order of the Public Service Commission.)

| | |
|---|---|
| $1.95/kW secondary | $1.94/kW secondary |
| $.79/kW primary | $.75/kW primary |
| (for each kW | (for each kW |
| of contract demand) | of as-used demand) |

Thus, as a result of the PSC decision, the energy rate for backup users was reduced by almost 75% and a new, more attractive supplemental structure was approved.

*One state in which partial requirements service has been the subject of commission review is Pennsylvania. The article below is included as an example of the context in which rate issues are decided. Drawn-out proceedings, intervenors, and contentious litigation are common. Our Philadelphia Electric Story is fair warning as to what could await you if you become involved in rate cases.*

## THE DEBATE OVER PHILADELPHIA ELECTRIC STANDBY RATES

**By Laurel W. Glassman**
**(Senior Attorney, Sutherland, Asbill & Brennan)**

The debate concerning appropriate rates for backup and maintenance power (collectively referred to as "standby" power) purchased by industrial cogeneration facilities has increasingly become the focus of attention at a number of state utility commissions. The Pennsylvania Public Utility Commission presided over such a debate in Pennsylvania Public Utility

Commission v. Philadelphia Electric Company, Docket No. R-850290 (decided October 9, 1986; final order on compliance tariff issued August 13, 1987), but, based on recent events in Pennsylvania, it appears that the Commission's work is far from over.

In the Philadelphia Electric case, the utility (PECO) had filed an Auxiliary Service Rider (ASR) to its tariff which reflected PECO's position that the normal service rate and applicable riders was the appropriate basis for pricing not only firm supplementary power (power supplied by the utility in addition to that which is supplied by cogeneration), but for firm backup power (power supplied by the utility to replace generation ordinarily provided by cogeneration equipment during periods of unscheduled outage) as well.

The ASR also reflected PECO's position that the normal service rate was the appropriate basis for pricing firm maintenance power (power supplied by the utility to replace the generation from cogeneration facilities when those facilities are scheduled out of service for maintenance), except that firm maintenance power charges under the ASR were prorated in accordance with the number of days that such service was supplied. This pro-ration was the only discount offered by PECO. Customers taking firm maintenance power, however, would not have avoided allocation of generation plant costs and common transmission plant costs. A customer taking firm maintenance power for a month would have paid the same charges as a full requirements customer, even though such usage occurred during a non-seasonal peak and was the only usage by that customer during the year. (Firm maintenance power used off-peak – PECO defined its peaks as diurnal rather than seasonal – would have been billed under PECO's Night Service Rider, under which the customer would have been charged for distribution costs, but no production or transmission costs except for a minimum billing demand of 25 kW.)

With regard to interruptible backup and maintenance power, under the ASR the normal service rate would have applied, except that the monthly billing demand based on the minimum contract demand or on the ratchet demand would not have been applicable to the interruptible load. In addition, under the ASR the demand and energy charges for the interruptible load were to be prorated in accordance with the numbers of days that interruptible load was used by the customer. The ASR also included a separate $/kW transmission and distribution reservation charge.

PECO's underlying position was that the full normal rate should be the starting point for sales of all power to cogenerators because, in its view, the demand characteristics for those customers were comparable to those of low load factor full requirements customers. This view was strongly opposed by the Commission Trial Staff and by a number of industrial firms participating in the proceeding, who asserted that there is a fundamental difference in the cost characteristics of customers taking standby power and the cost characteristics of the majority of other customers who form the basis for the normal service rates.

The administrative law judge and the full Commission agreed with Trial Staff and the industrials, finding that the effect of the Ashís firm and interruptible backup and maintenance power provisions was to assess excessive charges for these types of power and to discourage cogeneration development in Pennsylvania. By refusing to recognize "the actual costs imposed on the PECO system by the cogenerators" and by "insist[ing] on treating cogenerators simply as any other industrial customer," the utility had "erect[ed] obstacles to the development of cogeneration in its service territory. Instead of attempting to work with potential cogenerators for the benefit of all ratepayers, [PECO] has attempted to hold back the progress of cogeneration." (Recommended Decision, at 7).

In light of this finding, the administrative law judge recommended, and the full Commission adopted, a rate design wholly different from the one proposed by PECO. With regard to firm backup power, the administrative law judge found that "the only conclusion to be reached on this record is that backup demands do not bear peak responsibilities in any way similar to those imposed by full requirements customers." (Recommended Decision, at 36). (The evidence in the case supported a random forced outage rate for cogenerators of between 3%-8%.) He therefore determined that charging full requirements rates for firm backup service was inappropriate. Ultimately, it was determined that the firm backup rate should be comprised of two components: 1) a monthly minimum demand charge of $3.35/kW, based on PECO's 22% planning reserve margin for generating and transmission capacity; and 2) a flat $0.0652/kWh developed by dividing the test year claimed booked revenue by total booked kWh, plus the appropriate annual Energy Cost Rate (ECR) and State Tax Adjustment Surcharge (STAS).

With regard to firm maintenance power, the Commission confirmed the administrative law judge's decision setting the rate at a flat $.0375/kWh, plus ECR and STAS, for power taken during the eight off-peak months of October through May. The $0.0375/kWh was equivalent to the then-current energy "tail block" rate otherwise applicable to customers in the high-tension class. Given the expected off-peak nature of scheduled maintenance power, it was determined that the rate for such power should not recover peak production and transmission costs. Accordingly, the rate was designed to recover the costs of energy and some additional recovery of distribution and other non-energy costs. For scheduled maintenance power taken during the four on-peak months of June through September, the rate would be the same as the full normal rate, including demand charges and ratchets.

The Commission also confirmed the administrative law judge's decision to use the same $0.0375/kWh energy "tail block" rate, plus ECR and STAS, for both interruptible maintenance power and interruptible backup power. The rate was designed to recognize that these forms of interruptible power do not contribute to PECO's cost of meeting peak capacity requirements. however, the rate was intended to include some allocation of demand charges for generation, transmission and distribution.

PECO did not appeal the Commission's final decision. However, portions of the decision – including the $3.35/kW minimum monthly demand charge for firm backup power – have been appealed to the Pennsylvania Commonwealth Court by one of the industrial participants. Scott Paper Company v. Pennsylvania Public Utility Commission, Docket No. 1987 C.D. 2137 (filed September 11, 1987).

On May 1, 1987, the Pennsylvania Commission instituted a generic proceeding for the purpose of obtaining comments from a broad range of interests on a series of issues, regarding the Commission's current cogeneration regulations and whether those regulations should be fundamentally restructured. 17 Pa. Bull. 1776 (May 2, 1987). In this proceeding, which is not expected to conclude until the end of 1987, seven public utilities sponsored joint revisions to the cogeneration regulations. These revisions would, if adopted, amend Section 57.35 of the current regulations and, most importantly, Section 57.35(d). Section 57.35(d) currently requires maintenance power rates to include "all energy costs and a demand or capacity charge not in excess of a charge required to recover the appropriate transmission plant and full distribution plant costs." It also requires on-peak scheduled maintenance service to carry the full demand or capacity charge stated in the utility's filed tariff.

The utilities have recommended that Section 57.35(d) be deleted, and replaced with the following language: "Rates for sales to qualifying facilities shall be fully compensatory, and as such, they shall include charges required to recover production, transmission, distribution and all other costs incurred to provide such service." This new language is significant because, although ambiguous, it could be interpreted to mean that standby customers must pay generation and bulk transmission demand charges, as well as local transmission and distribution demand charges for firm off-peak scheduled maintenance power, as well as for interruptible backup and maintenance power, even though, as found in the Philadelphia Electric case, such power bears no responsibility for peak capacity costs. The utilities have thus far presented no expert testimony in support of the result which would be achieved by such a change in the regulations.

The Coalition of Ratepayers and In-State Power Producers (CRISSP) has sponsored testimony contrary to the utilities' position, and which would seek to preserve – and strengthen – the result achieved in Philadelphia Electric. CRISSP has argued in favor of retaining Section 57.35(d) in its entirety, but adding language which would make clear that the full demand charge would only apply to firm maintenance power, and only when such power cannot be scheduled during other than the utility's cost-causative peak periods. CRISSP has advocated that customers taking scheduled maintenance service (who do not also take backup service) would pay the on-going local transmission and distribution charge for their specified maintenance demand. To the extent that customers also contracted for backup service, no additional charge would be applicable. With notice requirements and other scheduling limitations on pre-arranged maintenance service, there should not, in CRISSP's view, be any reservation charge for scheduled maintenance service.

With regard to backup power, expert testimony sponsored by CRISSP supports use of the concept of equivalent forced outage rates (the ratio between the hours that a cogeneration facility is out of service as a result of a forced outage and the total hours in the period considered) to establish the charges for backup service. The equivalent forced outage rate of the most reliable facilities for which backup service is being provided would be the basis for calculating a minimum payment. For example, assuming that the appropriate unit cost of generation capacity for a utility is $10/kW/month, and that the equivalent forced outage rate of the most reliable facility is 5%, the minimum charge per kW of capacity would be $0.50/kW/month ($10 x 5%). All customers requesting backup service would pay $0.50/kW /month on an on-going basis.

A daily pro-ration methodology would then be used to ensure that less reliable facilities pay a commensurately higher amount. For example, assuming that a particular month contains 20 days that have on-peak hours, a customer requiring backup service during four days of that month would pay $0.50 x 4, or a total of $2.00/kW. Although this approach penalizes to some extent facilities with equivalent forced outage rates below 5%, it offers the dual benefits of providing ongoing compensation to the utility, and matching the charges paid for backup service with the reliability of the facilities being backed-up. The daily pro-ration concept thus accurately tracks the additional reserve obligation imposed on the utility system by facilities having varying forced outage characteristics.

Given the fact that the focus of the generic proceeding before the Commission is on issues pertaining to purchases from qualifying cogeneration facilities rather than on sales to such facilities, and that debate on methodologies for determining separate rates for firm and interruptible backup and maintenance power has not truly been joined, the Commission is not expected to amend its regulations to incorporate the methodology proposed by the CRISSP group. However, the CRISSP proposal does provide a basis for standardizing the methodology for developing standby rates in future proceedings in Pennsylvania?

## Philadelphia Electric PARTIAL REQUIREMENTS SERVICE TARIFF

Philadelphia Electric Company SUPPLEMENT No. 37 to Electric-PA. P.U.C. No. 26 FIFTH REVISED Page No. 52 Superseding 2nd, 3rd & 4th Revised Pages No. 52 Issued September 16, 1987 Issued by J.H. Austin, Jr., President Effective June 27, 1986

## STANDARD RIDERS- AUXILIARY SERVICE RIDER

### Applicability

Service to customers, including but not limited to qualifying facilities of small power producers and cogenerators as defined in the Public Utility Regulatory Policies Act, whose electrical requirements are partially or wholly provided by facilities not owned by the Company and when such facilities operate in parallel with the Company, will be supplied only under the provisions of this rider.

### Extent of Supply

The maximum firm supply available from the Company will be defined by contract except for customers served on Rates R, R-H and GS without demand measurement.

### Parallel Operation

The Customer shall not commence initial operation of any other source of supply in parallel with the Company's service until written permission is given by the Company for such parallel operation. Written permission is not necessary for reestablishing parallel operation, but the Customer shall notify the Company when resuming any parallel operation after an outage. The Company shall have the right to inspect the Customer's installation in accordance with Tariff Rule 9.3.

### Type of Supply

The following types of power supply are available:

Supplementary Power supply is available to add to alternative generating capacity whether or not owned by the Customer. All power provided pursuant to this Rider shall be Supplementary Power unless it is provided within the definition of Back-up Power or Maintenance Power.

**Back-up Power supply** is available to replace alternative generating capacity whether or not owned by the Customer during a forced outage of all or part of such generating capacity. Back-up Power shall be limited to 15% of the hours in any twelve-month period after which any additional power utilized shall be billed as Supplementary Power. The Customer must orally notify the Company immediately when Back-up Power is used, and within one business day after the forced outage giving rise to the need for Back-up Power, shall furnish the Company with a letter verifying the outage, specifying the time at which the outage commenced, the reason for the outage, and providing the best estimate possible of its duration. Oral and written notice shall also be provided to the Company within one business day following the conclusion of the forced outage. The Company may require verification of the cause of such forced outage. The foregoing 15% limitation on the number of hours in which Back-up Power may be received shall not apply during the following periods, nor shall such periods be taken into account in determining whether Back-up Power use in any subsequent period has exceeded such limitation: (a) in the case of an alternative generating facility with rated capacity of 1 MW or less, the three-month period commencing on the date such facility is first operated in parallel with the Company's service; and (b) in the case of an alternative generating facility with rated capacity in excess of 1 MW, the six-month period commencing on the date such facility is first operated in parallel with the Company's service.

**Maintenance Power** is available to replace alternative generating capacity whether or not owned by the Customer during periods of scheduled maintenance. Maintenance Power will be supplied on a scheduled basis in one of the following manners: (a) Upon mutual agreement, at any time. (b) Upon at least 60 days written notice and not more than 180 days written notice by the Customer, the Company will advise the Customer, within 30 days of the receipt of the request, of the availability of the requested Maintenance Power, for power required for a period of more than 48 hours duration. If the power is unavailable during the requested period, the Company will provide Maintenance Power within 30 days prior or subsequent to the beginning of the requested period and will so inform the Customer. (c) Upon 360 days written notice by the Customer, the Company will provide Maintenance Power during the requested period, unless the cumulative total of all such power requested during such time period will exceed 5% of the Company's operable generating capacity, in which case the provisions of (b) above will apply. (d) For Maintenance Power required for a period of 48 hours or less duration, at a demand of 50 MW or less, the Company will supply such power on at least 30 days written notice. (e) The Company in its sole discretion may refuse to schedule firm Maintenance Power during the months of June through September except that Maintenance Power as defined in (d) above will be made available during June through September as long as it can be scheduled during off-peak hours.

Maintenance Power will be limited to no more than 120 days in any twelve-month period, and no more than 60 consecutive days, after which any additional power utilized shall be billed as Supplementary Power. The foregoing limitations on the number of days in which Maintenance Power may be received shall not apply during the following periods, nor shall such periods be taken into account in determining whether Maintenance Power use in any subsequent period has exceeded such limitations: (a) in the case of an alternative generating facility with rated capacity of 1 MW or less, the three-month period commencing on the date such facility is first operated in parallel with the Company's service; and (b) in the case of an alternative generating facility with rated capacity in excess of 1 MW, the six-month period commencing on the date such facility is first operated in parallel with the Company's service. The supply of Maintenance Power will be terminated when generating capacity from which the Customer is supplied is returned to operation as indicated by the recorded demands on the Company's metering equipment, or upon notification to the Company by the Customer, or upon the expiration of the maximum maintenance period, whichever occurs first.

### Interruptible Power for Back-Up or Maintenance

Customers with a minimum of 1,000 kW of interruptible Back-Up or Maintenance Power may contract for interruptible supply. When a

Customer contracts for interruptible supply, such supply shall be interrupted when, in the sole judgment of the Company, any production, transmission or distribution capacity limitations exist. The Customer shall interrupt such load after a minimum of thirty minutes prior notice by the Company. The first time that the Customer is notified by the Company to interrupt service and the Customer fails to interrupt, a penalty of $24 per kilowatt shall be applicable to each kilowatt of demand that has not been interrupted. The second time that the Customer fails to interrupt, the same penalty shall be applicable and interruptible supply will not be available to the Customer for a period of three years.

### Rate and Billing

All monthly bills for service on this rider shall include one application of the Customer Charge of the applicable rate. All other Demand and Energy Charges of the applicable rate shall be modified as set forth below.

### Supplementary Power

Billing shall be under the provisions of the applicable rate and riders.

### Firm Back-Up Power

Demand Charge: $2.67 per kilowatt of demand specified in the contract for back-up supply. This charge shall include energy use equal in cost to the total monthly demand charge. Energy Charge: For service at: High Tension Voltage: 7.82 cents per kilowatt-hour Primary Voltage: 9.83 cents per kilowatt-hour Secondary Voltage: 11.67 cents per kilowatt-hour

### Interruptible Back-Up Power

Demand Charge: None.
Energy Charge:
For service at: High Tension Voltage:3.56 cents per kilowatt-hour
Primary Voltage:3.64 cents per kilowatt-hour
Secondary Voltage:5.14 cents per kilowatt-hour

### Firm Maintenance Power

June through September: Same as Supplementary Power.
October through May: Same as Interruptible Back-Up Power.

### Interruptible Maintenance Power

Same as Interruptible Back-Up Power.

STATE TAX ADJUSTMENT CLAUSE, FEDERAL TAX ADJUSTMENT CREDIT* and ENERGY COST RATE FACTOR apply to this rider. [*Applicable effective May 1, 1987.]

## CHARGES FOR STANDBY SERVICE IN CALIFORNIA

**Robert T. Howard**
**Vice President**
**Barakat, Howard & Chamberlin, Inc.**

### Introduction

Over the past several years, California's major utilities have endeavored to revise standby rate schedules for electric power in order to mitigate the impacts of cogeneration bypass by large commercial and industrial customers. Recently implemented or current proposals by Pacific Gas and Electric, Southern California Edison, and San Diego Gas and Electric increase standby charges significantly. These changes will affect the economics of natural gas-fired cogeneration projects in the state.

In California, utilities have generally provided standby service to backup customer generation facilities. In concept, standby charges recover the cost of facilities and reserve capacity made available by the utility in the event of outages of the customer's own generation.

When standby rates applicable to cogeneration were first implemented in California, they were designed with the assumption that most of the electricity generated by the customer would be sold to the utility. The tariffs were less stringent and generally provided encouragement to the cogeneration facilities to operate the facilities in a manner that maximized the benefit of sales to the utility system. For example, both Southern California Edison Company's (SCE) and San Diego Gas & Electric Company's (SDG&E) tariffs provided a waiver of some portion or all the standby charge when the customer's generator attained an 85 percent capacity factor in the company's on-peak period.

Through these provisions, standby charges evolved with a broader purpose than merely the recovery of fixed cost for facilities provided to backup the customers' own generation facilities. These incentives included performance incentives to operate the facility during the hours when the utility's power costs were the highest. Such provisions encouraged operation of the facility when it was of most benefit to the utility by providing the cogenerator an incentive through adjustment of the standby charge.

With more direct competition from natural gas-fired cogeneration, new tariffs are being designed with a more single-minded focus: to recover the fixed cost of the interconnections with the utility system. As noted in SCE's recent rate case exhibit, "Full-requirements customers should not be expected to carry the burden of fixed costs in excess of marginal costs for facilities dedicated to serving partial-requirements customers."

The remainder of this article will summarize the major components of existing standby charges and discuss how standby rates are changing in California.

**Summary of the Major Components of Existing Standby Charges**

Standby tariffs consist of more than rates; they include a number of other important terms and conditions, generally designed to reflect utility policies toward customer-owned generation options, including the following:

- Rates
- Definition of contract demand
- Waivers for forced outages and scheduled maintenance
- Performance incentives
- Interruptible service waivers

The rates include the monthly charge per kW for the contracted standby demand and a definition of the minimum charge. The charges normally apply in level addition to all charges and provisions of the otherwise applicable rate schedule.

The definition of the contract demand generally is determined under one of two options: unrestricted or partial operation of the customer's plant. The first option applies if the utility stands ready to supply the entire connected load or established demand of the customer's plant. The charge is applied to the entire connected load or the customer's generating capacity, whichever is less. In the second case, the customer contracts for a predetermined maximum demand to supply a portion of the load normally served by the customer's plant. In this case the customer is required to install, at his/her own expense, a circuit breaker that will break the connection with the utility's service when the demand exceeds the contract demand.

Several California utilities have allowed adjustments of standby charges for forced outages and scheduled maintenance. For example, SDG&E's current rates allow for a standby charge adjustment when a customer can show what part of the billing demand on the customer's bill for service was caused by a forced shutdown. This means that the standby charge for that portion of the demand would be waived in the event of a forced outage. With respect to scheduled maintenance, both SCE and Pacific Gas and Electric Company (PG&E) have provided that the added demand created by scheduled maintenance outages of cogeneration facilities would be ignored for the purposes of determining demand charges under the regular service schedule during the months of February, March, or other months determined acceptable to the utilities.

Current standby rates for SCE and SDG&E both provide performance incentives to standby customers. As discussed above, one provision common to both utilities is the capacity factor reward, whereby some or all of the standby charge is waived if the customer's plant attained a capacity factor of 85 percent or better during the utility's on-peak period. In addition, San Diego Gas and Electric offers a reward for the avoidance of on-peak demand. When on-peak usage is avoided, the standby charge is reduced by $2.50 per kW per month if the customer has not imposed on-peak demands on the utility during the "ratchet" period. Thus, when both the capacity factor condition and the avoidance of on-peak demand are met, the standby customer would fully waive the standby charge.

In SCE's case the customer can waive the standby charge in its entirety if the capacity factor condition is met. In SDG&E's case the customer is required to meet both the capacity factor and the on-peak demand condition before the entire standby charge is waived.

Finally, SCE and SDG&E have provided waivers for customers electing to take interruptible standby service in their currently effective rate structures. In the SCE case, the standby charge is waived. In SDG&E's case, the on-peak charges on the applicable time-of-use schedule are waived for the interruptible portion of the customer's load.

### Changing Standby Charges in Response to Increased Competition

The standby charges discussed above have been among the least restrictive in the nation. They are undergoing significant changes as the threat of direct competition from cogeneration increases. The proposed standby policies of utilities will have a significant impact on new projects and affect the economics of existing plants. The CPUC has recently approved revised standby rates for PG&E. SCE and SDG&E are proposing to revise their standby structures to increase the cost of standby through various proceedings now before the CPUC.

The CPUC staff has acknowledged that the revision of standby rates is an important issue. For example, in reviewing PG&E's proposal to significantly increase standby charges, the CPUC staff noted: The Public Staff Division (PSD) is aware of the possible threat of uneconomic bypass to become a serious issue for PG&E and agrees that a comprehensive review and possible revision of standby rates is appropriate.

Edison's proposed standby charges are designed to recover a significant portion of embedded distribution costs dedicated to serve standby customers at sub-transmission, primary, and secondary service voltage levels. These charges are intended to recover the costs of facilities that are essentially dedicated to servicing individual customers year-round – even during those months of the year when customers use little or no Edison-supplied power. SDG&E proposes changes to make their schedules "consistent with PG&E's new Schedule S, adopted by Commission Decision 86-12-091, including the elimination of performance incentives and waivers of demand charges."

SDG&E's proposed changes to standby charges are made in conjunction with a proposed revision to its retail level tariffs that would simultaneously increase the fixed charges collected from all customers for standard service. Thus, SDG&E proposes to collect a standby charge in addition to a fixed contract demand charge.

Under the current proposals, the utilities have based the revised charges on the principle of cost recovery of distribution facilities. In addition, the utilities argue for the elimination of the various waivers in the Original tariff because they have experienced very little, if any, fixed cost savings related to distribution plant investments when outages are scheduled in the off-peak hours or when higher capacity factors are achieved in the on-peak periods. In support of this position, the utilities argue that distribution facilities must be sized to serve standby customers' highest expected demands, even if these occur only once or twice a year. If the waivers are retained, the utilities argue that they will not recover the cost of these facilities.

These changes represent a significant departure from the structure of earlier standby charges. Earlier structures were in place to encourage the development of efficient, small power production to be sold to the utility on a non-discriminatory basis. The present structures will be more restrictive and will result in higher costs where interconnections are either required or necessary for the on-site facility.

**Strategic and Regulatory Implications**

Changes in standby charges are clearly part of a strategy to reduce the competitive threat of cogeneration alternatives. While it is natural for utilities to respond to the competition in their respective markets, the long term implications of more restrictive standby requirements are significant. On one hand, the utilities must consider the potential impacts on the utility system of providing sufficient resources to meet the reliability requirements of customers. On the other hand, significant energy efficiency gains may be lost through major restructuring of standby tariffs. The pricing for standby services, like any other service, should be cost based and not unduly discriminatory. However, standby charges may not be the most appropriate mechanism for pricing in a more competitive environment.

As an alternative, it would be more appropriate to make further distinctions in utility customer classes and develop rate structures accordingly. For example, the industrial customer class could be split into full requirements and partial requirements customer groups. As a separate class of service, the market structure, pricing, and cost of service considerations can be clearly and directly related to the operational requirements of this market. With such an approach it is more likely that any energy efficiency gains can be preserved while ensuring that utilities recover costs associated with the requirements of serving this market.

## RATE DESIGN FOR BACK-UP, MAINTENANCE, AND SUPPLEMENTAL POWER FOR SELF-GENERATORS

**Steven L. Walton**
**Manager, Pricing**
**Utah Power & Light Company**

Presented at:
*NERA Marginal Cost Working Group Meeting*
*October 14-16, 1987*

Our design of a rate for back-up, maintenance, and supplemental power might be viewed as an exercise in problem solving, to quote

Cicero, "A problem well stated is half solved." The fundamental step is to clarify the design objective. This can be done by identifying a set of critical problems which must be solved to meet the design goal. Identifying the critical problems focuses effort where it will be most effective. Successful inventors and entrepreneurs are successful because of an ability to define and solve such critical problems.

In rate design there are often conflicting goals which must be resolved to produce a practical tariff. Progress in resolving disputes in a rate proceeding often seems to founder because the critical problems are not clearly defined. There is often common ground for agreement, but because of poor problem definition, discussion never focuses on the true difficulties.

### Critical Problems

In approaching the back-up rate question, our design goal was to produce a rate schedule for back-up power, maintenance power, and supplementary power for facilities which generated either all or part of their own power requirements. We have several customers who either have or are in processes of installing cogeneration plants, and who want to receive these auxiliary services. Our previous "standby" rate schedule, designed in 1976, was not adequate. The dimensions of our design problem were further defined by the regulations issued by the Federal Energy Regulatory Commission (FERC) to implement PURPA. The FERC lists the following services which utilities must provide to a qualifying facility (QF) upon request: (1) supplementary power, (2) back-up power, (3) maintenance power, and (4) interruptible power. The rates for these services are to be based upon costing principles which do not "discriminate against any qualifying facility to the extent that such rates apply to the utility's other customers with similar load or other cost-related characteristics." Further, rates for back-up and maintenance power are not to be based upon the assumption that all QFs are simultaneously out of service or that QF outages occur at the time of system peak. Rates must also recognize the usefulness of coordinating maintenance between QFs and utility units.

It is clear that the rates for QFs must recognize the special characteristics of self-generators' operations and that the same principles used for other customers are to be applied to pricing QF services. Establishment of a traditional rate schedule depends upon a cost of service study, using defined usage patterns, to establish the revenue target for the class. Data drawn from billing histories is then used to design a rate which will collect the appropriate revenue. However, since there is not an existing class of QFs, there is no data upon which to base a cost of service study for QF auxiliary services. An additional difficulty is posed by the fact that the rate schedule is to be generic, yet a very wide range of installations with differing performance characteristics may request service under the schedule.

A designer learns more from a failed design than from a successful one. The evaluation of flaws in an early design are used to improve later models. Our first auxiliary services rate design was not very successful, judging by customer acceptance. We filed a proposed rate design that assumed a given level of availability and specified particular modes of operation most likely to occur. Intervenors filed testimony which challenged the proposed rates. Evaluating their information, together with our own data allowed us to sharpen our definition of the critical problems. The three critical problems identified were:

1. What will be the availability of the units for which back-up/maintenance services will be requested, i.e., how after and for how long will service be rendered.

2. What utility facilities are required to provide access to auxiliary services, to what degree will they be utilized, and how is the cost to be recovered.
3. How are back-up/maintenance power and supplemental power to be distinguished since each will be charged at different rates, i.e., how are services to be "unbundled"?

**Designing a Rate**

We now turned our attention to solving our critical problems. Since the rates offered must use the same cost basis as used for other system customers, we began with the unit costs used to design the Commercial/Industrial rates in our last approved tariff. The energy rate was set equal to the last energy block in the respective general service schedules. This energy rate contains both energy costs and a portion of the demand costs. The additional demand costs in both the earlier energy blocks and in the demand charge were used to produce the back-up rates.

Most ordinary customers have fairly steady demand patterns from day to day or week to week. However, the demand of self-generators as seen by a utility will be erratic since unit outages are not predictable. Therefore the rate design for the demand charges must incorporate pricing features that can follow the customer's use.

**Ability**

The most important of the three critical problems identified is the availability issue. In our first design, we assumed a 70% availability factor for cogenerators. This fixed availability was used to determine the portion of the generation costs which should be assigned to a back-up charge. The intervenors presented data showing a range of availabilities for various units. Many were in the 90% range, and some were claimed to have 99+% availability. Yet in checking with other utilities who presently serve self-generators, we found that some operators have availabilities of 50% or less.

The solution to uncertain availability is to make the price structure follow the actual usage, rather than assuming a level of usage based on someone's hypothesis of availability. We did this by converting the monthly demand charge to a daily charge for demands occurring on weekdays during on-peak hours. The daily demand charge provides an economic efficiency benefit; customers who achieve high unit availability will not be subsidizing poor operators who cannot keep their units on line. A protracted argument over availability was also avoided by addressing the problem directly through changing the billing demand.

Associated with the availability issue are charges to be made in maintenance periods. The intervenors asserted that rates should be lower during maintenance periods. The FERC regulations clearly indicate that some concession must be made when maintenance is scheduled with the utility. At the same time, notification of outages the day before they occur does not allow the utility any opportunity to change its plans and realize a savings.

The daily demand charge provides a resolution for maintenance outages which are truly planned. Since the daily demand charge applies only in on-peak hours, minor repairs can be done on weekdays without incurring a back-up demand charge by using the 8 hours between 11 p.m. and 7 a.m. In addition each week a period of 56 hours is available, from 11 p.m. Friday to 7 a.m. Monday, when more substantial work can be done. Should maintenance run over the planned period, the daily charge rate makes it possible for the self-generator to know how much each additional day of delay will cost. In order to accommodate annually scheduled maintenance, the daily demand charge was eliminated for one 30 day period or two 15 day periods a year. Advance scheduling of these annual maintenance periods is required by September 1 of the previous year. A monthly fee is added to a customer's bill if they do not want to coordinate maintenance scheduling with the Company.

**Required Facilities For Access to Utility Auxiliary Services**

In order for a self-generator to obtain auxiliary services from a utility, an interconnection must be in place. At the actual point of interconnection, it is obvious that facilities not shared with other customers should be paid for by the self-generator. However, deciding what the self-generator's responsibility is for the rest of the transmission and distribution system requires an understanding of those systems. In general, the electric power delivery systems, transmission and distribution, are poorly understood by customers. Many feel that if they have no use in a particular month, they should not pay anything; yet the systems must be able to handle the full load of the self-generator whenever the unit goes off-line. The costs of both the transmission system and the distribution are related to the maximum power demanded and not to the amount of energy delivered over time.

Our studies of the marginal cost of the distribution system show that distribution design requirements are not reduced by the low load factor of the self-generator. Our distribution system is a set of radial networks which contains no parallel paths. This design strategy lowers cost and simplifies design. Sufficient capacity must be present in each radial leg of the system to handle the full potential load which may be experienced. As a practical matter, since conductors only come in discrete sizes, this practice adds very little to system cost initially, but it minimizes the need to rebuild or reconductor circuits later. After a circuit is in service, work must often be done "hot," i.e., with the conductor energized, which dramatically raises labor costs.

Distribution system customers are not assigned the cost of specific equipment, but receive an average system cost as the distribution component in their rates. Since there is no difference between back-up and regular customers in distribution cost, the same cost was assigned to the back-up/maintenance rate as to regular rates. However, for the back-up rate, the distribution system cost was set out as a separate "Facilities Charge" which is levied against the back-up demand for which the customer contracts. No transmission costs were assigned to the facilities charge for the customers taking back-up service at distribution voltages, instead it was folded into the daily demand charge.

Transmission design practice is more complex than the distribution system practice described above. In most cases parallel paths exist in the transmission network over which a customer or company substation can be served. Service under the back-up rate schedule is limited to loads of 15 MW or less. These are smaller loads on the scale of transmission transactions, which can "run in the shadows," particularly on the higher voltage 230 kV and 345 kV facilities and perhaps even on 138 kV facilities. A 15 MW load however, is substantial on 46 kV and 69 kV facilities which make up the "sub-transmission" network. The facilities charge for transmission customers was set to recover the cost of only the sub-transmission lines and substations. More work needs to be done in this area. It is our intention in the future to examine the unique interconnection and line costs associated with serving customers from the transmission system in order to better quantify the appropriate facilities charges for customers served from the transmission system.

### Unbundling of Services

Most utilities services are bundled, i.e., all components of the service are covered by a single price structure. The FERC regulations require utilities to provide supplemental power if requested. The supplemental needs of the self-generator are to be provided under the standard rate schedules. This requires that an unbundling take place with separate prices for the back-up and supplemental services. The customer can then choose those services needed for a given load. Short of painting the electrons as they come and go, a single meter at the utilities point of delivery cannot tell a back-up kW from a maintenance kW from a supplemental kW. At least one more meter is needed to determine the operating level of the customer's generator. The use of daily demand charges requires that load profile metering be used for both meters. Even with two meters, some sort of rule is necessary to assign demand and energy into either the back-up/maintenance or the supplemental "buckets" since they have different pricing rules. By time sequencing the readings of the load profile meters, it is possible to use simple rules to assign the demand and energy from each 15 minute interval to the appropriate bucket. The customer charge includes the cost of the first metering set. The actual cost of additional metering is collected using a fixed charge multiplier. This is done to accommodate the wide range of physical arrangements which could occur at the self-generator's site, without unduly burdening all customers with high metering costs.

### An Example Problem

The following example will demonstrate some of the features of the back-up/maintenance rate schedule and show how the separation of services takes place. A copy of the rate schedule and a metering and billing example are attached as Exhibit "A." Suppose that a customer with a total load of 2,000 kW installs a 1,000 kW cogeneration facility. Service from the utility is delivered at 46 kV to a customer owned substation. Metering would be placed on the incoming utility connection and on the terminals of the customer's generator. Since our example customer has taken advantage of the coordination of maintenance by submitting a maintenance plan, the maintenance charge of $0.90/kW will not be assessed. Let us consider four billing intervals during a month. For simplicity this example uses one hour intervals instead of the actual 15 minute intervals.

**UTAH POWER & LIGHT COMPANY BACK-UP RATE EXAMPLE**

| Hour | Company Meter | Generator Meter | Total Load | Back-Up Power | Supp. Power |
|---|---|---|---|---|---|
| 9 | 600 | 1,000 | 1,600 | 0 | 600 |
| 10 | 1,600 | 0 | 1,600 | 1,000 | 600 |
| 11 | 2,000 | 0 | 2,000 | 1,000 | 1,000 |
| 12 | 1,500 | 500 | 2,000 | 500 | 1,000 |
| Energy, kWh | | | 404,000 | 2,500 | 401,500 |
| Demand, kWh | | | | 1,000 | 1,000 |
| Facilities Charge = (.98)(1,000) = | | | | | $980 |
| Back-Up Demand Charge = (.34)(1,000) = | | | | | $340 |
| Supplemental Demand Charge = (9.52)(1,000) | | | | | $9,520 |
| Energy Charge = (.031083)(404,000) | | | | | $12,557 |
| | | | | Total Load | $23,397 |

Only supplemental service is provided in Hour 9; the Utility measures a demand of 600 kW and the generator meter shows 1,000 kW. In Hours 10 and 11, the generator is off line, so 1,000 kW of back-up is provided, with the rest of the power being supplied as supplemental. In Hour 12 the unit comes back on at half power, and 500 kW of back-up power is supplied. In this example, the customer generator remains on line for the rest of the month at full load. No further back-up service is rendered but supplemental energy is supplied throughout the month. The calculation of the combined charges is shown with the exception of the customer charges.

**Final Notes**

By carefully examining the critical problems in designing a practical back-up rate, we were able to establish that a facilities charge and daily demands charges could be used to both cover the cost of the required facilities for the utility and charge a self-generator a price which follows the performance of a given installation. The daily demand feature avoided endless argument over the availability which would be appropriate for class-wide application to self-generators seeking auxiliary services. By using profile metering, we were also able to unbundle the services on a 15 minute interval basis, and cover supplemental service under the normal general service tariffs.

A point not made earlier was that the proposed rate schedule is not limited to QFs but is available to any self-generator not engaged in resale of services. Some have interpreted this as a lack of concern for possible penetration of cogeneration into our retail sales. This is not correct. A high back-up rate is not necessarily a deterrent to truly economic projects, instead a high back-up rate causes truly viable projects to buy diesel generation for back-up. The competitive forum is in the general service rates for the customers who are candidates for self-generation as an alternative to full utility services. We do not see an advantage to restricting this rate schedule to QFs.

# Appendix H-A Analysis of Standby Rate Tariffs

## ANALYSIS OF STANDBY RATE TARIFFS

Standby tariffs, by definition, aren't used very much. Due to this fact, the most important component of a standby rate is the "demand" or "reservation" charge. These charges vary widely among utilities, and are generally the make or break criteria for on-site generation facilities.

In this appendix we excerpt materials originally published in a survey, "Utility Standby Rates," published by the Power Marketing Association. The original survey is available for $195 from PMA, and includes the complete text of over 100 utility standby tariffs. To obtain a copy, fax your request to PMA at (703) 641-9265, call (703) 641-0613 or e-mail centaur@erols.com. Be sure to note your shipping and billing address.

**Alabama Power Co. / Alabama Public Service Commissioncentaur@erols.com**

Service available under Rates LPL, LPM, & RTP with following modifications:

Maximum integrated fifteen-minute capacity during each service period.

Projects to one month.

Demand Charge:

| | |
|---|---|
| May-September | $ 1.25 |
| October-April | $ 1.15 |

**Arizona Public Service Co. / Arizona Corporation Commission**

Fifteen minutes

Standby Monthly Reservation Charge: $ 2.25 per kW

*Proposed to change 7/1/96*

**Arkansas Power and Light**

Standby Service Charge:

$2.50 per kW of standby capacity less the excess of any minimum charges otherwise billed under the standard rate schedule but not less than zero.

Standby Capacity:

Excess of contract capacity over Customer's monthly demand for supplementary power during the current month.

Monthly Demand:

The kW as shown by the readings of the demand meter for the 15-minute period of greatest use during the month.

**Arkansas Power & Light Co. / Arkansas Public Service Commission**

Fifteen minutes

Standby service charge: $2.50 per kW

**Baltimore Gas & Electric Co. / Maryland Public Service Commission**

Backup Service: Rates for Backup Service are a percentage (the "factor") of the Controlling Schedule's Production & Transmission Demand Charge, plus the Distribution Demand charge. The percentage is subject to revision monthly as follows:

Factor

50% Base Level, plus (2%) decrement per month without purchase of Market Service on type B days, plus 1% increment per purchase of Market Service on Type B days.

In no event will the Factor for a given month be less than 20% or greater than 100%.

Demand is billed at a daily rate determined by dividing the Production & Transmission Demand Charge of the Controlling Schedule by 21.8, except that such charges are: reduced by 30% on Type A days; increased by 100% on Type B days.

**Black Hills Power & Light Co.**

Emergency Rate = 8.17 cents per kWh

Demand Charge:

a. Firm Contract Demand = $10.50 per kW

Interruptible Demand = $5.25 per kW

**Black Hills Power & Light Co. / South Dakota Public Utilities Commission**

Demand Charge:

$10.50 per kW of firm contract demand

$ 5.25 per kW of interruptible demand

Interruptible Demand shall be the maximum average load during any fifteen-minute period of use during the month

**Blackstone Valley Electric Company**

Secondary Distribution Voltage:

Demand Charge:

kW used during Peak Hours = $7.15 per kW

Billing Demand in kilowatts for each month will be the higher of the max. 15-minute period during peak hours in the month or 100 kilowatts.

Primary Distribution Voltage:

Demand Charge:

kW used during Peak Hours = $6.56 per kW

Billing Demand in kilowatts for each month will be the higher of the max. 15-minute period during peak hours in the month or 100 kilowatts.

**Blackstone Valley Electric Co. / Rhode Island Public Utility Commission**

Demand Charge kW during Peak Hours:

Primary $6.56 per kW

Secondary $7.15 per kW

The Billing Demand in kW for each month will be the higher of the maximum demand in any fifteen-minute period during Peak Hours in the month or 100 kW

**Carolina Power & Light Co. / North Carolina Utilities Commission**

Standby Service:

Contract Demand

| | Less than 1000 kW | | Greater than 1000 kW | |
|---|---|---|---|---|
| Transmission | | | | |
| without transformation | | | $1.74 | |
| with one transformation | | | $2.17 | |
| Distribution | | | | |
| without transformation | | | $1.50 | $ 2.34 |
| with one Transformation | | | $2.22 | $ 2.66 |
| | *Summer* | | *Non-summer* | |
| | *Summer* | | *Non-summer* | |
| Reservation Charge: | | | | |
| per kW of Contract Demand | | | | |
| | $2.09 | $ 1.59 | $3.87 | $ 3.16 |
| Daily Demand Charge: | | | | |
| per kW Maximum On-Peak | | | | |
| | $0.47 | $0.36 | $0.86 | $0.71 |

Standby times Number of days used during month

**Central Illinois Light Co. / Illinois Commerce Commission**

Demand Charge:

$5.44 per kVA summer months; $3.45 winter months

Determination of Maximum Monthly Demand:

Maximum Demand shall be greatest average load in kVA for any fifteen-minute period.

**Central Power & Light Co. / Public Utilities Commission of Texas**

Monthly Rate

For transmission voltage service the Minimum Monthly Charge shall be $ 2.36 per kW or kVA of Standby Contract Capacity.

Primary voltage service

$3.31

Secondary voltage service

$3.53

Demand charge per kW or kVA of Standby Contract Capacity will be 1.75% of the demand charge of the Rate Schedule up to seven days following the seven days during periods other than scheduled maintenance the demand charge shall be the demand charge in the Rate Schedule multiplied by the number of additional days or portion of days times 3.29% per kW or kVA or the higher created demand during such period 15 minute period of maximum use during the current month.

**Central Power & Light Co. / Public Utilities Commission of Texas**

As Available Standby Power

Monthly Rate

The Minimum Monthly Charge shall be $ 1.08 per kW of Contract Capacity.

During periods when service is furnished monthly demand charge will be Rate Schedule times 1.75% up to seven days after seven days Rate Schedule times number of additional days or portion of days times 1.65%.

**Cleveland Electric Illuminating Co. / Public Utilities Commission of Ohio**

Monthly Rates

Capacity Reservation Charge

| *Transmission* | *Sub-transmission* | *Distribution* |
|---|---|---|
| $3.50 kW | $4.50 kW | $5.50 kW |

Demand Charge

| *Transmission* | *Sub-transmission* | *Distribution* |
|---|---|---|
| $.85 kW day | $.90 kW day | $.95 kW |

Determination of daily Demand Charge shall be maximum daily Back-Up Demand times appropriate above daily rate. The billing period charge for Back-Up Demand shall be sum of billing periods daily demand charges.

**Commonwealth Edison Co. / Illinois Commerce Commission**

Required Facilities Charge:

$2.99 per kW per month

Demand Charge

*For Standby Service for firm load that would otherwise be supplied by the company under the otherwise applicable rate:*

**Summer Months**

first 10,000 kW $15.16
over 10,000 kW $ 6.29

**All Other Months**

first 10,000 kW $13.41
over 10,000 kW $ 6.03

for interruptible load

**Summer Months**

first 10,000 kW $0.70
over 10,000 kW $0.14

**Other Months**

first 10,000 kW $0.79

over 10,000 kW $0.16

30 minute maximum demand for month

**Connecticut Light & Power Co. / Connecticut Department of Utility Control**

Monthly Rate:

The customer shall be billed for service , in accordance with the applicable general service tariff available to the customer for the size of service taken based on the Back-up Contract Demand Level except as modified below.

Production/Transmission Demand or contribution Charge:

Customer shall pay the greater of:

A. D times (1-(1-K/2080)6) per kw of production/Transmission in demand, where D equals the applicable general service rate's Production/Transmission Demand Charge, or if absent, then the general service rate's total charge less $4.00 after July 1, 1993, and $4.50 after July 1, 1994. Where K equals the sum of the back-up/standby loads taken in each on-peak hour of the latest six months of December, January, February, June, July, August divided by the contracted back-up/standby demand, or

B. the Production/Transmission Demand Charge of $1.00 per kw of back-up Contract Demand.

**Consolidated Edison Company of New York / New York Public Service Commission**

Back-up Service Rate I

Contract Demand Charge (per kw per month of the contract demand):

*Primary Distribution*

| | |
|---|---|
| First 10kw | $4.55 |
| Next 90kw | $5.45 |
| Over 100kw | $5.81 |

*Secondary Distribution / Low Tension*

| | |
|---|---|
| First 10kw | $4.56 |
| Next 90kw | $5.52 |
| Over 100kw | $5.92 |

As-used Demand Charge (per kw per month of the maximum demand on an as-used basis, ratcheted during the summer billing period):

| | *Summer Billing Period* | |
|---|---|---|
| | *High Tension* | *Low Tension* |
| *Primary Distribution* (Monday-Friday 8:00A.M.-10:00P.M.) | | |
| First 10kw | $6.68 | $6.68 |
| Next 90kw | $9.39 | $9.39 |
| Over 100kw | $10.45 | $10.45 |
| *Secondary Distribution* | | |
| First 10kw | | $7.22 |
| Next 90kw | | $10.13 |
| Over 100kw | | $11.34 |

Rate II (Customer's whose demands exceed 900kw for two consecutive months.)

Contract Demand Charge:

| | *High Tension* | *Low Tension* |
|---|---|---|
| *Primary Distribution* (all hours-all day's) | | |
| First 10kw | $4.55 | $4.55 |
| Next 90kw | $5.45 | $5.45 |
| Over 100kw | $5.81 | $5.81 |
| *Secondary Distribution* | | |
| First 10kw | $4.56 | |
| Next 90kw | $5.52 | |
| Over 100kw | $5.92 | |

As-used Demand Charge (per kw per month of the maximum demand on an as-used basis, ratcheted during the summer billing period)

| | | |
|---|---|---|
| *Primary Distribution* (Monday-Friday 8:00A.M.-10:00 P.M.) | | |
| First 10kw | $6.68 | $6.68 |
| Next 90kw | $9.39 | $9.39 |
| Over 100kw | $10.45 | $10.45 |
| *Secondary Distribution* | | |
| First 10kw | | $7.22 |
| Next 90kw | | $10.13 |
| Over 100kw | | $11.34 |

**Consumers Power Co. /**
**Michigan Public Service Commission**

Monthly demand charge of contracted maximum standby capacity $0.33 per kW of the on-peak billing demand shall be the highest 15-minute kW standby demand created during on-peak period.

**COGENERATION AND SMALL POWER PRODUCTION RATE CG**

Optional Standby Rate:

Standby service contract of one year minimum contracted maximum standby demand in kW

Capacity Charge:

On-Peak Billing Demand Charge:

$0.39 per kW per day for the highest on-peak billing demand each day service is used.

Maximum Demand Charge:

*Secondary & Primary:*

$ 1.85 per kW per month contracted maximum demand.

*Sub-transmission:*

$0.97 per kW

*Transmission:*

$0.52 per kW

The maximum demand shall be the highest 15-minute kW demand created during current month.

**Detroit Edison Company**

Parallel Operation and Standby Service Rate

Monthly Generation Reservation Fee = $0.86 times the standby contract capacity in kW per month.

Non-Generation Charge: (per kW of standby contract capacity)

a. $1.51 per kW at transmission voltage

b. $2.36 per kW at sub-transmission voltage

c. $3.25 per kW at primary voltage

d. $3.80 per kW at secondary voltage

Demand Charge:

a. Daily highest 30 minute reading during on peak hours.

b. Daily on peak demand charge for back-up power = $0.90 per kW per day.

c. Daily on peak demand charge for maintenance = $0.48 per kW per day.

Auxiliary Or Standby Service Rates

Monthly demand charge = $0.33 per kW of contracted maximum standby capacity regardless of whether standby energy is used that month.

If operator uses on peak kWh for standby energy equal to 88.9 hours times the operator's contracted standby demand, the on-peak billing demand charge shall also be paid for all subsequent purchases of standby energy (excluding maintenance energy) during that contract year.

Cogeneration and Small Power Production Purchase Rate CG

An operator may opt out of the above auxiliary classification and agree to contract for standby service for a minimum of 1 year. A contracted maximum standby demand and, if applicable, a contracted maximum supplemental demand shall be established by mutual agreement.

Capacity charge:

a. On-peak Billing Demand Charge = $0.39 per kW per day for the highest on-peak billing demand occurring each day in which standby service is utilized.

b. Maximum Demand Charge
   - Secondary and Primary service = $1.85 per kW per month.
   - Sub-transmission service = $.97 per kW per month.
   - Transmission service = $.52 per kW per month.

**Detroit Edison Co. /**
**Michigan Public Service Commission**

Monthly Reservation Fee:

$0.86 per kW per month

Monthly Contract Capacity Charge based on peak period capacity

| | |
|---|---|
| Transmission: | $ 1.51 per kW |
| Sub-transmission: | $ 2.36 per kW |
| Primary: | $ 3.25 per kW |
| Secondary: | $ 3.80 per kW |

30-minute integrated reading during on-peak hours of demand meter which measures total load

Daily on-peak demand charge for back-up power: $0.90 per kW

Daily on-peak demand charge for maintenance: $0.48 per kW

**Duke Power Co.**

Parallel Generation

Rate: On-Peak Demand Charge per on-peak month

a. Transmission = $13.24 per kW
b. Distribution = $15.77 per kW

Standby Charge per month = .95 per kW.

**Duke Power Co. /**
**North Carolina Utilities Commission**

Rate:

Charges

On-Peak Demand Charge per on-peak month:

| | |
|---|---|
| Transmission: | $14.26 per kW |
| Distribution: | $16.94 per kW |
| Standby Charge per month: | $ 1.03 per kW |

Determination of demand charges:

The on-peak demand for billing purposes shall be the largest of the following:

1. the maximum integrated thirty-minute demand during the on-peak month for which the bill is rendered
2. seventy percent of contract demand
3. 30 kW

Demand charges do not apply in the months of October, November, April or May.

**El Paso Electric Co. /**
**Public Utilities Commission of Texas**

Monthly Rate

General Service Rate which is applicable to this class of service, will be applied when the bill for service rendered during any billing month, calculated at the base rate of the General Service Rate, exceeds the minimum specified for standby

Monthly Minimum

$16.75 per kW of installed capacity, or measured thirty minute demand, whichever is greater or stipulated in writing

**Florida Power Corp. /**
**Florida Public Service Commission**

Determination of Capacity:

Specified Standby Capacity for the current billing period shall be the greater of :

(1) mutually agreed upon Specified Standby Capacity

(2) the maximum 30-minute kW standby power requirement established in the current billing month or

(3) the maximum 30-minute kW standby power requirement established in any of the twenty three preceding billing months/applies to all 3 schedules.

FIRM STANDBY

(Rate Per Month)

Standby Service Charges:

Distribution Capacity

$ 1.50 per kW times the specified standby capacity

Note : no charge applicable to a customer who has provided all the facilities for interconnection to the company's transmission system/ applies to all schedules Generation & Transmission Capacity

The charge shall be the greater of :

1. $0.835 per kW times the specified capacity,
2. the sum of the daily maximum 30-minute kW demand of actual use occurring during On-Peak Periods times $0.398/kW times the appropriate monthly factor :

| *Billing Month of:* | *Factor* |
|---|---|
| March, April, May, October | 0.80 |
| June, September, November, December | 1.00 |
| January, February, July, August | 1.20 |

same rates as interruptible & curtailable.

**Florida Power & Light Co. /**
**Florida Public Service Commission**

Contract Standby Demand

| | |
|---|---|
| Below 500kW | 500-2000kW |
| 2000kW & Above | All Levels |

Distribution Demand Charge

| | |
|---|---|
| $ 2.14 per kW | $ 2.52 per kW |
| $ 2.21 per kW | |

Reservation Demand Charge

| | |
|---|---|
| $0.80 per kW | $0.79 per kW |
| $0.79 per kW | $0.76 per kW |

Daily Demand Charge for each daily maximum On-Peak Demand

| | |
|---|---|
| $0.38 per kW | $0.37 per kW |
| $0.37 per kW | $0.36 per kW |

Contract Standby Demand:

This Contract Standby Demand will not be less than the maximum load actually served by the customers generation during the current month or prior 23 month period less the amount specified as the customers load which would not have to be served by the Company in the event of an outage of the customer's generation equipment

**Georgia Power Co.**

Back-up Service

Monthly Rate

Firm Standby Reserve Charge = $2.13 per kW of firm standby capacity.

Local Facilities Charge = $0.76 per kW of total standby capacity

**Georgia Power Co. /**
**Georgia Public Service Commission**

Monthly Rate

| | |
|---|---|
| Firm Standby Reserve Charge | $2.13 per kW |
| Local Facilities Charge | $0.76 per kW |

**DETERMINATION OF NORMAL BILLING DEMAND.**

The highest 30-minute kW measurement, for the purpose of determining the normal billing demand for supplementary power as calculated under the provisions of either the Power & Light Rate or the Governmental Rate, shall be based on the greater of

(1) the maximum measured demand during the time standby service is not being taken, or

(2) the maximum measured demand during the time standby service is being taken, less the standby power demand times the standby demand adjustment factor, for the applicable on-peak and off-peak periods.

**Gulf Power Co. /**
**Florida Public Service Commission**

Contract Demand:

| 100-500kW | 500-7500kW | Above 7500kW |
|---|---|---|

Local Facilities Demand Charge:

| | | |
|---|---|---|
| $1.69 per kW | $1.25 per kW | $0.52 per kW |

On-Peak Demand Charge up to NC:

$2.45 per kW $7.27 per kW $7.73 per k

Reservation Charge per kW of BC:

$1.01 per kW $1.01 per kW $1.00 per kW

Daily On-Peak in excess of NC:

$0.47 per kW $0.47 per kW $0.47 per kW

**DETERMINATION OF STANDARD BILLING DEMAND:**

On-Peak Demand-the kW billing demand for billing purposes shall be the customers maximum integrated fifteen minute demand to the nearest kW during each service month as measured during the hours designated on-peak but not to exceed the NC.

**Gulf States Utility Co.**

Standby Service

1. Reserved standby power = The greater of either the amount contracted for or the maximum 30-minute standby service demand during the previous 12 months.
2. Monthly Standby Service Billing Demand = Sum of Daily Standby Service Demands.
3. Daily Standby Service Demands =
   a. Maximum metered demand less,
   b. The greater of (1) the current month's billing demand for firm or interruptible power or (2) Maximum 30 minute demand during non-standby/maintenance, less
   c. The amount of reserved standby power, less
   d. Scheduled Maintenance Billing Demand

**Houston Lighting & Power Co. / Public Utilities Commission of Texas**

Monthly Rate

| | *Distribution* | *Transmission* |
|---|---|---|
| Monthly Charge: | | |
| not in addition to firm: | $465 | $2,634 |
| in addition to firm: | $420 | $420 |

Standby Charge

Charge per Standby Kva:

| *Standby kVA Amount* | *Distribution* | *Transmission* |
|---|---|---|
| 0-15,000 Kva | $ 2.15 | $ 1.70 |
| 15,001-59,000 Kva | $ 2.43 | $ 1.99 |
| 59,001 Kva and larger | $ 2.72 | $ 2.28 |

In the event the customers Monthly Bill is based upon a period of less than 27 days or more than 33 days, the Kva prices & facilities charge expressed in item "(1) Rate" above, shall be adjusted by multiplying by a ratio determined by dividing the actual number of days by 30 days.

**Idaho Power Co.**

Standby and Supplemental Service Rates.

(Only applicable to customers with a uniform large power service agreement, not cogenerators or small power producers.)

Definitions:

Standby Billing Demand: Total Average kW supplied during 15 minutes of maximum use over billing period (monthly), adjusted for power factor, less supplementary contract demand.

Monthly Charge for Standby Service:

*Primary*

1. Standby Reservation Charge = $1.48 per kW of standby capacity.
2. Standby Demand Charge = 3.85 per kW of standby billing demand.

*Transmission*

1. Standby Reservation Charge = $0.61 per kW of standby capacity.
2. Standby Demand Charge = 3.73 per kW of standby billing demand.

**Idaho Power Co. / Idaho Public Utilities Commission**

Monthly Charge

Standby Reservation Charge:

| | |
|---|---|
| Primary | $ 1.48 per kW |
| Transmission | $0.61 per kW |

Standby Demand Charge:

| | |
|---|---|
| Primary | $ 3.85 per kW |
| Transmission | $ 3.73 per kW |

Excess Demand Charge:

$0.50 per kW times the sum of the daily Excess Demands recorded during

the billing period, plus

$5.00 per kW for the highest Excess Demand recorded during the billing Period.

This charge will not be prorated.

Standby Billing Demand is equal to the total average kW supplied during the 15-consecutive-minute period of maximum use during the Billing Period, adjusted for Power Factor, less Supplementary Contract Demand, but not less than zero.

**Illinois Power Co. / Illinois Commerce Commission**

Distribution Capacity Charge:

Customer served from supply line voltage below 138 kV shall be billed, for each billing period, a charge of $ 1.75 per kW for each kW of distribution.

Standby Demand Charges:

Customers Supply Line Voltage:

| | 12.47 kV & below | 34.5kv,69kv & 138kv |
|---|---|---|
| Summer Season | $14.00 per kW | $12.00 per kW |
| Winter Season | $ 8.00 per kW | $ 7.45 per kW |

Standby Billing Demand is the maximum integrated kW Standby Demand delivered during any 15 minute period in the On-Peak Period of the billing period, but shall not be less than zero.

**Indiana Michigan Power Co. / Indiana Utility Regulatory Commission**

Back-up & Maintenance Service:

Customer's with COGEN/SPP facilities having a total design capacity of more than 10kw shall be required to purchase back-up service to replace energy from COGEN/SPP facilities during maintenance and unscheduled outages of its COGEN/SPP facilities. The back-up capacity in kw shall be initially established by mutual agreement for electrical capacity sufficient to meet the maximum back-up requirements which the Company is expected to supply. Whenever the back-up capacity so established is exceeded by the creation of a greater actual maximum demand, then such greater demand becomes the new back-up capacity.

A monthly service charge of $1.526 per kw of back-up capacity shall be paid by customers served under demand metered rate schedules. Whenever back-up & maintenance capacity is used and the customer notifies the Company in writing prior to the meter reading date, the back-up contract capacity shall be subtracted from the total metered demand during the period specified by the customer for billing demand purposes. After 1900 hours of use during the contract year, the total metered demand shall be used as the billing demand each month until a new contract year is established.

**Jersey Central Power & Light / New Jersey Board Of Public Utilities**

Standby Rate per kw:

$3.09 for service classifications GS & GST
$2.32 for service classifications GP
$1.16 for service classification GT

Billing Demand KW:

Billing Demand is equal to MM-AG but not less than zero.

**Kansas Gas & Electric Co. / State Corporation Commission of Kansas**

Net Monthly Bill Back-up Service:

The rates shall be as stated in and total charge calculated according to the rate schedule under which a customer contracts for this service.

Back-up Service capacity shall be greater of

(1) the maximum capacity recorded by Company meter less a customer's contracted maximum Supplementary Service capacity, at times other than scheduled maintenance period's, or

(2) the maximum capacity recorded by Company meter less the sum of the customer's contracted maximum Supplementary Service capacity plus contracted Maintenance Service capacity scheduled for use during the hour in which maximum capacity is recorded. In no event shall Back-up Service capacity be less than zero or more than the maximum amount specified in the customer's Electric Service Agreement.

**Long Island Lighting Co. / New York Public Service Commission**

Rate II-Back-up & Maintenance Services:
Service charge(per installation per month):

| | *Backup & Maintenance only* | *Backup, Maintenance & Supplemental* |
|---|---|---|
| Secondary voltage(7kw & less): | $35.04 | $26.85 |
| Secondary (above 7kw ): | $48.18 | $30.96 |
| Primary | $94.71 | $77.85 |

Demand Charge(per kw per month):

Secondary: $2.97

Primary: $2.34

As-used demand charge(per kw per month):

Secondary: $2.97

Primary: $2.34

**Louisville Gas & Electric Co. / Kentucky Public Service Commission**

Rate:

Electric service actually used each month will be charged for in accordance with the provisions of the applicable rate schedule; provided, however, that the monthly bill shall in no case be less than an amount calculated at the rate of $5.61 per kw applied to the Contract Demand.

**Madison Gas & Electric / Wisconsin Public Service Commission**

Standby, Maintenance, And Supplemental Power

For standby, maintenance, and supplemental, the customer will purchase the energy under the applicable retail rate schedule. Maintenance of parallel generation equipment shall be scheduled with the Company. Demand charges paid by the customer during the maintenance period will be prorated based on the number of days the equipment is out for maintenance.

**MidAmerican Energy Co. / Iowa Utilities Board**

Rider No. 1 to Electric Large General Service-Standby and Supplementary Service

Standby service supplied under this rider is for no more than 10,000 kw.

Net Monthly Rate:

Standby Service Charge:

Summer Price

$1.55 per kw for all kw of Standby Capacity

Winter Price

$1.15 per kw

This rate is subject to Tax Adjustment.

Demand:

(1) The kw as shown by or computed from the readings of the Company's demand meter for the 15-minute period of customer's greatest use during the month, determined to the nearest kw, but not less than the minimum billing demand specified by the price schedule to which this Rider is applicable.

**Mississippi Power & Light Co. / Mississippi Public Service Commission**

Monthly Rate:

Service billed under applicable general service rate schedule plus,

Monthly Standby Charge: $3.79 per kw

When customer's potential maximum demand without respect to source of power is in excess of the customer's total capacity, the standby capacity shall be .3 times the capability of the customer's largest generating unit plus .05 times his total capacity.

When customer's potential maximum demand without respect to source of power is less than the customer's total capacity, the standby capacity shall be .35 times the customer's load without respect to the source of power.

**Mississippi Power Co. / Mississippi Public Service Commission**

Monthly Rate For Secondary Service:

| | |
|---|---|
| Reservation Charge | $ 1.25 per kW |
| Local Facilities Charge | $ 1.05 per kW |
| Billing Demand | $ 7.00 per kW |

**DETERMINATION OF MONTHLY BILLING DEMAND**

Customers kW billing demand each month shall be the average kW required from Company during the fifteen minute period of customers greatest use in month as measured by a suitable meter and rounded to the nearest whole kW.

Monthly Rates for capacity exceeding 5,000 kW:

| | |
|---|---|
| Reservation Charge | $ 1.25 per kW |
| Local Facilities Charge | $ 1.05 per kW |

Daily Demand Charges:

$0.40 per kW daily billing demand for the months of May through October during on-peak hours.

$0.26 per kW daily billing demands for the months of November through April on-peak.

**Montana Power Co. / Montana Public Service Commission**

Monthly Rates

Standby Reservation Charge:

| | |
|---|---|
| Transmission | $ 1.16 per kW |
| Substation | $ 2.64 per kW |
| Primary Distribution | $ 2.64 per kW |
| Secondary Distribution | $ 2.64 per kW |

Daily Demand Charge:

| | *Winter* | *Summer* |
|---|---|---|
| Transmission | $0.33 kW | $0.12 kW |
| Substation | $0.32 kW | $0.09 kW |
| Primary Distribution | $0.25 kW | $0.10 kW |
| Secondary Distribution | $0.28 kW | $0.11 kW |

Standby Reservation Demand Charge:

The monthly Standby Reservation Demand Charge is based on the Standby Contract Demand times the applicable Reservation Demand Charge for each month of the year.

**Montana-Dakota, State of Wyoming**

Contract Demand Charge = $10.38 per kW

**DETERMINATION OF CONTRACT DEMAND:**

Greater of the Max 15-minute measured demand in the current month or the contract demand in kilowatts. If customer exceeds his contract demand in any one month, such measured demand shall become the new contract demand for the next 12 months.

**Montana-Dakota Utilities Co. / Wyoming Public Utilities Commission**

Base Rate: $24.00 per month

Contract Demand Charge: $10.38 per kW

**DETERMINATION OF CONTRACT DEMAND:**

The demand in kW for billing purposes shall be the greater of either the maximum 15-minute measured demand in the current month or the contract demand in kW. Measured demand will be determined to the nearest one tenth kW. The company will require the customer to contract for additional standby and supplementary capacity if the customer exceeds his contract demand in any one month. Such measured demand shall become the new contract demand commencing with the month in which measured and thereafter for the eleven succeeding months, after which the customer and company re-determine full service requirements.

**Ohio Edison Co. / Public Utilities Commission Of Ohio**

Backup Capacity

| *Voltage Level* | *Reservation Charge* | *Daily Backup Power* |
|---|---|---|
| Secondary Voltage per kW | $4.41 per month | $.89 per day |
| Primary Voltage per Kva | $4.06 per month | $.87 per day |
| 23 & 34.5 kV per Kva | $3.55 per month | $.85 per day |
| 69 kV per Kva | $3.20 per month | $.83 per day |
| 138 kV per Kva | $2.56 per month | $.81 per day |

Demands created by daily backup power will not be used demand ratchet calculations in future months

Scheduled Maintenance Power

Power used during periods of agreed upon scheduled maintenance shall be billed on an on-peak daily basis based upon one-half the demand charges for Daily Backup Power

**BILLING DEMAND FOR SUPPLEMENTARY SERVICE**

The billing demand for the month for Supplementary Service shall be calculated per the Otherwise Applicable Rate Schedule with the exception that during periods of Daily Backup or Maintenance Power, each half-hour demand during the outage shall be reduced by the lesser

(1) contract demand for Backup or Maintenance Power

(2) amount of Backup or Maintenance Power actually used during that half hour as determined by interval metering of the customers on-site generation

**Nevada Power Company**

<u>Standby Service for Secondary Distribution Voltage:</u>

Facilities Charge = $1.97 per kW of Contract Demand per bill.

<u>Demand Charge:</u>

Summer On-peak Period = $5.70 per kW
Summer Mid-peak Period = $0.87 per kW
Summer Off-peak Period = $0.00 per kW
All Other Periods = 0.70 per kW

<u>Standby Service for Primary Distribution Voltage</u>

Facilities Charge = $2.00 per kW of Contract Demand per bill.

<u>Demand Charge:</u>

Summer On-peak Period = $5.19 per kW
Summer Mid-peak Period = $0.79 per kW
Summer Off-peak Period = $0.00 per kW
All Other Periods = $0.63 per kW

<u>Standby Service at Transmission Voltage</u>

Facilities Charge = $1.57 per kW of Contract Demand per bill.

<u>Demand Charge:</u>

Summer On-peak Period = $6.14 per kW
Summer Mid-peak Period = $1.02 per kW
Summer Off-peak Period = $0.00 per kW
All Other Periods = $0.57 per kW

**Northern Indiana Public Service Company**

<u>Back-up Power Capacity Charge (billed monthly):</u>

Monthly Base Cost: .003945 per kW per month times the number of hours selected for each month of the contract period for each kW of the net demonstrated capability of each generating unit for the hours of use selected for the back-up per year.

**Northern States Power Company (Minnesota)**

<u>Standby Service Rates:</u>

Demand Charge per month per kW
of contracted standby capacity = $3.25

Primary Voltage Service = $2.30

Transmission Transformed Voltage Service = $1.50

Transmission Voltage Service = .90

**Pacific Gas & Electric Co. /**
**California Public Utilities Commission**

Standby Reservation Charge ($/kw/month/ applied to 85% of the Reservation Capacity):

Secondary voltage: $2.55
Primary voltage: $2.55
Transmission voltage: $0.35

**Pacific Power, State of Wyoming except Wyoming West Service Territory**

For Contract Capacity of less than 1,000 kW.

<u>Demand Charge:</u>

Winter = $2.73 per kW of billing demand. (Nov.-Apr.)

Summer = $2.23 per kW of billing demand.

Determination of Demand: Max 15-minute period of use during the month.

<u>For Contract Capacity of more than 1,000 kW:</u>

<u>Demand Charge:</u>

On-peak Period Demand (Monday – Friday 7:00 a.m. to 11:00 p.m.)

Winter = 3.45 per kW of billing demand. (Nov.-Apr.)

Summer = 2.95 per kW of billing demand.

Determination of On-peak Billing Demand:

Greatest of:

1. Max 15-minute period of use measured during month, or
2. Average of the two greatest monthly measured On-peak Period kW established during the respective Summer or Winter months of the 12-month period which includes and ends with the billing month, or
3. 500 kW

**Pacific Power / Wyoming Public Service Commission**

Demand

The rate in kW at which electric energy is generated, transferred or used. Demand Power measurements are calculated based on the average integrated usage over consecutive fifteen minute periods of time. Demand Power measurements may be based on any one such fifteen minute period in a Billing Period, or on the period of greatest use during each day Demands may be adjusted for Power Factor. Abnormally high demands established as a result of extraordinary conditions existing on the Company's system or as a result of accidents resulting in temporary separation of Power Co. and Customers system shall be excluded from the determination of demand.

Back-up Facilities Charges:

| | |
|---|---|
| 138kv or higher | $2.68 per kW |
| 46kv or 69kv | $2.95 per kW |
| Less than 46kv | $3.23 per kW |

The Facilities Charge applies to the kW of Back-up Contract Demand

**Pacific Power (Wyoming West Service Territory)**

Back-up Power Charges:

138 kV or higher = $0.36 per kW Day
46 kV or 69kV = $0.39 per kV Day
Less than 46 kV = $0.43 per kV Day
During Scheduled = One half per kW Day Charge Maintenance

The power charge for back-up Power is billed on a per day basis and is based on the Max 15-minute period of back-up power used during each day.

**Pacific Power & Light / California Public Utilities Commission**

Standby Charge:

Fifty percent of the applicable Generation & Transmission Demand Charge of Schedule A-36 shall be applied to the kw by which customer's Contract Capacity or Total Load Demand, as provided by when the contract exceeds the customer's monthly measured demand.

**Pacific Power & Light Co. / Oregon Public Utilities Commission**

Partial Requirements 1000 kW or greater:

Monthly Billing:

Electric Service Charge:

The Electric Service Charge shall be computed in accordance with the Basic Demand Effective Energy including adjustments and Minimum Charges of Schedule 48T of this Tariff, and adjusted in accordance with the Primary Voltage Metering and Delivery Adjustments therein.

Standby Charge

Fifty percent of the applicable Demand Charge of Schedule 48T shall be applied to the kW by which Customers Contract Capacity or Total Load Demand, as provided by contract, exceeds the Billing Demand On-Peak Periods Billing Demand

The On-Peak Period kW shown by or computed from the readings of Company's demand meter for the 15-minute period of customers greatest use during the month, determined to the nearest kW.

Partial requirements Less Than 1000 kW Electric Service Charge

In accordance with Schedule 25 of this tariff.

Standby Charge

Fifty percent applicable Demand of Schedule 25.

Billing Demand

Greatest 15-minute period of use during the month.

**Pacific Power & Light / Washington Utilities & Transportation Commission**

Standby Charge:

Fifty percent of the applicable Demand Charge of Schedule 36 or 48T shall be applied to the kw by which customer's Contract Capacity or Total Load Demand, as provided by contract, exceeds the Billing Demand.`

**Pacific Power & Light Co. / Washington Utilities & Transportation Commission**

Partial Requirements 1000 kW & Over
Standby Charge:

Fifty percent of the applicable Demand Charge of Schedule 48T shall be applied to the kW by which customers Contract Capacity or Total Load Demand, as provided by contract, exceeds the Billing Demand.

Special Conditions

The contract for service shall specify customers selection from stated alternatives of service provisions by which the magnitude of Company's service and of the kW applicable to the standby charge is determined from

(a) customers Total Load Demand including any coincident power supplied by customers on site generation

(b) a lesser Contract Capacity expressed as a fixed total number of kW

In the absence of a currently applicable service contract providing for Total Load Demand billing or for a stated kW of Contract Capacity,

The Contract Capacity shall for billing purposes this schedule be deemed to be the average of the monthly measured demands of the most recently completed calendar year and shall not thereafter be increased Partial Requirements Less Than 1000 kW.

Standby Charge:

Fifty percent of applicable Demand Charge of Schedule 36
Same Special Conditions as for greater than 1000 kW

**PECO Energy Company**

Firm Back-up Power: Demand Charge = 3.00 per kW of demand specified in the contract for back-up supply. This charge shall include energy use equal in cost to the total monthly demand charge.

**Pennsylvania Electric Company**

Back-up Service:

Secondary Voltage = \$3.43 per kW
Primary Voltage = \$2.99 per kW
Transmission Voltage = \$1.67 per kW
Interruptible credit = \$0.67 per kW

During any month that the customer requires back-up due to an unscheduled outage, the customer will be required to pay for the back-up capacity at the above charges plus all capacity in excess of the specified back-up capacity, as well as all energy used. During any month in which the company is not required to provide back-up power, the customer will pay the above charges for back-up capacity.

**Pennsylvania Power & Light Company**

Back-up Power Capacity Charge:

480 Volts or less = $1.72 per kW
12,000 Volts = $1.67 per kW
69,000 Volts or more = 1.21 per kW

Determination of kW Demand: Average number of kilowatts supplied during the 15-minute period of maximum use during the billing period. (monthly)

Back-up Capacity Charge:

The demand charge per kVA of Back-up Capacity shall be initially established by multiplying the current demand charge in the Other wise applicable rate schedule by 0.30. The level of this multiplier may then be revised downward for subsequent contract periods to reflect the reliability of the individual self generator.

**Portland General Electric Company**

Standby Rate: 1.0 cents per kWh

**Potomac Electric Power Co. / District of Columbia Public Service Commission**

Standby Service:

PRODUCTION & TRANSMISSION RESERVATION CHARGE:

A monthly reservation charge of $0.50 per kw of contract demand shall be billed by the Company for standing ready to provide standby service. The contract demand shall be the maximum capacity for which the Company stands ready to serve.

**Potomac Electric Power Co. / Maryland Public Service Commission**

USAGE CHARGES:

When the Company is called upon to provide standby service, the demand and energy usage shall be billed under the schedule which would be normally applicable for the customer, including any fuel rate. The maximum demand charge for schedule "GT" will be credited by an amount equal to the facilities charge paid under this schedule, but not to exceed the amount of the respective demand charge.

**Public Service Company of Colorado**

Secondary Standby:
Back-up Service:

Capacity Reservation Charge = $5.05 per kW of Contract Reservation Demand.

Back-up Service Demand Charge = $4.71 per kW

Monthly Minimum: The Service and Facility Charge plus the applicable Capacity Reservation Charge.

Transmission Standby:

Back-up Service:
Capacity Reservation Charge = $2.64 per kW of Contract
Reservation Demand.

Back-up service Demand Charge = $4.52 per kW

Monthly Minimum: The Service and Facility Charge plus the applicable Capacity Reservation Charge.

Primary Standby:

Back-up Service:
Capacity Reservation Charge = $4.66 per kW of Contract
Reservation Demand.

Back-up service Demand Charge = $5.28 per kW

Monthly Minimum: The Service and Facility Charge plus the applicable Capacity Reservation Charge.

**Public Service Co. of New Mexico / New Mexico Public Service Commission**

Backup Power:

Rates for backup power will not contain demand ratchets or power factor penalties.

The rates for backup power may contain a monthly reservation fee, not to exceed 10 percent of the monthly demand charge contained in the retail rate schedule. Such a reservation fee will not be charged while a QF is taking backup power. Reservation fees are stated as per kw charge. Monthly reservation fees are computed by applying the per kw charges times the demand.

**Public Service Co. of Oklahoma / Public Service Commission of Oklahoma**

Company has no tariff for provision of standby/ backup service.
Rates, terms, and conditions for providing such service are determined by special contract, and are set in order such that the Company recovers the cost of service.

**Puget Sound Power & Light Co. / Washington Utilities & Transportation Commission**

Service shall be billed under the applicable general service rate, no separate standby rate applies.

**San Diego Gas & Electric Co. / California Public Utilities Commission**

Monthly Rates:

Standby Charge:

$3.44 per kw (Secondary)
$3.13 per kw (Primary)
$0.25 per kw (Primary Substation)
$0.25 per kw (Transmission)

Contract Demand:

If it is determined by the utility that the customer uses standby service in excess of his contract demand in any billing month, such increased demand shall become the new Contract Demand for the 12 months beginning with that month.

**Sierra Pacific Power Co. / Nevada Public Service Commission**

Fixed Standby Demand Charge

| | |
|---|---|
| For each kw of contract demand | $3.95 |
| Additional Variable Demand Charge | $3.95 |

For each kw of maximum total demand in excess of the contract demand

Variable Demand Charge

| | |
|---|---|
| For each kw Winter On-peak Billing Demand | $2.73 |
| Winter Mid-peak Billing Demand | $1.57 |
| Summer On-peak Billing Demand | $3.59 |

The demand for any billing shall be defined as the maximum measured fifteen minute average kw load in the billing period.

### Savannah Electric and Power Company

Monthly Standby Rates:

Basic Charge:

Transmission – 115 kV = $2.35 per kW of Contract Capacity.

Transmission – 46 kV = $2.40 per kW of Contract Capacity.

Distribution – Primary Voltage = $2.45 per kW of Contract Capacity.

Distribution – Secondary Voltage = $2.45 per kW of Contract Capacity.

Facilities Charge: (Only applies if Company provides a step down transformation)

Transmission – 115 kV = $0.24 per kW of Contract Capacity.

Transmission – 46 kV = $0.21 per kW of Contract Capacity.

Distribution – Primary Voltage = $0.17 per kW of Contract Capacity.

Distribution – Secondary Voltage = $0.17 per kW of Contract Capacity.

The contract capacity amount shall be automatically readjusted whenever the highest measured 30-minute demand in any billing period exceeds the previously established contract capacity and become equal to the highest measured 30-minute demand during the billing period. If such readjustment occur in the peak months of June – September then it shall cause a retroactive adjustment to the contract capacity for the preceding eleven months which shall be included in the current bill. If it exceeds contract capacity by more than 20% then a one time adjustment of six times the basic charge rate shall be added to the bill.

### South Carolina Electric and Gas Company

Standby Service Rate:

Demand Charge per kW of Contract Demand = $3.76

### Southern California Edison Co. / California Public Utilities Commission

Monthly Rate:

Standby charge:

$3.85 per kw (service below 2kv)
$3.40 per kw (2kv to 50kv)
$0.40 per kw (service above 50kv)

### St. Joseph Light & Power Co. / Missouri Public Service Commission

Rate:

The base rate shall be Schedule 311 or 411 on file by the Company, whichever schedule would be applicable to the class and type of service if the customer were purchasing his entire requirements for such type or class of service from the Company.

Minimum:

The net minimum monthly charge shall be $6.42 per kw of Reserved Capacity.

### Southwestern Electric Power Co. / Public Utilities Commission Of Texas

Monthly Rate:

During periods when Company provides As-Available Standby Power in accordance with the provisions of this rate, the kW to be used for billing determinations shall be based on the Customers 15-minute period of maximum use minus the kW of As-Available Standby Power requested.

| *Secondary Service* | *Primary Service* | *Transmission Service* |
|---|---|---|
| $1.62 | $1.25 | $0.30 |

Per kW of As-Available Standby Power Requested

**Tampa Electric Co. / Florida Public Service Commission**

Charges For Standby Service

Demand Charge:

$2.66 per kw of Standby Demand (Local Facilities Reservation Charge)

plus the greater of:

$0.87 per kw-Month of Standby Demand (Power Supply Reservation Charge) or

$0.34 per kw-Day of Actual Standby Billing Demand (Power Supply Demand Charge).

**Texas-New Mexico Power Co. / Public Utilities Commission Of Texas**

Demand Charge:

The per Kva charge of supplier multiplied by the amount of excess firm demands billed by Supplier to Company, and attributable to customers described in paragraph 10, if any plus $.14 per Kva

Reserve Charge:

The per Kva charge by Supplier to Company for Reserve Kva multiplied by the Reserve Kva billed by Supplier to Company and attributable to customer, if any, plus , where such charge is made, $.14 per Kva. 10. In the event one or more Standby Service customers have forced outages of generation or other power sources for which Standby Service is purchased from Company pursuant to this agreement and the sum of Standby Service requests exceeds the total Reserve Kva obtained by Company from Supplier and available at the given point of service between Company and Supplier for all Standby Service customers such that Company incurs a billing peak for firm wholesale electric service from its Supplier, the following shall apply:

Any Excess Firm Demand Costs calculated in accordance with this paragraph 10 which occur during the defined on-peak season of Supplier and subsequently contribute to the establishment of a ratcheted demand payment obligation from Company to Supplier, shall continue to be born by the customer(s) determined to have contributed to the establishment of the demand ratchet only to the extent and for the period that the ratcheted demand payment exceeds the amount that Company would otherwise have been required to pay supplier. This continuing payment obligation shall remain in effect so long as Company is billed by Supplier for the ratcheted demand payment as described herein.

**Texas Utilities Electric Co. / Public Utilities Commission Of Texas**

Monthly Rate:

| | *Secondary* | *Primary* | *Transmission* |
|---|---|---|---|
| Demand Backup (On-Peak ) | $3.05 per kW | $3.05 | $3.04 |
| Demand Backup (Off-Peak ) | $7.37 per kW | $6.67 | $5.43 |

On-Peak kW is the highest 15-minute kW recorded during the Company's on-peak hours of the current month.

Off-Peak kW is the highest 15-minute kW recorded during the Company's off-peak hours of the current month.

**Toledo Edison Co. / Public Utilities Commission of Ohio**

Monthly Net Rate

Capacity Reservation Charge:

$3.10 (transmission)
$4.00 (sub-transmission)
$4.85(distribution)

Back-up Demand

*Method A:*

The Demand Charge found in the appropriate, applicable Full Service Tariff shall be used to calculate the billing period charge for Back-up Demand.

The Capacity Reservation Charge for the billing period shall be reduced to the extent that Back-up Demand, under method A, is used during the billing period.

Method B

The billing period Demand Charge for Back-up Power shall be determined by use of daily demand, and will use the appropriate following rate based on service voltage:

$ .75 per kva per day (transmission)
$ .80 per kva per day (sub-transmission)
$ .85 per kva per day (distribution)

**Utah Power and Light, State of Utah**

Monthly Charges:

Facilities Charge:

Secondary Voltage = $2.63 per kW of Contract Demand

Primary Voltage = $2.12 per kW of Contract Demand

Transmission Voltage = $1.42 per kW of Contract Demand

**Utah Power and Light, State of Idaho**

Less than 2,300 Volts:

Power Rate for May – October, inclusive = $11.85 per kW of Contract Demand

Less than 2,300 Volts:

Power Rate for Nov. – April, inclusive = $8.58 per kW of Contract Demand

More than 2,300 Volts:

Power Rate for May – October, inclusive = $9.48 per kW of Contract Demand

More than 2,300 Volts:

Power Rate for Nov. – April, inclusive = $7.14 per kW of Contract Demand

**Virginia Electric & Power Co. / Virginia State Corporation Commission**

30-Day Rate

Distribution Demand Charge

| | |
|---|---|
| First 5000 kW | $0.613 per kW |
| Additional kW | $0.463 per kW |
| RKVA Demand Charge | $0.15 per rkva |

Standby Service Demand Charge:

| *Contract Available Hours* | *Applicable Rate Per kW of Standby Demand* |
|---|---|
| 175 | $0.76 |
| 350 | $1.53 |
| 525 | $2.29 |
| 700 | $3.06 |

Determination Of Distribution Demand

The Customer shall contract for Distribution Demand only where the service voltage is less than 69kv. Service voltage is defined as the voltage associated with facilities which the Company would normally provide for the service required by the Customer.

Initially Distribution Demand shall be established by mutual agreement. In case the maximum Measured Demand at any time exceeds the Distribution Demand, Distribution Demand shall be increased by such excess demand. When the Customer's power factor is less than 85 percent, a minimum Distribution Demand of not less than 85 percent of the customers maximum Kva demand may be established. Distribution Demand may be changed by mutual agreement as to the amount of change and term of agreement; however, in no case shall Distribution Demand be reduced below the maximum Measured Demand during the preceding eleven billing months.

**Western Massachusetts Electric Company**

Distribution Demand Rate:

13,000 Volts or above = $3.86 per kW
13,000 Volts or less = $4.83 per kW

Production/Transmission Demand Rate for Firm Back-up Power:

D x [1 – (1 – K/H)squared]; where

D equals the applicable general service demand rate minus the applicable Distribution Demand rate, and

K equals the sum, over the most recent twelve months, of the Back-up/standby demand in each on-peak hour, divided by the applicable Firm Back-up Contract Demand, and

H equals the number of on-peak hours in the most recent twelve billing months which fall in the current period.

Production/Transmission Demand Charge for firm Back-up Power, is the greater of:

A. The production/Transmission Demand Rate for Firm Back-up Power times the Back-up Standby Demand for the current billing month, or
B. $2.00 times the firm Back-up Contract Demand for the current period.

The monthly Charge shall be the sum of:

1. The Administrative Fee, plus
2. The Customer Service Charge, plus
3. The Distribution Demand Charge, plus
4. The Production/Transmission Demand Charge for Firm

Back-up Power, plus

5. The Energy Charge.

**Wisconsin Power & Light Co. / Public Service Commission of Wisconsin**

This standby & maintenance rider is available to any customer on Rate Schedule cp-2

Standby Demand Charge:

The greater of: $0.75 per kW (Reservation Demand Charge) of contract standby demand per month or $0.32 per kW of daily demand per month.

Measured Maximum Demand:

The measured maximum demand in any billing month shall be that demand in kW necessary to supply the average kW in 15 consecutive minutes of greatest consumption of electricity during each billing month. Such measured maximum demand shall be determined from readings of permanently installed meters or, at the option of the company, by any standard methods of meters. The demand meter shall be reset to zero at the beginning of each billing month.

**Wisconsin Public Service Corporation / Public Service Commission of Wisconsin**

Small Commercial & Industrial Standby Service

Where service is made available to loads which can be served by a source of power other than the company's (excluding emergency standby maintained in the event of failure of Company's supply), billing shall be at the normally applicable rate, but the monthly minimum demand charge for standby service shall be not less than $1.50 per kw of contracted demand (subject to primary discounts). The delivery of service portion of the primary discount, if applicable, shall be applied to the contract demand for purposes of determining the monthly minimum demand charge.

# Appendix H-B Standby Rate Demand Charges Quick Reference

This is a quick reference to reservation fees for standby services. In order to keep it simple, we have done such things as taken the "middle" rate, when a company has different rates for different voltage levels. We've calculated a weighted average when there are different summer and winter rates, and used a few other similar methods to provide a good idea of a company's reservation charges. Other charges apply if one were to actually use power under the standby rates. Obviously, the precise terms of the tariff may have a dramatic impact on an individual customer. Complete tariffs appear in the PMA survey "Utility Standby Rates," published by the Power Marketing Association. The original survey is available for $195 from PMA, and includes the complete text of over 100 utility standby tariffs. To obtain a copy, fax your request to PMA at (703) 641 9265, call (703) 641 0613 or e-mail centaur@erols.com. Be sure to note your shipping and billing address.

| Company | Demand Charge (Reservation Only) |
|---|---|
| Alabama Power Co. | $1.25 |
| Arizona Public Service Co. | $2.25 |
| Arkansas Power & Light Co. | $2.50 |
| Baltimore G&E | 50% of regular rate |
| Black Hills Power & Light Co. | |
| Blackstone Valley Electric Co. | $6.56 |
| Carolina Power & Light Co. | $5.85 |
| Central Illinois Light Co. | $4.11 |
| Central Light & Power | Regular rate minus $4.50 |
| Cleveland Electric Illuminating | $4.50 |
| Commonwealth Edison Co. | $2.99 |
| Connecticut Light & Power | Regular rate minus $4.50 |
| Consolidated Edison | $5.81 |
| Consumers Power Co. | $1.85 |
| Detroit Edison Co. | $2.36 |

| Company | Demand Charge (Reservation Only) |
|---|---|
| Duke Power Co. | $6.62 |
| El Paso Electric Co. | $16.75 |
| Florida Power & Light | $3.00 |
| Florida Power Corp. | $2.335 |
| Georgia Power Co. | $2.13 |
| Gulf Power Co. | $2.26 |
| Gulf States Utilities | 95¢ |
| Houston Lighting & Power | $1.99 |
| Idaho Power Co. | $1.48 |
| IES | $6.34 |
| Illinois Power Co. | $1.75 |
| Indiana Michigan Power Co. | $1.52 |
| Jersey Central Power & Light | $3.09 |
| Long Island Lighting Co. | $2.34 |
| Louisville Gas & Electric Co. | $5.61 |
| Maine Public Service Co. | $2.50 |
| Metropolitan Edison | $2.41 |
| MidAmerican Energy Co. | $1.55 |

| Company | Demand Charge (Reservation Only) | Company | Demand Charge (Reservation Only) |
|---|---|---|---|
| Mississippi Power & Light | 35% of regular rate | Public Service New Mexico | 10% of regular rate |
| Mississippi Power Co. | $2.30 | Puget Sound P&L | $3.94 |
| Montana Power Co. | $2.64 | San Diego Gas & Electric | $3.13 |
| Montana-Dakota Utilities | $10.38 | Savannah Electric & Power Co | $2.40 |
| Nevada Power | $5.19 | Sierra Pacific Power Co. | $3.95 |
| NIPSCO | $2.88 | South Carolina Electric & Gas | $3.76 |
| Northern States | $3.25 | Southern California Edison Co. | 40¢ |
| Ohio Edison Co. | $3.55 | St. Joseph Light & Power Co. | $6.42 |
| Pacific Gas & Electric Co. | $2.55 | Tampa Electric Co. | $3.53 |
| Pacific Power | $1.05 | Texas Utilities Electric Co. | $1.90 |
| Pacific Power & Light | 50% regular rate | Toledo Edison Co. | $4.00 |
| PECO Energy | $3.00 | Utah Power & Light Co. | $2.12 |
| Pennsylvania Electric Co. | $2.31 | VEPCO | $3.06 |
| Pennsylvania Power & Light | $1.72 | West Penn Power Co. | $3.12 |
| Pennsylvania Power Co. | 30% regular rate | Western Massachusetts Electric | $3.86 |
| PEPCO | 50¢ | Wisconsin Power & Light Co. | $1.50 |
| Public Service Co. Colorado | $5.05 | | |
| Public Service Electric & Gas | $3.58 | | |

# Section I

## Conclusion: So You Have Decided to Explore the Cogeneration Option

# Section I

## Conclusion: So You Have Decided to Explore the Cogeneration Option

Cogeneration is not a new way of generating electricity. As recently as the 1920s, large industrial users accounted for a large percentage of the nation's electrical generation. What is different about cogeneration today is that it occurring in the context of a competitive generation sector. Consequently, the decision on whether to proceed with cogeneration now is a much more complex task than it used to be.

The process by which you decide to explore the cogeneration option will differ depending on the circumstances.

### ON-SITE GENERATION

If your company is interested primarily in on-site generation with minimal off-site sales, you need to pull together a team of experts on the economic, financial and regulatory issues involved in the construction of a cogeneration plant to determine whether on-site generation makes sense in your particular situation.

Among other things, this team should be responsible for:

- Doing an analysis of the costs to self-generate;
- Comparing the costs of self-generation with that of utility service;
- Identifying the legal and regulatory issues involved in the decision to self-generation; and
- Investigating the costs and charges of the local utility to determine whether the local utility can reduce your electric rates.

If the team's preliminary analysis is that self-generation is a viable option, then you want to begin talking with the local utility to determine whether it will offer you a rate break to keep you on its system.

At the same time the team pulled together to do the preliminary analysis should be refining its work by contacting prospective suppliers and coming up with more complete cost estimates. Not surprisingly, the interest of the local utility in negotiating a rate break often is a direct function of its perception of the customer's willingness to proceed with the cogeneration option.

If no agreement with the local utility has been reached with the local utility at the time the review team has refined its analysis, the customer should proceed to the next step: issuing an RFP and starting negotiations with the local utility as to appropriate rates for partial requirements service.

While the optimal result in a region in which utilities have surplus capacity is not a cogeneration project, sometimes it is necessary to construct one due to institutional problems. For example, merely having a cogeneration facility in some cases qualifies a customer for lower-cost treatment than other customers, even if the facility is not used regularly.

### COGENERATION WITH SIGNIFICANT OFF-SITE SALES

An industrial exploring self-generation with significant off-site sales should undertake many of the same steps discussed with regard to the construction of a unit for on-site generation. Here, however, the analysis is complicated by the need to find a home for the excess kilowatts.

The first step in this case is identifying prospective electric power customers. You need to determine who is buying and who is selling, and the actual prices, terms and conditions they have found agreeable.

Second, you need to package the power product. Once prospects are identified, you must determine if there are power products they require in which you can add value. If this appears likely, you should begin development of the package, obtaining letters of intent from prospective short-term capacity providers, backup suppliers, surplus capacity customers, and supplementary power suppliers to develop a winning product for the customer, in which your ability to provide top-quality generating capacity is the touchstone.

Third, you need to finalize the package. This involves closing the deal, obtaining signed contracts for the other participants in the package committing them to a position in the product, contingent upon final acceptance by the customer.

One should note it may be necessary to bring actions against a recalcitrant local service provider who is unwilling to provide transmission services or to negotiate acceptable terms for service. Sometimes the local utility will fold after one action is commenced, with nothing more than a simple filing, but one should be prepared to see these actions through to the end.